Quantitative Chemical Analysis by Electrolysis

Quantitative Chemical Analysis by Electrolysis

Contributors

Sergei Grokhovsky, Irina Il'icheva et al.

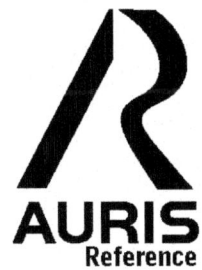

AURIS
Reference

www.aurisreference.com

Quantitative Chemical Analysis by Electrolysis

Contributors: Sergei Grokhovsky, Irina Il'icheva et al.

Published by Auris Reference Limited

www.aurisreference.com

United Kingdom

Quantitative Chemical Analysis by Electrolysis

ISBN: 978-1-78154-898-1

British Library Cataloguing in Publication Data
A CIP record for this book is available from the British Library

Printed in the United Kingdom

Exclusively distributed by CBS Publishers & Distributors Pvt. Ltd.

Sales & Distribution Rights only for India, Pakistan, Bangladesh, Sri Lanka, Nepal and Bhutan.This book is not to be sold outside these territories.

Contents

List of Abbreviations

CAEC	Carbon Assisted Electrolysis Cell
CCS	Carbon Capture and Storage
CCU	Carbon Capture and Utilization
CD	Circular Dichroism
CE	Chain Elongation
CEM	Cationic Exchange Membrane
CEM	Complete Electrode Model
DAS	Dendrite Arm Space
DCFC	Direct Carbon Fuel Cell
EC	Electrolysis Cell
EIT	Electrical Impedance Tomography
FESEM	Field Emission Scanning Electron Microscopy
FISH	Fluorescence In Situ Hybridization
FTIR	Fourier Transform Infrared Spectrometer
GT	Gas Turbine
HER	Hydrogen Evolution Reaction
HOR	Hydrogen Oxidation Reaction
NTIRE	Non Thermal Irreversible Electroporation
OUT	Operational Taxonomic Units
PAGE	Polyacrylamide Gel Electrophoresis
RDA	Redundancy Discriminant Analysis
SEE	Synergistic Electrolysis and Electroporation
SRT	Solids Retention Time
TSS	Total Suspended Solids
VFA	Volatile Fatty Acids
XRD	X-Ray Diffraction

List of Contributors

Irina Il'icheva
Engelhardt Institute of Molecular Biology, Russian Academy of Sciences, Moscow Russia

Dmitry Nechipurenko
Department of Physics, Moscow State University, Moscow, Russia

Michail Golovkin
Department of Physics, Moscow State University, Moscow, Russia

Georgy Taranov
Department of Physics, Moscow State University, Moscow, Russia

Larisa Panchenko
Department of Biology, Moscow State University, Moscow, Russia

Robert Polozov
Institute of Theoretical and Experimental Biophysics Russian Academy of Sciences, Puschino, Russia

Yury Nechipurenko
Engelhardt Institute of Molecular Biology, Russian Academy of Sciences, Moscow Russia

Martins Vanags
Institute of Solid State Physics, University of Latvia, Riga, Latvia

Janis Kleperis
Institute of Solid State Physics, University of Latvia, Riga, Latvia

Gunars Bajars
Institute of Solid State Physics, University of Latvia, Riga, Latvia

Arie Meir
Biophysics Graduate Program, University of California, Berkeley, California, United States of America

Boris Rubinsky
Department of Mechanical Engineering, University of California, Berkeley, California, United States of America

Michael K. Stehling
Inter Science GmbH, Biophysics, Luzern, Switzerland

Enric Guenther
Inter Science GmbH, Biophysics, Luzern, Switzerland

Paul Mikus
Inter Science GmbH, Biophysics, Luzern, Switzerland

Nina Klein
Institut fuer Bildgebende Diagnostik - Tumortherapy Center, R&D, Offenbach, Germany

Liel Rubinsky
Inter Science GmbH, Biophysics, Luzern, Switzerland

Boris Rubinsky
Inter Science GmbH, Biophysics, Luzern, Switzerland

Ruyao Wang
Institute of Material Science and Engineering, Donghua University, Shanghai,
P.R.China

Weihua Lu
Institute of Material Science and Engineering, Donghua University, Shanghai,
P.R.China

Stephen J. Andersen
Laboratory of Microbial Ecology and Technology (LabMET), Ghent University, Coupure Links 653, Building A, Room A0.092, B 9000 Ghent, Belgium

Pieter Candry
Laboratory of Microbial Ecology and Technology (LabMET), Ghent University, Coupure Links 653, Building A, Room A0.092, B 9000 Ghent, Belgium

Thais Basadre
Laboratory of Microbial Ecology and Technology (LabMET), Ghent University, Coupure Links 653, Building A, Room A0.092, B 9000 Ghent, Belgium

Way Cern Khor
Laboratory of Microbial Ecology and Technology (LabMET), Ghent University, Coupure Links 653, Building A, Room A0.092, B 9000 Ghent, Belgium

Hugo Roume
Laboratory of Microbial Ecology and Technology (LabMET), Ghent University, Coupure Links 653, Building A, Room A0.092, B 9000 Ghent, Belgium

Emma Hernandez Sanabria
Laboratory of Microbial Ecology and Technology (LabMET), Ghent University, Coupure Links 653, Building A, Room A0.092, B 9000 Ghent, Belgium

Marta Coma
Laboratory of Microbial Ecology and Technology (LabMET), Ghent University, Coupure Links 653, Building A, Room A0.092, B 9000 Ghent, Belgium
Centre for Sustainable Chemical Technologies, University of Bath, Claverton Down, Bath BA2 7AY, UK.

Korneel Rabaey
Laboratory of Microbial Ecology and Technology (LabMET), Ghent University, Coupure Links 653, Building A, Room A0.092, B 9000 Ghent, Belgium

Guangzai Nong
Institute of Light Industry and Food Engineering, Guangxi University, Nanning 530004, China

Zongwen Zhou
Institute of Light Industry and Food Engineering, Guangxi University, Nanning 530004, China

Shuangfei Wang
Institute of Light Industry and Food Engineering, Guangxi University, Nanning 530004, China

Hyun Sic Park
Department of Chemical and Biomolecular Engineering, Yonsei University, 262 Seongsanno, Seodaemun-gu, Seoul 120-749, Korea

Ju Sung Lee
Department of Chemical and Biomolecular Engineering, Yonsei University, 262 Seongsanno, Seodaemun-gu, Seoul 120-749, Korea

JunYoung Han
Proton Conductors Section, Department of Energy Conversion and Storage, Technical University of Denmark, Kemitorvet 207, Kgs. Lyngby DK-2800, Denmark

Sangwon Park
CO2 Sequestration Department, Korea Institute of Geoscience and Mineral Resources (KIGAM), 124 Gwahak-ro, Yuseong-gu, Daejeon 305-350, Korea

Jinwon Park
Department of Chemical and Biomolecular Engineering, Yonsei University, 262 Seongsanno, Seodaemun-gu, Seoul 120-749, Korea

Byoung Ryul Min
Department of Chemical and Biomolecular Engineering, Yonsei University, 262 Seongsanno, Seodaemun-gu, Seoul 120-749, Korea

Bruce C.R. Ewan
Chemical & Biological Engineering Department, University of Sheffield, Mappin Street, Sheffield S1 3JD, UK

Olalekan D. Adeniyi
School of Engineering and Engineering Technology, Federal University of Technology, Gidan Kwano Campus PMB 65, Minna, Niger State, Nigeria

Kafoumba Bamba
Laboratoire de Thermodynamique et de PHysico-Chimie du Milieu, UFR-SFA, Université Nangui Abrogoua, Abidjan, Côte d'Ivoire

Nahossé Ziao
Laboratoire de Thermodynamique et de PHysico-Chimie du Milieu, UFR-SFA, Université Nangui Abrogoua, Abidjan, Côte d'Ivoire

Romdhane Ben Slama
Unit of Research: Environment, Catalysis & Processes Analysis, National School of Engineers of Gabes, University of Gabes, Gabes, Tunisia.

Preface

Quantitative chemical analysis, deals with the determination of the amount or percentage of one or more constituents of a sample. A variety of methods is employed for quantitative analyses, which for convenience may be broadly classified as chemical or physical, depending upon which properties are utilized. Chemical methods depend upon such reactions as precipitation, neutralization, oxidation, or, in general, the formation of a new compound. Quantitative Chemical Analysis by Electrolysis offers consistently modern portrait of the tools and techniques of chemical analysis, incorporating real data, spreadsheets, and a wealth of applications, all presented in a witty, personable style that engages students without compromising the principles and depth necessary for a thorough and practical understanding. First chapter focuses on the developed protocols for gel data treatment that helped us to quantitatively describe the observed phenomenon of sequence-specific ultrasonic cleavage of DNA. Water electrolysis with inductive voltage pulses is presented in second chapter. The primary goal of third chapter is to explore the hypothesis that changes in pH during electrolysis can be detected with Electrical Impedance Tomography (EIT). The goal of fourth chapter is to expand on earlier studies with small animals and use the pig liver to establish SEE treatment parameters of clinical utility. In fifth chapter, we have discussed the structural heredity of alloys upon remelting. In sixth chapter, we fermented thin stillage to generate a mixed VFA extract without chemical pH control. The objective of seventh chapter is to develop an effective means to remove the water pollutants by recovery of both lignin and sodium hydroxide from black liquor, based on electrolysis. Eighth chapter proposes a method to fixate $CaCO_3$ stably by using relatively less energy than existing methods. Ninth chapter demonstrates about carbon-assisted water electrolysis. In tenth chapter, we test for the first time both ceramic components (61 mol% ZrB_2-αSiC and 61 mol% TiB_2-αSiC) as cathode rotating disk electrode RED for hydrogen evolution reaction HER and hydrogen oxidation reaction HOR in PEMFC. In last chapter, the goal is to produce hydrogen by water electrolysis, using photovoltaic solar energy. Although occasionally there may be certain imperfections with these old texts, we feel they deserve to be made available for future generations to enjoy.

Chapter 1

QUANTITATIVE ANALYSIS OF ELECTROPHORESIS DATA – APPLICATION TO SEQUENCE-SPECIFIC ULTRASONIC CLEAVAGE OF DNA

Sergei Grokhovsky[1], Irina Il'icheva[2] , Dmitry Nechipurenko[3] , Michail Golovkin[3] , Georgy Taranov[3] , Larisa Panchenko[4] , Robert Polozov[5] and Yury Nechipurenko[1]

[1]Engelhardt Institute of Molecular Biology, Russian Academy of Sciences, Moscow, Russia

[2]Engelhardt Institute of Molecular Biology, Russian Academy of Sciences, Moscow Russia

[3]Department of Physics, Moscow State University, Moscow, Russia

[4]Department of Biology, Moscow State University, Moscow, Russia

[5]Institute of Theoretical and Experimental Biophysics Russian Academy of Sciences, Puschino, Russia

INTRODUCTION

The complete genomes of many different species are now being revealed in ever increasing pace. The impressive progress made in genome sequencing was largely attributed to development of high resolution denaturing polyacrylamide gel electrophoresis (PAGE). Next-generation sequencing platforms use new powerful technologies, providing gigabases of genetic information in a single run (Farias-Hesson et al., 2010). Nevertheless, scientific research often deals with situations when one needs to change the experimental conditions or the data analysis protocols, but commercial available devices and programs don't give such an opportunity. We have faced this problem during the research focused on the phenomenon of sequence specific ultrasonic cleavage of double-stranded (ds) DNA (Grokhovsky, 2006). The observed sequence dependence of DNA cleavage efficiency was quite surprising. It seems that sequence-specificity of ultrasonic cleavage reflects the local variations in DNA structural dynamics. Thus, ultrasound may provide a basis for developing a new method for studying sequence effects on local structural dynamics of DNA fragments.

It is generally accepted that recognition of various DNA binding sites by many types of transcription factors depends not only on the base pair sequence, but also on local variations in structural parameters of the DNA molecule. Among the important factors involved in these processes are conformational flexibility of sugar-phosphate backbone, geometry of DNA grooves and local bending propensities of the double helix.

Local conformational parameters of DNA are sequence-dependent but in many cases different DNA sequences might carry similar structural profiles (Travers, 2004; Parker et al., 2009). Besides, structural parameters of DNA are sensitive to temperature, ionic strength, pH and other factors and might also drastically change in presence of closely bound proteins and other ligands (Neidle et al., 1987; Belikov et al., 2005; Wells, 2009). Structural information for short double-stranded oligonucleotides have been obtained by various experimental methods, including crystal structure data sets (Olson et al., 1998; Sims & Kim, 2003; Svozil et al., 2008), NMR, and Fourier transform infrared and Raman spectroscopy (Heddi et al., 2010; Abi-Ghanem et al., 2010).

Nevertheless, elucidation of sequence effects on conformation and dynamics of longer DNA fragments remains a challenge. Hence, the development of new methods which would allow studing local structural properties in long double-stranded DNA fragments of several hundreds base pairs is of great importance. Currently there are some methods available for studying sequence-specific variations of DNA flexibility and grooves width along DNA fragments exceeding 100 base pairs in length. These methods are based on the analysis of DNA cleavage produced by various types of agents and irradiations. Cleavages with DNase I (Waring, 2006; Bullwinkle & Koudelka, 2011), hydroxyl radicals (Tullius, 1988; Van Dyke & Dervan, 1983) and laser (Spassky & Angelov, 1997; Vtyurina et al., 2011) or X-ray irradiation (Zubarev & Grokhovsky, 1991) are typically used for these type of research. The analysis of sequence-specific DNA cleavage is performed by PAGE. Cleavage patterns obtained by PAGE are further analyzed in order to correlate bands intensities with cleavage efficiencies of corresponding covalent bonds. Variations in cleavage intensities along DNA are attributed to heterogeneity of local structural parameters of the molecule. For example, DNase I-cleavage is sensitive to the geometry of the minor groove and the DNA stiffness that resists bending towards major groove. Thus, variations of phosphodiester bonds cutting efficiencies by DNase I reflect heterogeneity of these parameters along DNA (Hogan et al., 1989; Brukner et al., 1995). Recently we have developed a new method for studying sequence-dependent structural dynamics of extended DNA fragments (Grokhovsky, 2006, Il'icheva et al., 2009, Nechipurenko et al., 2009).

The approach is based on the analysis of ultrasound - induced DNA cleavage using PAGE. Sequence-specificity of ultrasonic cleavage of DNA is attributed to variations in local conformational flexibility of the sugar-phosphate backbone along the irradiated fragments (Grokhovsky et al., 2011). Data produced by this method is complementary to the information recovered using chemically- and DNase I-induced cleavage since ultrasonic cleavage represents a mechanochemical reaction and depends on the local dynamical properties of the DNA molecule. Using ultrasonic cleavage patterns of DNA by PAGE we show various effects which should be considered during the analysis of the electrophoresis data. In this chapter we will focus on the developed protocols for gel data treatment that helped us to quantitatively describe the observed phenomenon of sequence-specific ultrasonic cleavage of DNA. These methods can be used for the analysis of electrophoresis data obtained during various types of experiments.

METHODS AND EXPERIMENTAL PROTOCOLS

In order to address the sequence dependence of the DNA cleavage with different agents, fragments containing a radioactive or fluorescence label at 5'- or 3'-end of one of the DNA strands are commonly used. We used intense ultrasound treatment of fragments containing a radioactive label at the 3'-end of one of the DNA strands. Restriction fragments of DNA were generated by digestion of plasmid DNA by the restriction endonucleases. The fragments were 3'-end-labeled with [a- 32P]dATP in the presence of the unlabeled other dNTP and the Klenow fragment of Escherichia coli DNA-polymerase I. The DNA fragments were isolated by nondenaturing polyacrylamide gel in a 1-mm thick 5% gel with subsequent elution and precipitation (Maniatis et al., 1982).

Sonication of DNA Fragments

For sonication, 10 mL of DNA fragments (~104 Bq) in water were mixed with 10 mL of 0.2M NaOAc, pH 6.0, in the bottom of a thin-walled polypropylene microcentrifuge tube of 0.2 mL capacity. The final concentration of the fragments was 5–10 µg/mL (~10 mM base pair). The test-tube ends were located ~0.5 cm below the horn sonicator edge, which had a diameter of 12 mm. The ring and horn sonicator were placed in a vessel with water and crushed ice (Fig.1). Ultrasound was generated by a 300 W generator with a frequency of 44 or 22 kHz using the maximum power output. The sonication was adjusted in continuous operation mode at 1-min intervals, and after each interval the ring was turned 180o and fresh ice was added. The ultrasonic power exerted on the system was determined calorimetrically and exceeded 60 W. To obtain rough measurements of the chemical effects induced by cavitation inside the tubes, we used a test tube containing 0.05 M KI in 0.025% starch solution. After 8 min of irradiation, the coloration extent was equal to that obtained by adding ~0.1 mM of hydrogen peroxide. This yield is comparable to reported results obtained under normal temperature conditions and with ultrasound intensity exceeding 2 W/cm² (Margulis, 1984). It is worth noting that both low ultrasound frequency and low temperature conditions are known to increase the power of cavitational effects (Basedow & Ebert, 1977).

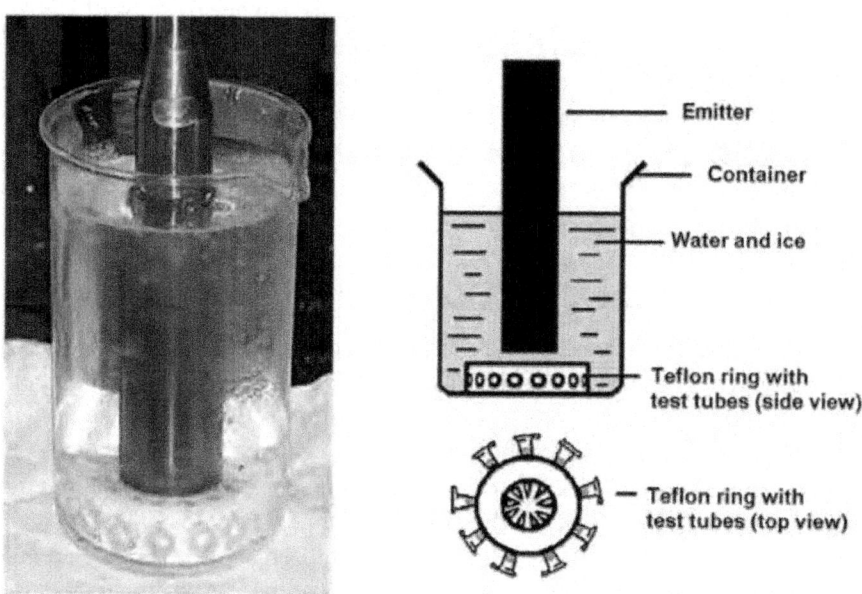

Figure. 1. The ring with test tubes and horn sonicator in a vessel with water and crushed ice. Ultrasound was generated by a 300 W generator UZDN-2T (Ukraine) with a frequency of 44 or 22 kHz using the maximum power output.

Separation of DNA Fragments in Nondenaturing Gel

After sonication, the sample was combined with an equal volume of 50% glycerol with 0.02% bromphenol blue. Aliquots (0.5 µl) were applied on nondenaturing polyacrylamide gel of 40 cm in length and 0.15 mm in thickness. Electrophoresis was carried out in 1×TBE at 1.3 kV (~30W) for 3 h (gel temperature~30°C) or at 300 V (~2W) for 18 h at +2°C. Gels were dried on a glass which was pretreated with γ-methacrylpropyloxysilane and were then exposed with a luminescent screen.

Separation of DNA Fragments in Denaturing Gel

To localize the DNA cleavage sites with single-nucleotide precision the sonication product was resolved by denaturing polyacrylamide gel. After sonication, the samples were combined with 180 mL of a solution containing 0.15 M NaCl, 50 mM Tris-HCl (pH 7.5), and 10 mM EDTA. The samples were then extracted with phenol. The DNA was precipitated with ethanol, washed with 70% ethanol, dried, and dissolved in 1 µL of 95% formamide (which contained 15 mM EDTA (pH 8.0), 0.05% bromphenol blue, and 0.05% xylencyanol FF). It was then heated for 1 min at 90oC, rapidly cooled down to 0oC, and applied on polyacrylamide gel containing 8 M urea (length: 40 cm; gradient width: 0.15–0.45 mm) (Kraev, 1988). Electrophoresis was carried out for 55 min (100 W, 2500 V) at 60–70oC. Afterward, the gel was fixed in 10% acetic acid and dried on a glass plate pretreated with g- methacrylpropyloxysilane. The dried gel was exposed to a luminescent screen and then scanned with a Cyclone Storage Phosphor System device (Packard BioScience). Cleavage pattern bands were assigned to particular nucleotide sequences of fragments by comparison with the lanes of "A+G" track DNA samples.

Impact of DNA Fragment Size, Ultrasound Frequency, PH and Ionic Strength on Ultrasonic Cleavage Patterns

Cleavage profiles of the 470-bp DNA fragment were obtained by nondenaturing gel (Fig. 2). The fragment contained AT clusters alternating with GC clusters. Gel was carried out at 2°C (Fig. 2a). A similar pattern was observed when gel was carried out at 30°C (Fig. 2b). This temperature makes it possible to detect double-strand breaks with cleavage sites located several nucleotides apart on the two DNA strands: such sticky ends melt during PAGE under these conditions.

Figure 3 shows the cleavage patterns of the 475- and 439-bp fragments obtained by denaturing gel. The fragments carried the label on different strands of the same sequence: one 3' end of each fragment was labeled. Gel reports cleavage for only one strand under such conditions. Several sites were revealed whose cleavage rate was considerably higher than the background level.

Preliminary analysis of the nucleotide sequence of these and other fragments demonstrated that DNA strands break more readily between cytosine and guanine in the 5'-CpG-3' sequence. As Fig. 2 shows, the fragment was cleaved preferentially at several sites, which corresponded to alternating GC pairs. Many double-strand breaks arose as early as within the first four minutes of sonication. Further sonication enhanced the cleavage pattern, but the ends of the fragment still remained noncleaved.

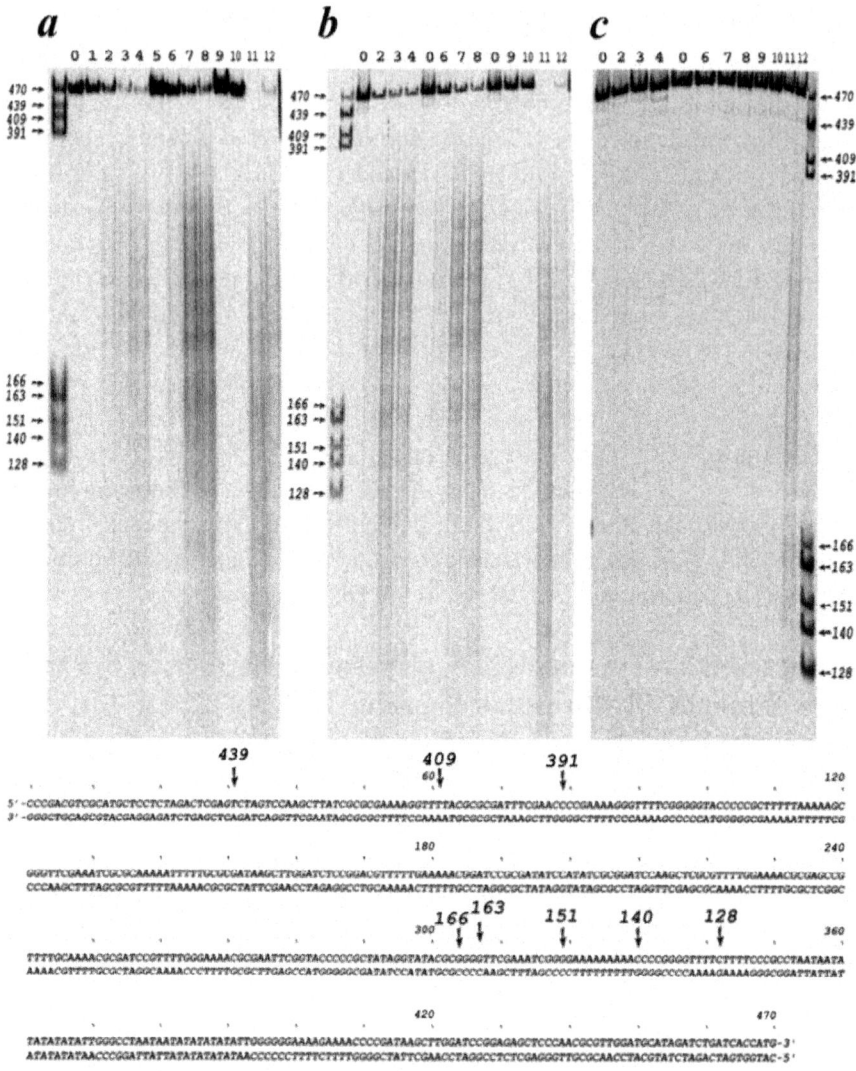

Figure. 2. Cleavage profiles of the 470-bp DNA fragment in nondenaturing 5% poly-acrylamide gel after sonication at 44 kHz. DNA was (a) sonicated at 0°C and resolved at 2°C, (b) sonicated at 0°C and resolved at 30°C, or (c) sonicated at 30°C and resolved at 30°C. The fragment sequence is shown at the bottom. Positions of labeled marker double-stranded DNA fragments of known sizes are shown with arrows on the electrophoretic patterns and on the sequence. The fragment was analyzed (lane 0) before and after sonication (lanes 1–4) in 5 mM NaOAc (pH 7.0) for 2, 4, 8, and 16 min, respectively; (lanes 5–8) in 0.5 M NaOAc (pH 7.0) for 2, 4, 8, and 16 min, respectively; (lanes 9,10), in 0.5 M NaOAc (pH 11.0) for 4 and 8 min, respectively; and (lanes 11,12) in 0.5 M NaOAc (pH 5.0) for 4 and 8 min, respectively.

The efficiency of cleavage slightly increased with ionic strength increasing from 5 mM to 0.5 M at pH 7.0. When pH was varied, cleavage was almost undetectable at pH 11.0, while its efficiency considerably increased at pH 5.0. This finding is explained by the fact that the double helix is partly unwound at alkaline pH, which increases the flexibility and the condensation of DNA. When sonication temperature was increased to 30°C, cleavage was almost completely suppressed (Fig. 2c). The fragment was cleaved to a significant extent at 30°C only at low pH. These findings suggest that the main contribution to cleavage DNA strands is made by hydrodynamic forces which arise when cavitation bubbles collapse and which depend on the water vapor pressure, decreasing with a decrease in temperature (Suslick & Price, 1999). The chemical processes generating radicals during cavitation play only a minor role, if any. The character of fragment cleavage was the same upon sonication at 22 and at 44 kHz.

Elucidation of the Terminal Groups Resulting from DNA Cleavage

Chemical cleavage of DNA with formic acid — diphenylamine reagent eliminates a purine from the cleavage site (Tate & Petersen 1975; Belikov & Wieslander, 1995). Thus, the bands seen in lanes "A+G" (on figs. 3 - lanes 1; fig. 4 - lanes 1 and 18; on fig. 7 - lanes 1 and on fig. 8 - lane 10) correspond to oligonucleotides lacking the terminal purine. 3'-endlabeled fragments contain the uncharged 3'-OH group at the 3' end and the phosphate group, which carries two negative charges, at the 5' end. When the 5'-terminal phosphate is removed with calf intestinal alkaline phosphatase, the electrophoretic mobility of DNA fragments changes (Fig. 3; lanes 1,2). The mobility of a fragment depends on its molecular weight, its total charge, and the gel density. The longer the fragment, the lower the contribution of the two terminal charges to the total charge and the weaker the dependence of the electrophoretic mobility on the fragment size. For example, the 20 base pairs (b.p.) fragment without phosphate at the 5' has overall negative charge of 19 e, while the terminal phosphate adds 2e which results in 21 e for total charge in presence of 5' phosphate. Thus, the relative electric charge difference for 20 b.p. fragments is 11%. The corresponding values for 40 b.p. and 90 b.p. fragments would be 41 : 39 - 5% and 91 : 89 - 2%, respectively. The length of the fragment is less affected with the presence of 5' phosphate. Experiment demonstrates that in denaturing 6% gel electrophoretic mobility shifts by about 1.5 steps in the region of 20-mer oligonucleotides, by 1 step in the region of 40-mer oligonucleotides, and by 0.5 steps in the region of 90-mer oligonucleotides. In 14% gel, similar shifts are observed in the regions of 14-, 25-, and 55-mer oligonucleotides, respectively. The bands observed after sonication of the DNA fragment exactly coincided with the bands observed after its chemical cleavage at purines (Fig. 3 and Fig.4). This result indicated that the products had phosphates at their 5' ends.

Evidence of Mechanochemical Nature of Observed Cleavage

Fig. 4 demonstrates the gel image obtained after sonication of the DNA fragments for various periods of time. This image represents the results of ultrasonic irradiation of three fragments that differed in initial length (311, 251, and 218 base pairs, respectively) but shared the same base pair sequences. The left part of the gel contains lanes that correspond to cleavage of the longest fragment (lanes 1–6). The central part represents lanes corresponding to cleavage of the middle-sized fragment (lanes 7–12), and the lanes at the right side of the gel demonstrate the cleavage patterns of the shortest fragment (lanes 13–17).

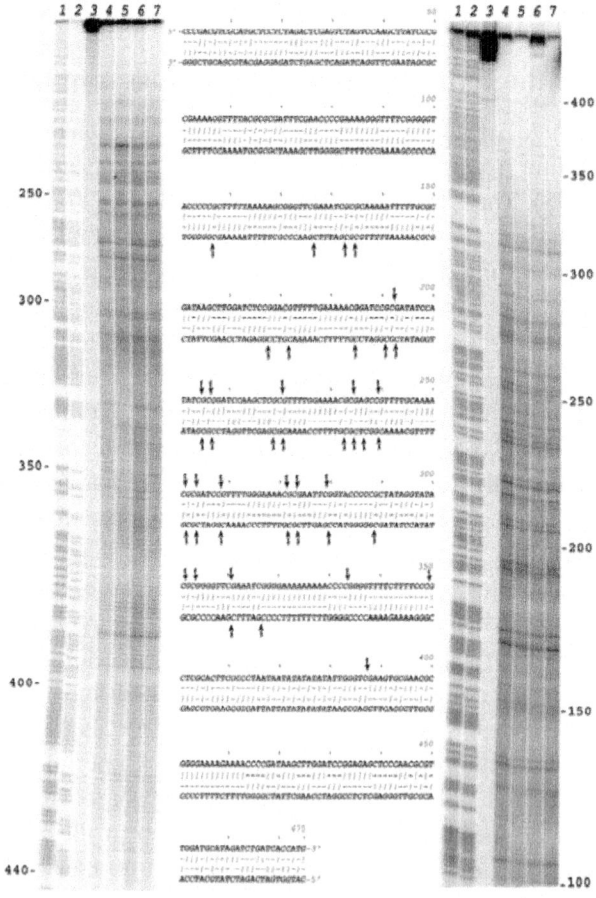

Figure. 3. Cleavage profiles of DNA fragments in denaturing 6% gel after sonication at 44 kHz for 20 min. Lane 1, products of chemical cleavage at purines with subse-

quent treatment with calf intestinal alkaline phosphatase; lane 2, product of chemical cleavage at purines by formic acid — diphenylamine reagent.; lane 3, the initial fragment without treatment; lane 4, the fragment sonicated in isolation; and lanes 5 – 7, the fragment sonicated in the presence of 1, 0.5, or 0.25 µM of Pt-bis-netropsin, respectively (for details see section 6.2.). The nucleotide sequence of the fragments is shown in the center. To simplify comparison with the bands seen on gel, purines are marked with slants for each strand. The sites with a cleavage rate far higher than the cleavage intensity baseline are indicated with arrows. The cleavage profiles shown on the left correspond to the cleavage of the upper strand, i.e. when this strand is radiolabeled, while the cleavage patterns on the right side represent the cleavage of the lower strand.

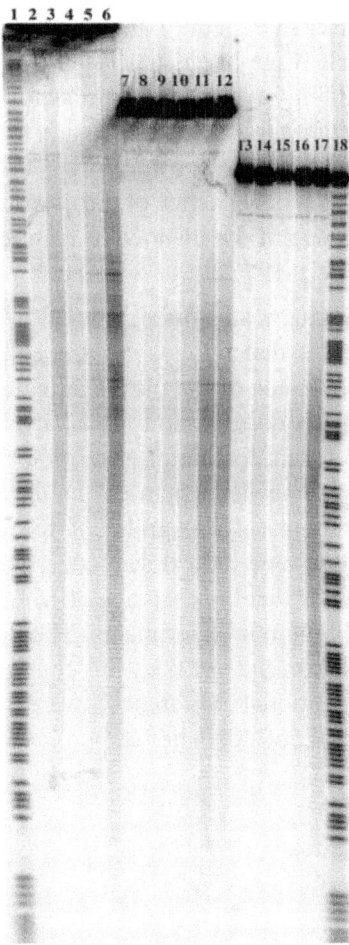

Figure. 4. Cleavage pattern of DNA fragments in 6% denaturing gel after irradiation with ultrasound (22 kHz). Lanes 1 and 18: Chemical cleavage at purines by formic acid — diphenylamine reagent. Lanes 2, 7, and 13: Sonication of fragments for 2 min.

Lanes 3, 8, and 14: Sonication of fragments for 2 min in the presence of 50% glycerol. Lanes 4, 9, and 15: Sonication of fragments for 4 min. Lanes 5, 10, and 16: Sonication of fragments for 8 min. Lanes 6, 11, and 17: Sonication of fragments for 16 min. Lane 12: Sonication of fragments for 16 min in the presence of 0.5 M tiourea.

It is clear that increasing the irradiation time from 2 to 16 min leads to a sufficient increase in overall cleavage intensity for all three types of fragments. Fig. 4 also demonstrates that the addition of tiourea has no visible effect on the cleavage patterns (lanes 11 and 12). The same result was obtained when others free radical scavengers (dithiothreitol and sodium ascorbate) were added to the irradiated solution (data not shown). On the other hand, adding 50% glycerol, which increased the viscosity of the solution by roughly 10-fold, led to a significant increase in cleavage intensities (lanes 3, 8, and 14). The cleavage patterns obtained by adding glycerol are similar to those obtained without it but with a longer irradiation time. Thus, increasing the viscosity of the solution leads to an overall increase in cleavage intensity but does not affect the relative intensities of cleavage. This dependence of the ultrasonic cleavage intensity on the solution viscosity is one of the distinctive features of a mechanochemical reaction (Basedow & Ebert, 1977).

Fig. 3 and 4 also demonstrates the positional effect (i.e., the damping of ultrasonic cleavage) at sites that are closer to the ends of the DNA fragments. Accordingly, the darkest bands of the cleavage patterns that give the highest values of cleavage intensity correspond to breakages at the central part of the DNA fragments. This relevant feature of ultrasonic cleavage patterns of DNA also supports the idea of mechanochemical nature of the cleavage process observed in our experiments. The significant role of the positional effect and the minor influence of free radical scavengers on the observed cleavage patterns lead us to conclude that the cleavage of DNA induced by free radicals in solution was negligible in our experiments. The distinctive features of free radical cleavage of DNA on the gel (i.e., the emergence of overall cleavage background with no positional preference) were observed only at higher temperature conditions (>25oC; data not shown).

What Physical Processes in Aqueous Solution under Sonication Lead to DNA Cleavage?

The ultrasonic cleavage of DNA reported here is most likely the result of hydrodynamic shearing stresses caused by the collapse of cavitation bubbles (Basedow & Ebert, 1977, Suslick & Price, 1999). Their collapse results in a drastic increase of local temperature and pressure (Margulis, 1984; Didenko, et al., McNamara et al., 1999). The critical size of the bubbles weakly depends on the sound frequency in a wide frequency range. Shearing forces that act on the DNA fragments are thought to originate from high-velocity gradients of water near the collapsing bubble. It is known that in the case of asymmetric collapse, the velocity of the microjets exceeds 100 m/s (Suslick & Price, 1999), whereas the theoretical value of the bubble's interface velocity in the case of symmetrical collapse might exceed 200 m/s. High-velocity gradients in the streaming solution may cause mechanical deformation of the molecule by friction forces. Thus, the observed cleavage of DNA fragments most likely represents a complex mechanochemical process, which includes mechanical deformation of the molecule before the actual chemical reaction takes place (Basedow & Ebert, 1977). Because cavitational flows are accompanied by turbulence, any mathematical treatment of the problem is restricted to highly simplified models.

To estimate the values of the shearing forces that act on DNA molecules in cavitating solution, we used the model proposed by Thomas (Thomas, 1959). This model is generally accepted for describing the degradation of polymers in cavitating solution. Our computation of cavitation bubble dynamics showed that in the final stage of the bubble's collapse, the radial velocity gradient calculated for water flow near the bubble's interface exceeded 10^7 s^{-1}. Calculations showed that such flow gradients are capable of producing stretching forces acting on a 200 base pair DNA fragment of >3 nN (unpublished results). Single-molecule studies of various polymers have shown that the rupture force of a single covalent bond is in the nanonewton range and depends logarithmically on the stretching rate (Bustamante, 2000).

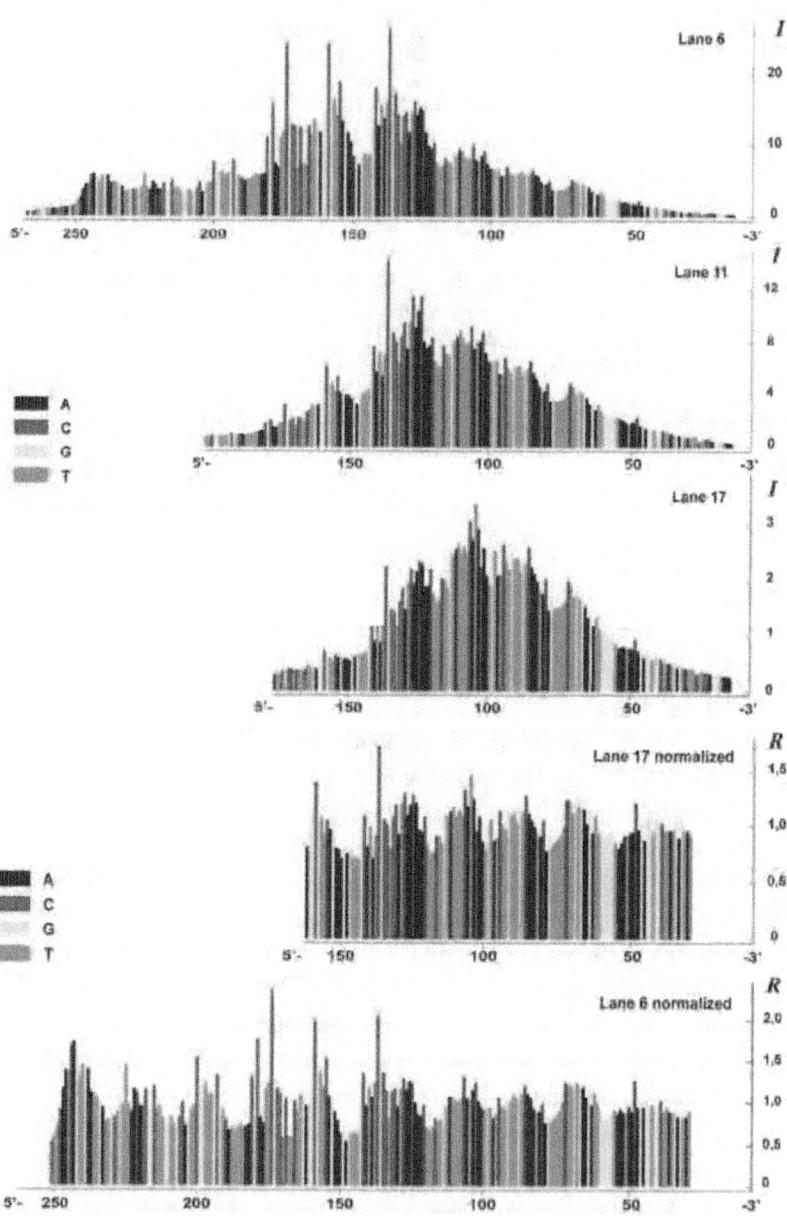

Figure. 5. Cleavage pattern of lanes obtained by computer digitization of the gel band densities. Histograms correspond to lanes in Fig. 4. The upper plot represents the profile of intensity of cleavage (I) for lane 6 and is followed by the same type of profiles built for lanes 11 and 17. The last two plots represent profiles of the relative intensity of cleavage (R) for lanes 17 and 6, obtained by the moving-average method, to demonstrate the normalization procedure.

We should note that single-molecule studies on double-stranded DNA mechanics have clearly demonstrated the existence of several stages upon DNA stretching, such as B/S-form transition and melting (Bustamante et al., 2000). Nevertheless, the timescale of these experiments is approximately seconds, whereas the impulsive stretching force in cavitation flow acts on timescales of several nanoseconds. Hence, we assume that drastic conformational changes in DNA, such as the B/S transition, do not occur in this case.

Results of the Gel Digitization

Fig. 5 presents the results of the gel digitization. Profiles of the intensity of cleavage (I) for several lanes of Fig. 4 are shown. Profiles of the relative intensity of cleavage (R) calculated for 2 lanes by normalization procedure (which eliminates the positional effect) are given below the intensity profiles. The values of the relative intensities of cleavage along with the corresponding local nucleotide sequences are used as the input data for a statistical analysis of sequence effects on ultrasonic cleavage of DNA.

We analyzed the cleavage patterns for 48 different radiolabeled DNA restriction fragments with lengths ranging from 100 to several thousand base pairs from l-phage DNA and plasmids pBR322, pUC18, and pGEM7(f+) (Promega), and their modified analogs, which contained different insertions of various eukaryotic functional genomic sequences (like promoters) into the polylinkers. For statistical analysis, we used the central parts of the gels where the bands were clearly separated. Because experiments with the same sequences showed some data scattering, these experiments were repeated two or three times. It should be noted that 1–2% of the greatest intensities, as well as 1–2% of the lowest intensities, were out of scope in the statistical treatment. The first group of excluded values (i.e., the greatest intensities) came from gel defects (an example of such a defect is clearly seen in the lowerleft corner of Fig. 3) or foreign fragments, whereas the second group (the lowest intensities) results from incorrect approximation of the overall band intensity value due to its curvature or overlap with neighboring band (such overlaps can also be seen in the upper part of the lanes shown in Fig. 3 and Fig.4). The nucleotide sequences and other supplementary materials are available at http://grok.imb. ac.ru/en/.

GEL DATA ANALYSIS

The first stage of gels analysis comprises computation of band intensities and their correlation with corresponding nucleotide sequence. The gel surface generally contains defects, strongly hindering data digitization and further processing. The nonhomogeneous thickness, air bubbles, and different amounts of salts in the deposited samples lead to the bending and deformation of the tracks and shift the bands on different tracks relative to one another. The labeled DNA fragment solution often has small contaminations of foreign fragments. Due to imperfect wells or well loading different lanes often have different total intensity of bands. Prolonged physical or chemical treatment of labeled fragments leads to the "double strike" effect, increasing the fraction of short labeled fragments in the reaction mixture. The mechanochemical origin of DNA cleavage leads to the damping of ultrasonic cleavage near the ends of the fragments. All these factors point to the difficulty of quantitative analysis of cleavage data for DNA with a definite nucleotide sequence. Therefore, it is important to seek adequate methods for analyzing the experimental data.

For primary PAGE data analysis we used the SAFA package (Das, 2005). This software was exploited to align the gel lanes, calculate the overall intensity of each band and correlate the band sequence with the corresponding nucleotide sequence. The calculation of overall band intensities produced by SAFA is based on several models (Shadle et al, 1997; Takamoto, et al., 2004) introduced earlier to account for such effects as band overlapping and asymmetric distribution of single band intensity along the gel which is better fit with Lorentz function rather than the Gaussian function. As far as band intensities are sensitive to various parameters of the experiment, it is important to calculate the normalized values of cleavage intensities. The strategy of normalization procedure depends on the observed properties of cleavage patterns: in the case of purely chemical cleavage – such as OH radical-induced cleavage – the positional preference of breakage along the molecule is attributed only to local sequence effects on DNA structure. In this case no general trends in cleavage patterns are observed provided that the cleavage is not efficient enough to produce "double strike" effect. In order to perform normalization of such cleavage pattern it is sufficient to validate the baseline of cleavage intensity which might be calculated by averaging the intensity values for a number of particular bands. In case of OHradical induced DNA cleavage analysis the baseline might be determined by averaging the cleavage intensity of the common sequences which flank the test DNA sequence (Greenbaum et al, 2007). Despite the success of this normalization method in the case of OH-radical induced cleavage, it is not suitable for the treatment of cleavage patterns possessing general trends of cleavage variation along the fragment.

Ultrasonic cleavage patterns demonstrate pronounced positional effect, i.e. the dependence of the band intensity value on its position in the gel. In order to analyze the sequencedependence of cleavage it is important to eliminate this general trend and operate with normalized values of cleavage intensities. This procedure might be performed using several approaches.

We have compared the efficiency of several methods listed below:

- no trend elimination and normalization of band intensity by dividing band intensity by the mean value of bands intensities calculated using all analyzed bands of the lane;

- using the moving average method with various window sizes;

- describing the positional effect in terms of asymmetric gauss functions;

- approximation of the trend with various degree polynomials (Nechipurenko et al, 2009).

As far as the basic goal of these approaches is retrieving the values of relative cleavage intensities for each band, the efficiency of each method might be characterized after calculation the mean value of cleavage for all bands which were analyzed. In the case of large sample size, comprising many gels and many various sequences, this value should be close to 1. Thus, the calculated mean value might be used to compare the methods listed above.

The comparison of these approaches performed during the analysis of DNA ultrasonic cleavage has shown that the best results are obtained with the moving average method.

Further we will focus on this method which shows good performance in case of low-term specificity observed for ultrasonic cleavage, hydroxyl radical cleavage, X-ray and laserinduced breakages of DNA. The intensities of cleavage in these experiments do not demonstrate such great variations which might be seen for DNAse I–induced cleavage, when the cleavage in particular sites is hundreds and even thousands times greater than in the other sites of DNA.

The absolute value of an individual band's intensity, or cleavage intensity, is further denoted as I, while the normalized value, i.e. the relative intensity of cleavage, is denoted as R. We normalized the band intensities by dividing their values by the local basic band intensity values, which were determined by using the moving-average method separately for each band. Thus, the array of R_n values was calculated using the formula:

$$R_n = (2m+1) \frac{I_n}{\sum\limits_{k=n-m}^{n+m} I_k},$$

where R_n denotes cleavage rate corresponding to band number n, and I_k denotes the intensity of band number k, while (2m+1) is the window size which is constant during analysis. The optimal number of adjacent bands used for the mean intensity calculation has been shown to be 31. Lowering this number results in an increased scattering of data points, whereas increasing the number of adjacent bands does not change the ratio of the obtained relative intensities of cleavage but does lead to a decrease in the number of analyzed data points.

STATISTICAL ANALYSIS OF THE SEQUENCE-DEPENDENT DNA ULTRASONIC CLEAVAGE

To study the relationship between the nucleotide sequence and the relative intensity of ultrasonic cleavage (R) of the central phosphodiester bond in all possible di- and tetranucleotides, we used analysis of variance, nonparametric methods (i.e., the KruskalWallis test and Brown-Mood test) and multiple-comparison methods (i.e., Tukey-Kramer test, and Dunn test) (Zar, 1999). We found that the nonparametric analysis yielded the same results as the parametric analysis.

The results of the statistical analysis of the 20,588 relative cleavage intensities for each of 16 dinucleotides are shown in Table 1 and on fig. 6. The sample mean values of the relative cleavage intensities (cleavage rates, $\bar{R}$), of the dinucleotides were found not to be significantly different from the corresponding values from our previous study (Grokhovsky et al., 2008) where the total length of the analyzed sequences was ~2500 nucleotides. The effect of the dinucleotide type on the cleavage rate was also shown to be statistically significant ($p \ll 0.05$).). The statistical results for relative intensities of ultrasonic cleavage (R) of the central phosphodiester bond in all 256 tetranucleotides are available at http://grok.imb.ac.ru/en/

In addition we found a significant difference between the cleavage rates at complementary dinucleotides. Therefore, cleavage of particular phosphodiester bond does not always result in cleavage of the opposite phosphodiester bond in the complementary strand. Importantly, we showed that the cleavage rates at dinucleotides d(CpC), d(CpT), d(CpA), and d(CpG) are significantly different from each other and all much higher than from the cleavage rates obtained for all other dinucleotides. The cleavage rate of dinucleotide d(CpG) was the highest among all 16 dinucleotides.

Table 1. Designations: N – the sample size; $\bar{R}$ - the sample mean; S - standard deviation; $S_{\bar{R}}$ - standard error of mean.

Type of di-nucleotide	N	$\bar{R}$	S	$S_{\bar{R}}$	The 95% confidence limits	
					Lower limit	Upper limit
AA	1636	0.919	0.129	0.003	0.913	0.926
AC	1076	0.913	0.128	0.004	0.905	0.920
AG	1028	0.900	0.124	0.004	0.892	0.907
AT	1374	0.904	0.119	0.003	0.898	0.910
CA	1265	1.160	0.209	0.006	1.149	1.172
CC	1141	1.007	0.144	0.004	0.999	1.015
CG	1230	1.444	0.334	0.010	1.426	1.463
CT	1077	1.130	0.198	0.006	1.118	1.142
GA	1153	0.970	0.133	0.004	0.962	0.978
GC	1317	0.954	0.146	0.004	0.947	0.962
GG	1168	0.922	0.145	0.004	0.914	0.931
GT	1101	0.952	0.126	0.004	0.944	0.959
TA	1065	0.973	0.120	0.004	0.966	0.980
TC	1173	0.912	0.131	0.004	0.904	0.919
TG	1305	0.979	0.126	0.003	0.972	0.986
TT	1672	0.932	0.127	0.003	0.926	0.938

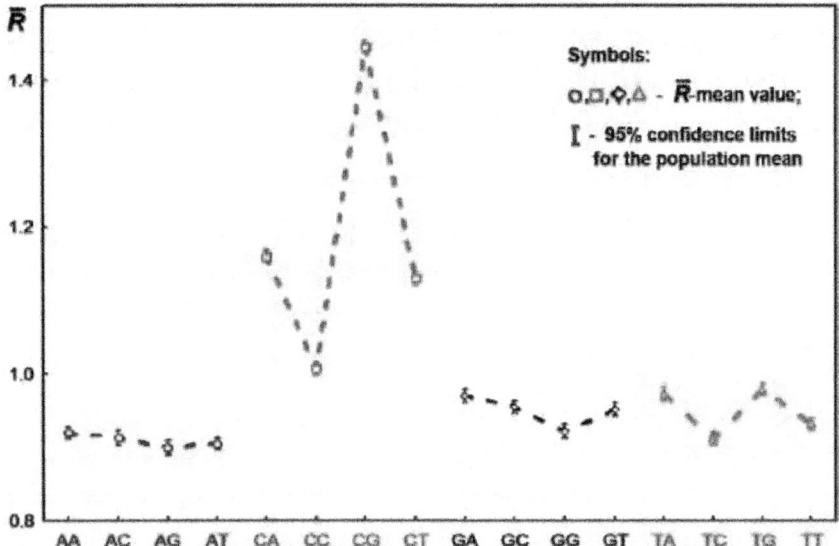

Figure. 6. The sample mean values of the relative intensity of cleavage for all di-nucleotides, and 95% confidence limits for the population mean.

As a result of our analysis we showed the significant dependence of nucleotide type on the ultrasonic cleavage rate at the 3'-position in all four groups of dinucleotides ($p \ll 0.05$). Moreover, the results of the statistical analysis of the relative cleavage intensities of the central phosphodiester bond in tetranucleotides showed the context dependence of cleavage rates in dinucleotides.

To estimate the contribution of the different analyzed DNA restriction fragments to the overall variability of ultrasonic cleavage rate we used two-level nested ANOVA. This study was performed for precisely resolved runs of bands in the cleavage patterns corresponding to cleavage of 140 DNA restriction fragments of known base pair sequences. The size of such runs varied from 100 to 250 base pairs. Two following factors were tested affecting the values of ultrasonic cleavage rate: the type of dinucleotide (the constant factor) and the type of fragment including the analyzed phosphodiester bond cleavage in the full length sequence of this fragment (the random factor).

The results of statistical analysis have led to the conclusion that the influence of both factors is statistically significant for dinucleotides of types CN, GN, TN (where N=A, C, G, T). Contribution of the random factor (DNA fragment) to the overall variability of cleavage rates is much smaller than the contribution of the constant factor, i.e. of the type of dinucleotide. For dinucleotides of type AN (N=A, C, G, T) only the random factor's influence was shown to be statistically significant.

Thus, sequence effects on conformational dynamics in any dinucleotide seem to propagate beyond mono and dinucleotide levels: further neighboring nucleotides might also influence the dynamics of sugar phosphate backbone.

ULTRASONIC CLEAVAGE OF NICKED DNA

Fig. 7 demonstrates the PAGE data obtained after ultrasonic irradiation of intact and nicked fragments of DNA. The initial length of fragments was 253 base pairs. Band numeration is given from 5'- to the 3'-end of the labeled strand. The positions of the nicks in complementary strand are given at the right side of the lane.

The results of analysis performed for over 20 cleavage patterns of nicked DNA made it possible to conclude that the intensity of ultrasonic cleavage near the nick is one order of magnitude higher than intensity of ultrasonic cleavage in the same sites of the intact dsDNA fragments.

If one chain of dsDNA is nicked the intensity of cleavage near the nicks is (in average) about 20 times higher than cleavage in the same sites of the intact dsDNA fragments (Fig. 7 a,b) (Il'icheva I. A. et al, 2009). At the same time, the cleavage rates in positions beyond the regions of the nick markedly grow weak even comparing to the sequence-specific cleavage of intact double-stranded DNA fragments (Fig. 7 c). Thus, the presence of the nick serves as an expressive structural indignation, which exceeds modulation of the structure caused by the base-pair sequence and is capable of absorbing mechanical stresses applied to the nearby sites of the molecule.

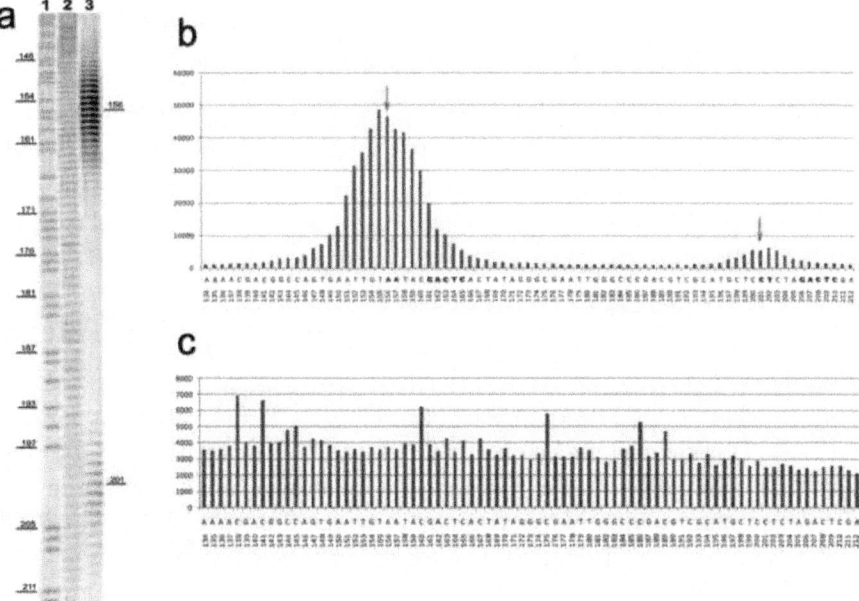

Figure. 7. Ultrasonic cleavage patterns of intact and nicked dsDNA a- Cleavage pattern of dsDNA: lane 1 corresponds to chemical cleavage of the dsDNA fragment by the purines; second lane represents ultrasonic cleavage pattern of primarily intact ds-DNA; lane 3 demonstrates cleavage pattern of twice nicked dsDNA. b- Band intensity data for third lane of the gel. The local maximums of cleavage intensities are disposed in front of the nicks. Both regions of cleavage enhancement spread about 10 b. p. around the nick and their amplitudes depend on the nick distance from the ends of dou-blestranded DNA fragment. c- Cleavage profile of the second lane corresponding to ultrasonic cleavage of intact dsDNA fragments without nicks shows sequence-specific cleavage of DNA.

ULTRASONIC FOOTPRINTING

Sequence-Specific Ligands Alter the Local Conformation of the DNA Double Helix

The structure of double-stranded DNA is not perfectly monotonous, but depends on the nucleotide sequence. The nucleotides differ in geometry, and their combinations show various deviations from the ideal helical structure: bends, turns, and changes in the widths of the minor and major grooves. Such features are of importance for DNA condensation and recognition by various proteins in the cell (Crothers, 1987; Tolstorukov et al., 2004). In addition, structural changes arise in DNA when the parameters of its aqueous environment are changed or various ligands are bound. A convenient model for studying the local parameters of the double helix is provided by low-molecular-weight sequence-specific ligands, which bind to particular DNA sequences (Gursky et al., 1983; Zimmer, Ch. & Wahnert, 1986; Bailly et al., 2005).

Pt-bis-Netropsin Changes the DNA Ultrasonic Cleavage Rate

The above data suggest that sonication can be used to probe the local conformation of the DNA double helix and for localization of different ligands on DNA. Previously we have shown that X-ray irradiation of restriction fragment complexes with a platinum(II)- containing ligand results in DNA cleavage at the location of the platinum atom (Grokhovsky & Zubarev, 1991). This effect is obviously due to the preferential adsorption of X-ray quantum by the atoms with large atomic weights, with subsequent emission of Auger electrons and generation of a multicharge positive ion. Pt-bis-netropsin have been used in those experiments (Grokhovsky et al., 1992) and its binding sites were localized on the DNA fragment with a known sequence. Fig. 8 demonstrates that the sites where the sugar– phosphate backbone was cleaved in both strands (long arrows) were detected in regions tightly bound with Pt-bis-netropsin and corresponded to the position of the platinum atom. Netropsin residues orient differently relative to the DNA helix and recognize two symmetrical consensus sequences, 5'-TTTT-3' (underlined). The orientation of the CO-NH groups of each residue of the netropsin within the complex coincides with the 5'-3'-direction of the AAAA tetranucleotide. (Grokhovsky et al., 1992). A scheme of the complex of Pt-bisnetropsin with DNA is shown on the right. However, it remained unclear why minor cleavage sites are detectable in a sequence of alternating AT pairs.

Also we have studied the DNA binding properties of a series of bis-linked netropsin and distamycin derivatives (the chemical structure of distamycin is very similar to the netropsin) in which two monomers were bridged by different dicarboxylic acid and peptide residues (Nikolaev et al., 1996; Surovaya et al., 1996, 2008). Using circular dichroism (CD) spectroscopy and DNase I footprinting studies it was found that bis-linked netropsin derivatives bind selectively to clusters of AT- base pairs and form several types of complexes with DNA. They exhibit strong preference for binding in the extended conformation to long clusters of AT-base pairs. In the complex, each bound bis-netropsin molecule covers approximately one turn of the DNA helix in such a way that both netropsin -like fragments are implicated in specific interaction with DNA base pairs. The observed preference of Pt-bridged bis-netropsin for binding to DNA regions with the sequence 5'-TTTTAAAA-3' and lower affinity to the site in which blocks of Ts and As are interchanged can be explained by the increased width of the minor groove in the DNA site with 5'-TpA -3' step.

In 1992 using NMR techniques Fagan & Wemmer have shown that the minor groove can accommodate not only a single distamycin molecule, but also side-by-side antiparallel binding of two distamycin molecules (Fagan & Wemmer, 1992). Further analysis of the binding of Pt-bis-netropsin with double-stranded oligonucleotides revealed complexes of sandwich type (Surovaya et al., 2001, 2002, 2008). In this case, two netropsin residues of a Ptbis-netropsin molecule are arranged as a parallel pin which forms a tight complex with a sequence of four alternating AT pairs (Fig. 8, left scheme).

Evidently, the increased width of the minor groove is needed for simultaneous accommodation of two netropsin-like fragments and cis-diamminoPt(II) group of the bisnetropsin molecule in the minor DNA groove. The cis-diammino-Pt(II) bridged bisnetropsin and bis-netropsins containing oligomethylene linkers can also bind in the parallelstranded hairpin form to shorter DNA regions with the sequence 5'-TATAT-3' . Molecular model building studies revealed that two parallel oligopyrrole carboxamide chains can be sandwiched in the minor DNA groove and form bifurcated hydrogen bonds with AT-base pairs. Parallel-stranded hairpin motifs extend the possible repertoire of hairpin polyamides that can be used for DNA sequence recognition and drug design. A head-to-tail bisnetropsin in which two monomers are bridged by a triglycine residue exhibits different DNA binding properties. In close similarity with Pt-bridged bis-netropsin it binds to a long AT-cluster in the extended conformation. However, it cannot form intramolecular parallel hairpin structure and binds to DNA in the form of dimer or hiher order associates stabilized by interaction between the halves of two bis-netropsin molecules bound at adjacent AT-rich sites on DNA (Grokhovsky, et al., 1998). Different complex geometries are characterized by distinctly different CD patterns and can be discriminated by CD spectroscopy. Footprinting and CD studies revealed that affinities and specificities shown by bis-netropsins to AT-rich regions on DNA depend on their nucleotide sequences, local DNA conformation and width of the DNA minor groove. Sonication can be used to localize the preferred positions of this netropsin derivatives on DNA.

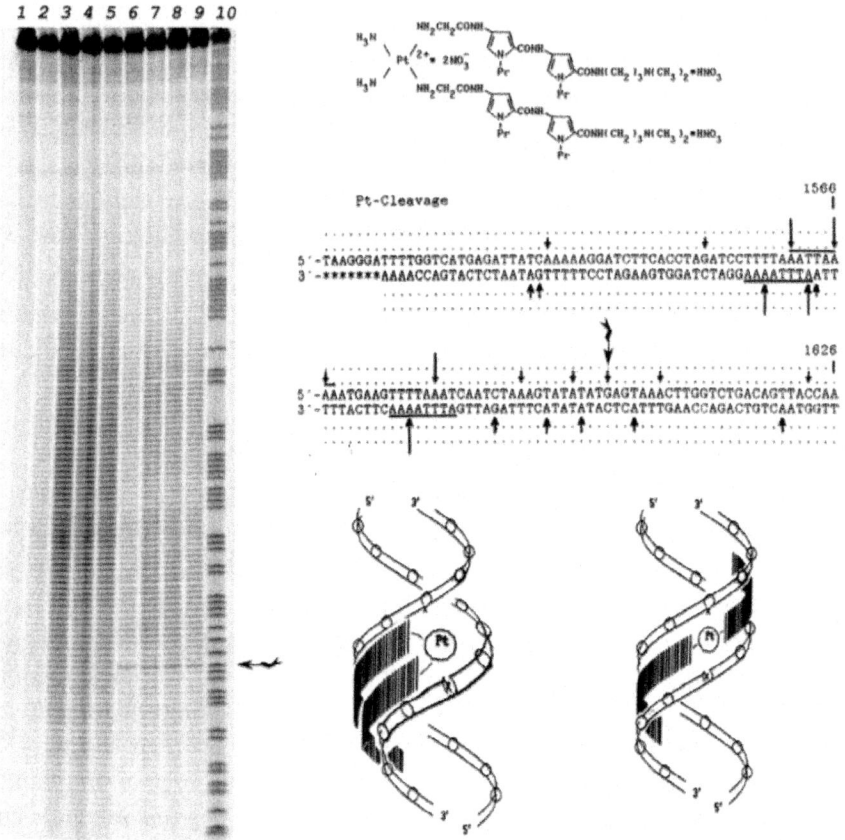

Figure. 8. Cleavage pattern of the 166-bp fragment in denaturing 6% gel after sonication at 44 kHz. The fragment was analyzed before (lane 1) and after (lanes 2–5) sonication for 2, 4, 8, and 16 min, respectively; ((lanes 6–9) - sonication in the presence of 0.5 μM Pt-bis-netropsin for 2, 4, 8, and 16 min, respectively; and ((lane 10) chemical cleavage at purines. Plain arrows indicate the sites of complex cleavage upon exposure to X-rays (Grokhovsky & Zubarev, 1991). A wavy arrow indicates the site where sonication-induced cleavage became more intense.

When a complex of Pt-bis-netropsin with DNA fragment was sonicated (Fig. 8), the average background cleavage was slightly decreased in sites where the ligand was bound to DNA in the extended conformation (Grokhovsky,

2006). A local increase in cleavage was observed at nucleotides adjacent to the sites where Pt-bis-netropsin pins were bound (wavy arrows). Similar regions where the cleavage rate decreased or increased depending on the conformation of Pt-bis-netropsin in complex with DNA sequences containing blocks of thymines or alternating adenines and thymines in one strand are detectable in Fig. 3 (lanes 5 – 7).

CONCLUSION

We have developed a new method for studying sequence-dependent structural dynamics of DNA fragments in solution. The approach is based on the analysis of ultrasound - induced DNA cleavage using high resolution denaturing polyacrylamide gel electrophoresis. Ultrasonic cleavage of DNA observed in our experiments represents mechanochemical reaction induced by cavitational processes in irradiated solution. It has been shown that the intensity of cleavage of sugar-phosphate backbone depends on the nucleotide sequence of irradiated fragments.

The computer methods for treatment and analysis of cleavage rate data have been developed along with several models for qualitative description of experimental effects. Based on recent data for structural and dynamical properties of DNA the interpretation of observed cleavage specificity has been offered.

As far as currently used methods for genome sequencing commonly use ultrasonic cleavage of the sample DNA and basically imply that this cleavage is non-specific, it is possible that the observed effect of sequence-dependence of DNA cleavage with ultrasound actually should be taken into account in order to avoid systematic errors during sequence assembly procedure.

Cleavage rates, i.e. the mean values of the relative intensities of cleavage of the central phosphodiester bond in all 16 dinucleotides and all 256 tetranucleotides, were determined by multivariate statistical analysis. We observed a remarkable enhancement of cleavage rates of phosphodiester bonds after deoxycytidine, which diminished in the following row of dinucleotides: d(CpG) > d(CpA) > d(CpT) >> d(CpC). The cleavage rates for all pairs of complementary dinucleotides were significantly different from each other. The effect of flanking nucleotides in tetranucleotides on cleavage rates of all 16 types of central dinucleotides was also statistically significant. The sequence-dependent ultrasonic cleavage rates of dinucleotides are consistent with reported data on the intensity of the conformational motion of their 5'-deoxyribose. The sequence specificity of ultrasonic cleavage is the result of sequence-dependent conformational dynamics, and is likely modulated by the intensity of the sugar ring $S \leftrightarrow N$ interconversion. Sequence effects on conformational dynamics in any dinucleotide seem to propagate beyond mono and dinucleotide levels. Local conformational motions in complementary strands are independent (Grokhovsky et al., 2011).

Hence, the relative intensity of ultrasound cleavage may serve as indicator of sequencespecific flexibility in both strands of DNA. Each complementary chain can be characterized independently by the cleavage rate, and the diversity of conformational dynamics in both complementary chains can be estimated. Such numerical evaluation may be useful for identifying promoter regions in the genome and assessing preferences for nucleosome positioning.

ACKNOWLEDGMENT

We thank Georgy Gursky and Victor Salyanov for their useful discussions about the paper. This work was supported by the Program of the Presidium of the Russian Academy of Sciences on Molecular and Cell Biology, and the Russian Foundation for Basic Research (projects 11-04-02001a and 12-04-01584a). 9.

REFERENCES

1. Bailly C., Kluza J., Martin C., Ellis T. & Waring M.J. (2005). DNase I footprinting of small molecule binding sites on DNA. Methods Mol. Biol., Vol. 288, pp. 319-342.

2. Belikov S., Wieslander L. (1995). Express protocol for generating G + A sequencing ladders. Nuclei Acids Res. Vol. 23, p. 310.

3. Belikov S. V., Grokhovsky S. L., Isaguliants M. G., Surovaya A. N. & Gursky G. V. (2005). Sequence-specific minor groove binding ligands as potential regulators of gene expression in Xenopus laevis oocytes. J. Biomol. Struct. Dyn., Vol. 23, pp. 193-202.

4. Basedow A. M. & E. B. Ebert. (1977). Ultrasonic degradation of polymers in solution. Advances in Polymers Science. A. Abe, A.-C. Albertsson & J. Genzer, editors. Springer, Berlin/Heidelberg., Vol. 22, pp. 83–148.

5. Brukner I., Sánchez R., Suck D. & Pongor S. (1995). Sequence-dependent bending propensity of DNA as revealed by DNase I: parameters for trinucleotides. EMBO J., Vol. 14, pp. 1812–1818.

6. Bullwinkle T.J. & Koudelka G.B. (2011). The lysis-lysogeny decision of bacteriophage 933W: a 933W repressor-mediated long-distance loop has no role in regulating 933W PRM activity. The Journal of Bacteriology, Vol. 193, pp. 3313-3323.

7. Bustamante C., Smith S.B., Liphardt J. & Smith D. (2000). Single-

molecule studies of DNA mechanics. Curr. Opin. Struct. Biol. Vol. 10, pp. 279–285.

8. Crothers D.M. (1987). Gel electrophoresis of protein–DNA complexes. Nature, Vol. 325, pp. 464 - 465.

9. Das R., Laederach A., Pearlman S. M., Herschlag D. & Altman R. B. (2005). SAFA: semiautomated footprinting analysis software for high-throughput quantification of nucleic acid footprinting experiments. RNA, Vol. 11, pp. 344–354.

10. Didenko Y.T., McNamara W.B. & Suslick K.S. (1999). Temperature of multibubble sonoluminescence in water. J. Phys. Chem. A. Vol. 103, pp. 10783–10788.

11. van Dyke M. W. & Dervan P. B. (1983). Methidiumpropyl-EDTA*Fe(II) and DNase I footprinting report different small molecule binding site sizes on DNA. Nucleic Acids Research, Vol. 11, pp. 5555-5567.

12. Fagan P. & D. E. Wemmer. (1992). Cooperative binding of distamycin-A to DNA in the 2:1 mode. J. Am. Chem. Soc. Vol. 114, pp. 1080-1081.

13. Greenbaum J.A., Pang B. & Tullius T.D. (2007). Construction of a genome-scale structural map at single-nucleotide resolution. Genome Res., Vol. 17, pp. 947–953.

14. Grokhovsky S.L. & Zubarev V.E. (1991). Sequence-specific cleavage of double-stranded DNA caused by X-ray ionization of the platinum atom in the Pt-bis-netropsin - DNA complex. Nucl. Acids Res., Vol. 19, pp. 257-264.

15. Grokhovsky S.L., Gottikh B.P. & Zhuze A.L. (1992). Ligands with affinity for certain DNA sequences: IX. Sythesis of netropsin and distamycin A analogs containing a sarcolysin residue or a platinum (II) atom. Bioorg. Khim. (Russ.), Vol. 18, pp. 570– 583.

16. Grokhovsky S. L., Nikolaev V. A., Zubarev V. E., Surovaya A. N., Zhuze A. L., Chernov B. K., Sidorova N. Yu., Zasedatelev A. S. & Gursky G. V. (1992) Specific DNA Cleavage by a Netropsin Analog Containing a Copper(II)-Chelating Peptide GlyGly-His. Molecular Biology (Russ). VoL 26, pp. 839-858.

17. Grokhovsky S. L., Surovaya A. N., Burckhardt G., Pismensky V. F., Chernov B. K., Zimmer Ch. & Gursky G. V. (1998). DNA sequence recognition by bis-linked netropsin and distamycin derivatives. FEBS Lett., Vol. 439, pp. 346–350.

18. Grokhovsky S. L. (2006). Specificity of DNA cleavage by ultrasound.

Molecular Biology (Russ), Vol. 40, pp. (276–283).

19. Grokhovsky S. L., Il'icheva I. A., Nechipurenko D. Yu., Panchenko L. A., Polozov R. V. & Nechipurenko Yu. D. (2008). Ultrasonic cleavage of DNA: quantitative analysis of sequence specificity. Biophysics (Russ.), Vol. 53, pp. 250–251.

20. Grokhovsky S. L., Il'icheva I. A., Nechipurenko D. Yu., Golovkin M. V., Panchenko L. A., Polozov R. V. & Nechipurenko Y. D. (2011). Sequence-specific ultrasonic cleavage of DNA. Biophysical J., Vol. 100, pp. 117-125.

21. Gursky G.V., Zasedatelev A.S., Zhuze A.L., Khorlin A.A., Grokhovsky S.L., Streltsov S.A., Surovaya A.N., Nikitin S.M., Krylov A.S., Retchinsky V.O., Mikhailov M.V., Beabealashvili R.S. & Gottich B.P. (1983). Synthetic sequence-specific ligands. Cold Spring Harbor Symp. Quant. Biol., Vol. 47, pp. 367–378.

22. Hogan M. E., Roberson M. W. & Austin R. H. (1989). DNA flexibility variation may dominate DNase I cleavage. Proc. Natl. Acad. Sci. USA, Vol. 86, pp. 9273-9277.

23. Il'icheva I. A., Nechipurenko D. Yu. & Grokhovsky S. L. (2009). Ultrasonic cleavage of nicked DNA. J. Biomol. Struct. Dyn., Vol. 27, pp. 391-398.

24. Kraev A.S. (1988). A simple system for phage M13 cloning and DNA sequencing with the use of terminators. Mol. Biol. (Russ.), Vol. 22, pp. 1164–1197.

25. Maniatis T., Fritsch E.F. & Sambrook J. (1982). Molecular Cloning: A Laboratory Manual. Cold Spring Harbor, N.Y.: Cold Spring Harbor Lab. Press.

26. Margulis M. A. (1984). Osnovi Zvukohimii. Himiya, Moscow. McNamara W.B., Didenko Y.T. & Suslick K.S. (1999). Sonoluminescence temperatures during multibubble cavitation. Nature 401, pp. 772–775.

27. Nechipurenko D. Yu., Golovkin M. V., Nechipurenko Yu. D., Il'icheva I. A., Panchenko L. A., Polozov R. V. & Grokhovsky S. L. (2009). Characteristics of ultrasonic cleavage of DNA. Journal of Structural Chemistry. (Russ.), Vol. 50, pp. 1007-1013.

28. Neidle S., Pearl L.H. & Skelly J.V. (1987). DNA structure and perturbation by drug binding. Biochem. J., Vol. 243, pp. 1-13.

29. Nikolaev V.A., Grokhovsky S.L., Surovaya A.N., Leinsoo T.A., Sidorova N.Yu., Zasedatelev A.S., Zhuze A.L., Strachan G.A., Shafer R.H. & Gursky G.V. (1996). Design of sequence-specific DNA-binding ligands

that use two-stranded peptide motif for DNA sequence recognition. J. Biomol. Struct. Dyn., Vol. 14, pp. 31-47.

30.   Parker S.C.J., Hansen L., Abaan H.O., Tullius T.D. & Margulies E.H. (2009). Local DNA topography correlates with functional noncoding regions of the human genome. Science, Vol. 324, pp. 389–392.

31.   Shadle S. E., Allen D. F., Guo H., Pogozelski W. K., Bashkin J. S. & Tullius T. D. (1997). Quantitative analysis of electrophoresis data: novel curve fitting methodology and its application to the determination of a protein-DNA binding constant. Nucleic Acids Res., Vol. 25, pp. 850–860.

32.   Spassky A. & Angelov D. (1997). Influence of the local helical conformation on the guanine modifications generated from one-electron DNA oxidation. Biochemistry, Vol. 36, pp. 6571–6576.

33.   Surovaya A.N., Burckhardt G., Grokhovsky S.L., Birch-Hirschfeld E., Gursky G.V. & Zimmer Ch. Hairpin polyamides that use parallel and antiparallel side-by-side peptide motifs in binding to DNA. J. Biomol. Struct. Dyn. (1997). Vol. 14, pp. 595- 606.

34.   Surovaya A.N., Burckhardt G., Grokhovsky S.L., Birch-Hirschfeld E., Nikitin A.M., Fritzsche H., Zimmer C. & Gursky G.V. (2001). Binding of bis-linked netropsin derivatives in the parallel-stranded hairpin form to DNA. J. Biomol. Struct. Dyn., Vol. 18, pp. 689- 701.

35.   Surovaya A.N., Grokhovsky S.L., Burkhardt H., Fritsche H., Zimmer K. & Gursky G.V. (2002). Effect of local DNA conformation in bis-netropsin binding to DNA. Mol. Biol. (Russ), Vol. 36, pp. 901–911.

36.   Surovaya A.N., Grokhovsky S.L., Bazhulina N.P. & Gursky G.V. (2008), DNA-Binding Activity of Bis-Netropsin Containing a cis-Diaminoplatinum Group between Two Netropsin Fragments. Biophysica (Russ.), Vol. 53, pp. 344-351.

37.   Suslick K.S. & Price G.J. (1999). Applications of ultrasound to materials chemistry. Annu. Rev. Mater. Sci. Vol. 29, pp. 295–326.

38.   Takamoto K., Chance M.R. & Brenowitz M. (2004). Semi-automated, single-band peakfitting analysis of hydroxyl radical nucleic acid footprint autoradiograms for the quantitative analysis of transitions. Nucleic Acids Res., Vol. 32, No. 15, e119.

39.   Tate W. P. & Petersen G. B. (1975). Stability of pyrimidine oligodeoxyribonucleotides released during degradation of deoxyribonucleic acid with formic acid — diphenylamine reagent. Biochem J., Vol. 147, pp. 439–445.

40.   Tolstorukov M. Y., Jernigan R. L. & Zhurkin V. B. (2004). Protein–DNA hydrophobic recognition in the minor groove is facilitated by sugar

switching. J. Mol. Biol., Vol. 337, pp. 65–76.

41. Thomas J.R. (1959). Sonic degradation of high polymers in solution. J. Phys. Chem. 63, pp. 1725-1729.

42. Travers A. A. (2004). The structural basis of DNA flexibility. Phil. Trans. R. Soc. Lond. A, Vol. 362, pp. 1423–1438.

43. Tullius T.D. (1988). DNA footprinting with hydroxyl radical. Nature 332: 6165, pp 663-664. Vtyurina N. N., Grokhovsky S. L., Filimonov I. V., Medvedkov O. I., Nechipurenko D. Yu., Vasiliev S. A. & Nechipurenko Yu. D. (2011). Cleavage of DNA fragments induced by UV nanosecond laser excitation at 193 nm. Biofizika (Russ.), Vol. 56, pp. 399–402.

44. Waring M. J. (2006) Sequence-Specific DNA Binding Agents (M.J. Waring, ed.), Royal Society of Chemistry, Biomolecular Science series. 258 pp.

45. Wells R.D. (2009). Discovery of the role of non-B DNA structures in mutagenesis and human genomic disorders. The Journal of Biological Chemistry, Vol. 284, pp. 8997-9009.

46. Zar J. H. (1999). Biostatistical Analysis. Prentice Hall, Upper Saddle River, NJ.

47. Zimmer Ch. & Wahnert U. (1986). Nonintercalating DNA-binding ligands: specificity of the interaction and their use as tools in biophysical, biochemical and biological investigations of the genetic material. Prog. Biophys. Mol. Biol. Vol. 47, pp. 31-112.

Chapter 2

WATER ELECTROLYSIS WITH INDUCTIVE VOLTAGE PULSES

Martins Vanags[1], Janis Kleperis [1] and Gunars Bajars[1]

[1] Institute of Solid State Physics, University of Latvia, Riga, Latvia

INTRODUCTION

The main idea of Hydrogen Economy is to create a bridge between the energy resources, energy producers and consumers. If hydrogen is produced from renewable energy sources (wind, solar, hydro, biomass, etc.), and used for energy production in the catalytic combustion process, then the energy life cycle does not pollute nature longer. With transition to Hydrogen Economy the Society will live accordingly to the sustainable development model, defined in the 1987 (Our Common Future, 1987).

Hydrogen is not available on Earth in free form; therefore the production process is representing a major part of the final price of hydrogen (Hydrogen Pathway, 2011). This is the main reason while research for effective electrolysis methods is very urgent. On our Planet the hydrogen is mainly located in compounds such as hydrocarbons, water, etc. and appropriate energy is needed to release hydrogen from them. In principle the amount of consumed energy is always greater than that which can be extracted from the hydrogen, and in the real operating conditions, the cycle efficiency does not exceed 50% (The Hydrogen Economy, 2004). The current problem is motivated to seek improvements to existing and discovering new technologies to produce hydrogen from the water – widely available and renewable resource on the Earth. Water electrolysis is known more than 130 years already, and different technologies are developed giving power consumption around 3.6 kWh/m³ - high temperature electrolysis, and 4.1 kWh/m³ - room temperature alkaline electrolysers and proton exchange membrane electrolysers (The Hydrogen Economy, 2004). Lower hydrogen production costs is for technologies using closed thermo-chemical cycles, but only in places where huge amount of waste heat is available (for example, nuclear power stations (The Hydrogen Economy, 2004). Nevertheless what will be the hydrogen price today, in future only hydrogen obtained from renewable resources using electricity from renewable energy sources will save the World, as it was stated in 2nd World's Hydrogen Congress in Turkey (Selected Articles, 2009). For Latvia the hydrogen obtained in electrolysis using electricity from renewables (wind, Sun, water) also would be the best solution to move to Hydrogen Economics (Dimants et all, 2011). That is because all renewables available in Latvia's geographical situation are giving non- stable and interrupt power, for which the storage solutions are necessary. Usage of hydrogen as energy carrier to be produced from electricity generated by renewables, stored and after used in fuel cell stack to generate electricity is the best solution (Zoulias, 2002). Efficient and stable electrolysers are required for such purposes. Smaller electrolysis units are necessary also for technical solutions were hydrogen is produced and used directly on demand, for example, hydrogen welding devices, hydrogen powered internal combustion cars (Kreuter and Hofmann, 1998).

DC power typically is used for electrolysis, nevertheless pulse powering also is proposed (see, for example, Gutmman and Murphy, 1983). Using a mechanically interrupted DC power supply (Brockris and Potter, 1952; Bockris et all, 1957) next interesting phenomena was noticed: immediately upon application of voltage to an electrochemical system, a high but short-lived current spike was observed. When the applied voltage was disconnect, significant current continues to flow for a short time. In 1984 Ghoroghchian and Bockris designed a homopolar generator to drive an electrolyser on pulsed DC voltage. They concluded that the rate of hydrogen production would be nearly twice as much as the rate for DC.

The Latvian Hydrogen Research Team is developing inductive pulse power circuits for water electrolysis cell (Vanags et all, 2009, 2011a, 2011b). The studies revealed a few significant differences compared to conventional DC electrolysis of water. New model is established and described, as well as the hypothesis is set that water molecule can split into hydrogen and oxygen on a single electrode (Vanags et all, 2011a). There has been found and explained the principle of high efficiency electrolysis. A new type of power supply scheme based on inductive voltage pulse generator is designed for water electrolysis. Gases released in electrolysis process from electrodes for the first time are analyzed quantitatively and qualitatively using microsensors (dissolved gases in electrolyte solution nearby electrode) and masspectrometer (in atmosphere evolved gases). The hypothesis of hydrogen and oxygen evolution on a cathode during the process of pulse electrolysis is original, as well as interpretation of the process with relaxation mechanisms of electrons emitted by cathode and solvated in electrolyte (Vanags, 2011b).

LITERATURE REVIEW

A Brief History of the Electrolysis of Water

Adriaan Paets van Troostwijk, 1752.–1837., and Johan Rudolph Deiman, 1743.–1808., while using Leyden jar and a powerful electrostatic generator noticed a gas evolution on the electrodes of water electrolysis cell as a result of spark overjumping in the electrostatic generator. The evolved gases displaced water out of the Leyden jar during the experiment and spark jumped into the collected gas mixture creating an explosion. The researchers decided that they have decomposed water into hydrogen and oxygen in a stoichiometric proportion 2:1; they published the results in 1789, which is considered to be the year of discovering the water electrolysis (Zoulias et all, 2002; De Levie, 1999). In more that hundred years, in 1902 there were more than 400 industrial electrolysers used all over the world, but in 1939 the first large water electrolysis plant was commissioned with the hydrogen production capacity of 10000 Nm³/h (Zoulias et all, 2002). The high-pressure electrolysers were produced in 1948 for the first time; in 1966 General Electric built the first electrolysis system with solid electrolyte and in 1972 the first solid oxide high-temperature electrolyser was built. However the development of electrolysis devices is in progress nowadays as well along with the development of proton exchange membrane, which can be used in the water electrolysers and fuel cells, along with the development of high-temperature solid oxide electrolysers likewise the optimization of alkaline electrolysers (Kreuter W, and Hofmann H (1998).

Direct Current Water Electrolysis

When dissolving acid in the water (e.g. sulfuric acid), molecules of water and acid dissociate into ions. The same happens if alkali (e.g. KOH) is dissolved in the water, the solution dissociates into ions, creating ionic conductor or electrolyte. There has been formed an ionic conductor, where a direct current will be passed through. The processes taking place on electrodes are in the case of sulfuric acid - the positive hydroxonium ions H_3O^+ (cation) move to the side of negative electrode. When cations reach the electrode, they receive missing electrons (Zoulias et all, 2002):

$$2H_3O^+ + 2e \Rightarrow H_2 + 2H_2O \tag{1}$$

Hydrogen is produced as gas from the medium, in its turn, water dissociates into ions again. The reaction on an anode or positive electrode in an alkaline medium is (Zoulias et all, 2002):

$$2OH^- - 2e \Rightarrow \frac{1}{2}O_2 + H_2O \tag{2}$$

Oxygen evolves as gas, but water dissociates into ions again. There are produced three parts of volume of gaseous substance in the process described - two parts of hydrogen and one - oxygen. In the case of alkaline electrolyte there are polarized water molecules, which have their hydrogen atoms oriented toward an electrode, near the cathode and dissociation reaction takes place:

$$2H_2O + e \Rightarrow 2OH^- + H_2 \qquad (3)$$

The first law of thermodynamics for the open system states that:

$$Q - W_s = \Delta H \qquad (4)$$

where Q is the amount of heat supplied to the system, W_s the amount of appropriate work performed by the system and ΔH is the change in system's enthalpy. The work done is electricity used up in electrolyser, therefore W_s is:

$$W_s = -nFE \qquad (5)$$

where:

n - The amount of transferred electrons;

F – Faraday constant: = 23,074 cal/volts g-equivalent;

E - Electric potential of the cell in volts.

Using equation (5), transform the expression (4), resulting:

$$E = \frac{\Delta H - Q}{nF} \qquad (6)$$

In an isothermal reversible process (without loss) Q is:

$$Q_{atg} = T\Delta S \qquad (7)$$

where T is absolute temperature and ΔS is system entropy change. From (6) and (7) the value of reversible reaction potential is obtained, where it is impossible to decompose water into hydrogen and oxygen in real time:

$$E_{rev} = \frac{\Delta H - T\Delta S}{nF} \qquad (8)$$

$(\Delta H - \Delta S)$ is the change in Gibbs free energy ΔG. At normal temperature and pressure ((25 °C temperature and 1 atm pressure) ΔH equals 68 320 cal/gmol and ΔG equals 56 690 cal/gmol. Therefore the reversible potential of the cell is:

$$E_{rev} = \frac{\Delta G}{nF} = \frac{56,690}{2(23,074)} = 1,23V \qquad (9)$$

Potential where Q equals zero and supplied energy transforms into chemical energy, is called thermo-neutral voltage (Oldham and Myland, 1993; Bockris and Potter, 1952):

$$E_{thermo} = \frac{\Delta H}{nF} = 1,48V$$

(10)

The voltage to split water in practical electrolysis devices is higher than thermo-neutral cell voltage due transformation into heat, which heats up the cell. Therefore industrial electrolyser requires additional cooling and the value of DC voltage is defined:

$$E = E_{rev} + loss$$

(11)

where the

$$loss = E_{anode} + E_{cathode} + E_{mt} + IR$$

(12)

In equation (12) E_{anode} – activation overvoltage of the anode; $E_{cathode}$ – activation overvoltage of the cathode; E_{mt} – overvoltage of the mass transfer and IR – ohmic overvoltage (includes resistance in an electrolyte, on electrodes, leads). Current density must be higher than 100 mA/cm² in industrial electrolysers, therefore voltage applied to individual cell partly transforms into the heat, becoming typical loss in DC water electrolysis.

It is possible to write an expression for the efficiency factor of the water electrolysis, calculated versus the thermo-neutral voltage, using relations above (Bockris and Potter, (1952):

$$\eta = \frac{\Delta H}{\Delta G + loss} = \frac{E_{thermo}}{E}$$

(13)

When ΔG is negative, the reactions are spontaneous and work has been done by releasing the energy. When ΔG is positive, for reaction to happen external work must be used. As for ensuring the reaction, work must be done, water electrolysis cell does not operate spontaneously. The reaction in fuel cells is spontaneous because of the catalyst and during reaction the energy is released (Salem, 2004).

Hydrogen Evolution Reaction (HER) is one of the most widely researched reactions in the electrochemistry. The studies of HER are carried out in different kind of systems and following each other processes is divided (Salem, 2004; Heyrovsky, 2006; Murphy, 1983; Bockris, 1957; El-Meligi, 2009; Sasaki and Matsuda, 1981; Noel and Vasu, 1990; Kristalik, 1965): Volmer electrochemical discharge step, Heyrovsky electrochemical desorption step, Tafel catalytic recombination step. Each of steps can be a reaction limiting step in a certain system during the whole reaction. This means, that each step can have different reaction rate, and the slowest step will determine the speed of reaction. Charge transfer may begin when the reagent is next to the electrode. Two most typical steps are charge transfer ending with the adsorption of hydrogen atom, and recombination of the adsorbed atoms with next desorption of H_2 molecule.

The general equation of the electrochemical reaction links current with potential (Noel and Vasu, 1990):

$$i_c = \vec{i} - \overleftarrow{i} = i_0 \left[\exp\left(-\frac{\beta\lambda\eta F}{RT} \right) - \exp\left(\frac{(1-\beta)\lambda\eta F}{RT} \right) \right]$$

(14)

β is symmetry factor (0, ½, 1 for process without activation, normal process and barrier-free process accordingly).

Interface Between Electrode And Electrolyte: Double Layer

When two equal electrodes (conductors) are immersed in an electrolyte, initially there is no measurable voltage between them. But when the current is caused to flow from one rod to the other by a battery, charge separation is naturally created at each liquid/solid interface and two electrochemical capacitors connected in series are created. Typical capacitors store electrical charge physically, with no chemical or phase changes taking place, and this process is highly reversible; the discharge-charge cycle can be repeated over and over again, virtually without limit. In electrochemical capacitor at electrode/electrolyte interface solvated ions in the electrolyte are attracted to the electrode surface by an equal but opposite charge in it. These two parallel regions of charges at interface form the "double layer" where charge separation is measured in molecular dimensions (i.e., few angstroms), and the surface area is measured in thousands of square meters per gram of electrode material (Miller and Simon, 2008).

Double-layer phenomena and electro-kinetic processes are the main elements of electrochemistry. It is considered that the behavior of the interface is and should be described in terms of a capacitor. It is a consequence of the "free charge" approach that in the case of a continuous current flow through the interface a strict distinction should be made between the so-called non-Faraday and Faraday currents. The former is responsible for charging of the double-layer capacitor, while the latter is the charge flow connected with the charge transfer processes occurring at the interface (Horányi, Láng, 2006). The state of an interface at constant pressure and temperature can be changed by changing the concentration of the components in the bulk phases, and by constructing an electrical circuit with the aid of a counter electrode and forcing an electric current through the circuit, which can be expressed as:

$$i = i' + i'',$$ (15)

Were i` is charging current of double layer, and i`` – charge transfer or Faraday current. Double layer charging current can be viewed as an ideal capacitor charging current equal to

$C\frac{\partial \eta}{\partial t}$, were η is overvoltage and C – double layer capacitance. Rewriting equation (15), next current equation is obtained:

$$i = C\frac{\partial \eta}{\partial t} + i''$$ (16)

which has caused much debates. Paul Delahais wrote in 1966 that this equation permits the decoupling of the non-Faraday from Faraday processes, but at the same time concludes that Faraday charge transfer and charging processes cannot be separated a priori in non-steady-state electrode processes because of the phenomenon of charge separation or recombination at the electrode-electrolyte interface without flow of external current. Charging behaviors as ideal polarized or reversible electrode represent only two limiting cases of a more general case (Delahay, 1966). Nisancioglu and Newman (2012) in their article even without going into the assumptions and basing only on the mass balance equation, obtained next current equation: $i=dq/dt + i''$ and showed that a priori separation of double-layer charging and Faraday processes in electrode reactions is the component mass balance for the electrode surface. Equation (1) is valid if the rate of change of concentration of the species, which take part in the electrode reaction, can be neglected at the electrode surface.

Water Splitting with the Pulse Electrolysis

There are different ways of water splitting first reviewed by Bockris et all (1985), that sharply differs from conventional DC water electrolysis. The most common could be: thermo chemical, sonochemical, photocatalytic, biological water splitting; water splitting under the magnetic field and centrifugal force of rotation; pulse electrolysis and plasma electrolysis. Regarding pulse electrolysis, Ghoroghchian and Bockris in 1956 already defined that the pulse electrolysis is more effective than conventional electrolysis. Many new patents on pulse electrolysis appeared in 1970-1990 (Horvath, 1976; Spirig, 1978; Themu, 1980; Puharich, 1983; Meyer, 1986; Meyer, 1989; Meyer, 1992a, 1992b; Santilli, 2001;Chambers, 2002) stating to be invented over-effective electrolysis (i.e. the current efficiency is higher than 100%). The water splitting scheme described in these patents initiated a huge interest, but nobody has succeeded in interpreting these schemes and their performance mechanisms up to now, and what is more important, nobody has succeeded to repeated patented devices experimentally as well.

In interrupted DC electrolysis the diffusion layer at the electrode can be divided into two parts: one part, which is located at the electrode surfaceis characterized with pulsed concentration of active ions, and another part is fixed, similar to the diffusion layer in case of DC. The concentration of the active ions in pulsing diffusion layer changes from defined initial value when the pulse is imposed, to a next value when it expired. The concentration of active ions in pulse may fall or cannot fall to zero. Time, which is necessary for active ion concentration would fall to zero, is called the transition time τ. The transition time is depending from pulse current ip and pulse duration T. If depletion in stationary diffusion layer is small, i.e. $c'e{\approx}c0$, were $c'e$ is concentration of ions in pulsed-layer outer edge, and $c0$ is bulk concentration, the transition time can be found from Sand equation (Bott, 2000):

$$\tau = \frac{\pi D c_0^2 (zF)^2}{4 i_p^2}$$

(17)

were F is Faraday's constant, z – charge number and D is its diffusion coefficient. As can be seen, transition time τ depends on the ion concentration in bulk volume $c0$ and pulse current density ip. Thickness of pulsed layer δ_p at the end of the pulse depends only from the density of pulse current:

$$\delta_p = 2 \left(\frac{DT}{\pi} \right)^{1/2}$$

(18)

With very short pulses extremely thin pulsating layer can be reached. This thin layer would allow temporary to impose a very high current densities during metal plating (more than $250 A/cm^2$, which is 10,000 times higher than currents in conventional electrolysis), which accelerates the process of metal electroplating (Ibl et all, 1978). The rough and porous surface is formed during metal plating with direct current, when the current value reached the mass transport limit. When plating is done with pulse current, pulsating diffusion layer always will be much thinner than the surface roughness, what means that in case the mass transfer limit is reached, the plated surface is homogeneous still and copy the roughness of substrate. This feature gives preference to pulse current in metal plating processes, comparing with conventional DC plating, because the highest possible power can used (current above mass transfer limit) to obtain homogeneous coatings in shortest times (Ibl et all, 1978).

Pulse electrolysis is widely investigated using various technologies (Hirato et all, 2003; Kuroda et all, 2007; Chandrasekar et all, 2008). In all of these technologies rectangular pulses are mostly used which have to be active in nature. Shimizu et all (2006) applied inductive voltage pulses to water electrolysis and showed significant differences with conventional DC electrolysis of water. The conclusion of this research is that this kind of water electrolysis efficiency is not dependant on the electrolysis power, thus being in contradiction to the conventional opinion of electrolysis.

We studied inductive voltage electrolysis and promoted the hypothesis that pulse process separates the cell geometric capacitance and double layer charging current from the electrochemical reaction current with charge transfer (Vanags (2009), Vanags (2011a, 2011b). To prove this we have done plenty of experiments proving double layer charging process separation from the electrochemical water splitting reaction. There are no studies about the usage of reactive short voltage pulse in the aqueous solution electrolysis; also no microelectrodes are used to determine the presence of the dissolved hydrogen and oxygen near the cathode in electrolysis process.

EXPERIMENTAL

Materials and Equipment

Materials, instruments and equipment used in this work are collected in Tables 1 and 2.

Inductive Reverse Voltage Pulse Generator

The inductive voltage pulses were generated in the electric circuit (Fig. 1) consisting of a pulse generator, a DC power source, a field transistor BUZ350, and a blocking diode (Shaaban, 1994; Smimizu et all, 2006). A special broad-band transformer was bifilarly wound using two wires twisted together. Square pulses from the generator were applied to the field transistor connected in series with the DC power source. The filling factor of pulses was kept constant (50%). To obtain inductive reverse voltage pulses, the primary winding of the transformer is powered with low amplitude square voltage pulses. In the secondary winding (winding ratio 1:1) due to collapse of the magnetic field induced in the coil very sharp inductive pulse with high amplitude and opposite polarity with respect to applied voltage appears. Pulse of induced reverse voltage is passed through the blocking diode, and the resulting ~1 µs wide high-voltage pulse is applied to the electrolytic cell. A two-beam oscilloscope GWinstek GDS-2204 was employed to record the voltage (i.e. its drop on a reference resistance) and current in the circuit.

MOSFET (IRF840) is used as semiconductor switch between DC power supply and ground circuit. Pulse transformer is a solenoid type with bifilar windings; length is 20 cm and a coil diameter of 2.3 cm and ferrite rod core. Number of turns in both the primary and the secondary winding is 75, so ratio is 1:1. Inductance of solenoid is approximately 250 μH. Super-fast blocking diode with the closing time of 10ns is included in the secondary circuit, to pass on electrolysis cell only the pulses induced in transformer with opposite polarity. Direct pulses are blocked by diode.

Table 1. Materials and equipment used in this work

	No	Name	Parameters	Producer
Chemicals	1.	KOH	99%	Aldrich
	2.	NaOH	99%	Aldrich
	3.	LiOH	99.9%	Aldrich
	4.	K_2C_O3	99.8%	Aldrich
	5.	H_2S_O4	95%	Aldrich
	6.	$(NH_2)_2CO$	98%	Aldrich
	7.	H_2O	0.1 μS	Deionised
Metals	1.	Stainless Steel (parameters Table 2)		316L
	2.	Tungsten	95%	Aldrich
	3.	Platinum	99.9%	Aldrich
Equipments and Instruments	1.	DC power supply Agilent N5751A	300V; 2.5A	Aligent echnologies
	2.	Frequency Generator GFG-3015	0 – 150 MHz	GW-Instek
	3.	Oscilloscope GDS-2204	4 beams, resolution 10 ns	GW-Instek
	4.	Power Meter HM8115-2	16A, 300V	Hameg Instruments
	5.	Water Deionization Crystal – 5	Water - 0.1 μS	Adrona Lab. Systems
	6.	Masspectrometer RGAPro 100	0 – 100 m/z units	Hy-Energy
	7.	X-ray fluorescence spectrometer EDAX/ Ametek, Eagle III		Ametek
	8.	Microsensors for dissolved gases H_2 and $_O2$	Resolution 0.1 μmol/l	Unisense, Denmark

Table 2. Composition of Stainless steel 316L used for electrodes (wt%)

Element	C	Si	P	S	Ti	Cr	Mn	Fe	Ni	Cu
Quantity, wt%	0.12	0.83	0.04	0.02	0.67	17.88	2.02	68.36	9.77	0.29

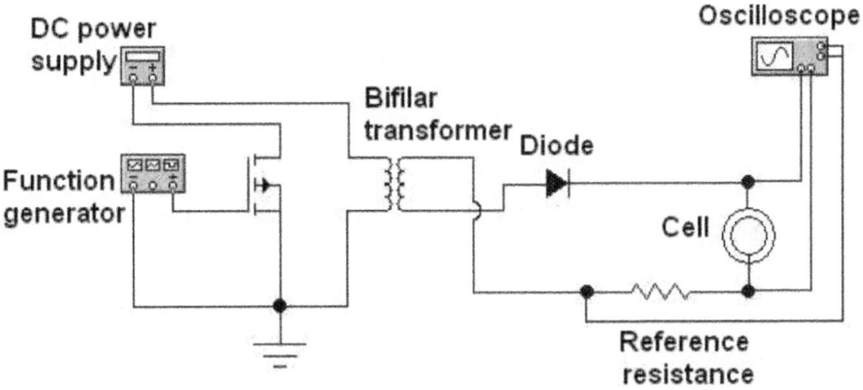

Figure 1. Experimental circuit for generation of inductive reverse voltage pulses.

Construction of Electrolyses Cells

Experiments in this chapter are divided into five parts. In the first part the gas evolution rate is explained and performance efficiency coefficients defined (current efficiency and energy efficiency). The second part examines kinetics of the inductive voltage pulse applied to electrolysis cell were electrolyte concentration and the distance between electrodes are changing. The third part describes the application of respiration microsensors to measure concentration of dissolved hydrogen gas directly to the cathode surface in an electrolysis cell, powered with inductive voltage pulses. The fourth experiment studied inductive voltage pulse kinetics in very dilute electrolyte solutions. The fourth experiment also noticed interesting feature in current pulse kinetics, therefore additional experiment is performed, to measure concentration of evolved hydrogen at the cathode with oxygen microsensor. This experiment devoted to fifth.

Amount of released gases during electrolysis was determined with volume displacement method (Fig. 2).

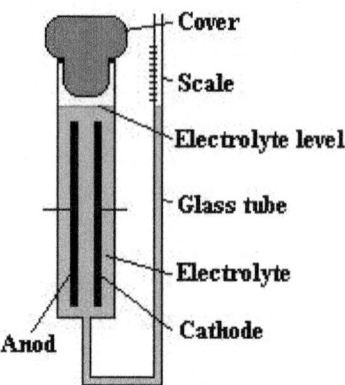

Figure 2. Principal scheme to determine the volume of released gases.

Electrolysis cell is in a separate chamber closed with a sealing cap. Glass tube bent in 180 degrees is attached to the bottom of the electrolysis chamber. The tube is graduated in units of volume above the level of the electrolyte. Gases arising in electrolysis process are pressing on electrolyte and the level in calibrated tube is increasing giving approximate volume of gases produced. In measured value of volume the 5% relative error is from different reasons; the biggest uncertainty is determined by the pressured gas generated during electrolysis - higher than atmospheric pressure. Gases are produced in electrolysis by volume 2/3 hydrogen and 1/3 oxygen. Knowing the mass of hydrogen generated in period *texp*, the charge necessary to produce such amount can be calculated and compared with consumed energy – result is current efficiency of particular electrolysis cell. Energy efficiency is calculated from consumed energy compared to what can be obtained from burning the produced amount of hydrogen at highest calorific value - 140 MJ/kg.

Self-made water electrolysis cell with movable electrode was used in experiments. It consists of a polyethylene shell with built in micro-screw from one side. Using stainless steel wire the micro-screw is connected to the movable electrode, situated perpendicular to the electrolyte cavity (diameter 40mm). Stainless steel stationary electrode with same area is situated against a moving electrode. SUS316L stainless steel plate electrodes with equal area (2 cm^2) were used in experiments. Before experiments the electrodes were mechanically polished and washed with acetone and deionized water. As an electrolyte KOH solution in water was used in different concentrations. At each electrolyte concentration the distance between electrodes was changed with micro-screw from 1mm to 5mm. During experiment the appropriate concentration of the electrolyte solution was filled and cell attached to an inductive voltage pulse generator. At each electrolyte concentration the current and voltage oscillograms were taken for 1 to 5 mm distance between electrodes (step 1 mm). Oscillograms were further analyzed calculating consumed charge, pulse energy, and in some cases - energetic factors.

To measure the concentration of dissolved hydrogen at the cathode during electrolysis, self-made cell was used (Fig. 3). Cell consists from three cameras connected with ion conducting bridges.

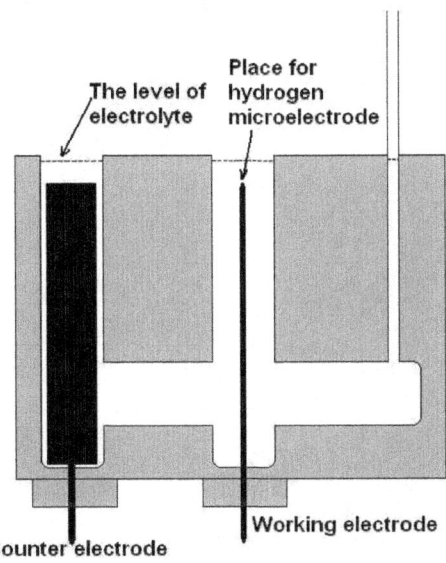

Figure 3. Three-camera electrolysis cell to measure concentration of dissolved hydrogen.

The first camera is for nickel plate counter electrode, second – for working electrode - smooth wires (diameter 0.5 mm, length 100 mm) of tungsten and platinum, but third camera was used for reference electrode in some specific experiments. Pt and W electrodes were cleaned before experiments, etching them 24 hours in concentrated alkali solution and rinsing with deionized water. The concentration of dissolved hydrogen was determined with respiration microsensor used typically in biological experiments (Unisense, 2011). The Unisense hydrogen microsensor is a miniaturized Clark-type hydrogen sensor with an internal reference electrode and a sensing anode. The sensor must be connected to a high-sensitivity picoammeter where the anode is polarized against the internal reference. Driven by the external partial pressure, hydrogen from the environment will pass through the sensor tip membrane and will be oxidized at the platinum anode surface. The picoammeter converts the resulting oxidation current to a signal. In our experiments sensor H2100 having the tip with diameter 110 μm was placed as closely as possible to cathode (<1 mm distance). Before experiments the microsensor was graduated in two points – zero H_2 concentration (Ar gas is bubbled through deionized water) and 100% or 816 mmol/l at 20 °C (H_2 gas is bubbled through deionized water – from Unisense, 2011 user manual). The experiment was carried out as follows: separate inductive voltage pulses was delivered to the cell and voltage and current oscillograms recorded. At the same time the concentration of the dissolved hydrogen was measured using microsensor.

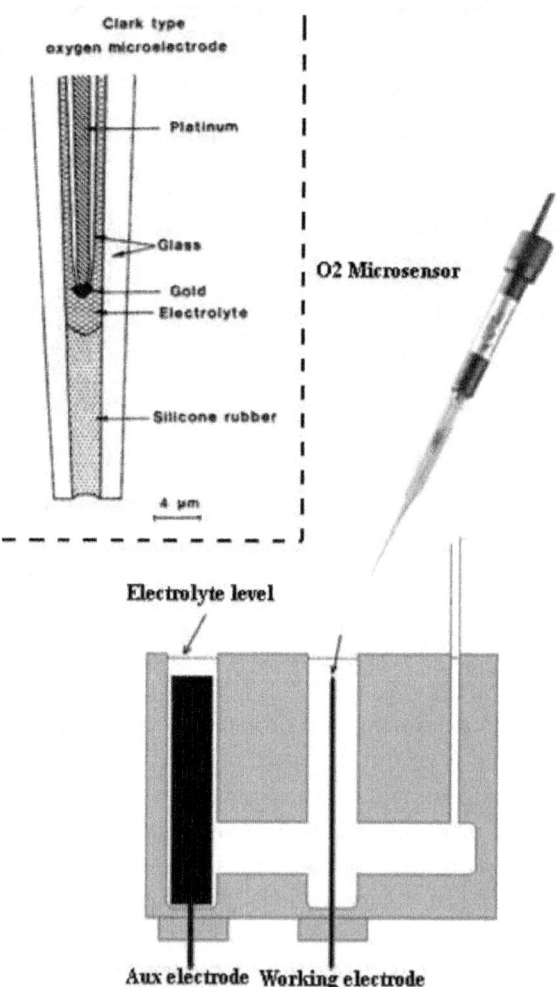

Figure 4. Water electrolysis cell to measure the concentration of dissolved oxygen.

The oxygen microsensor (also from Unisence,2011) was used to measure the concentration of dissolved hydrogen close to cathode during inductive pulse electrolysis (Figure 4). The oxygen micro-sensoris all Clark-type sensorbased on diffusion of oxygen through a silicone membrane to an oxygen reducing cathode which is polarized versus an internal Ag/AgCl anode. The flow of electrons from the anode to the oxygen reducing cathode reflects linearly the oxygen partial pressure around the sensor tip (diameter 100 μ) and is in the pA range. The current is measured by a picoammeter. To generate short inductive voltage pulses, the same circuit (chapter 3.2) is used. Cell is filled with deionized water, and the generator set in mode when expressed negative current peak is observed in oscillograms.

Concentration of oxygen is measured simultaneously with microsensor, previously calibrated in deionized water bubbled with oxygen.

Specific electrolysis cell was made for study the kinetics of inductive pulse electrolysis in diluted electrolytes (Figure 5). It was made from glass bowl with two separate electrode holders equipped with screws for electrodes from stainless steel 316L wires (diameter 2 mm, length 100 mm). Steel electrodes were cleaned before experiments, etching them 24 hours in concentrated alkali solution and rinsing with deionized water together with glass bowl of electrolysis cell. Very diluted electrolyte was prepared pouring in the cell 350 ml deionized water and adding drops of 5 M electrolyte from calibrated volume dropper (0.05±10% ml). Four electrolytes (KOH, NaOH, LiOH, H_2SO_4) were used in experiments and measurements were registered after each drop.

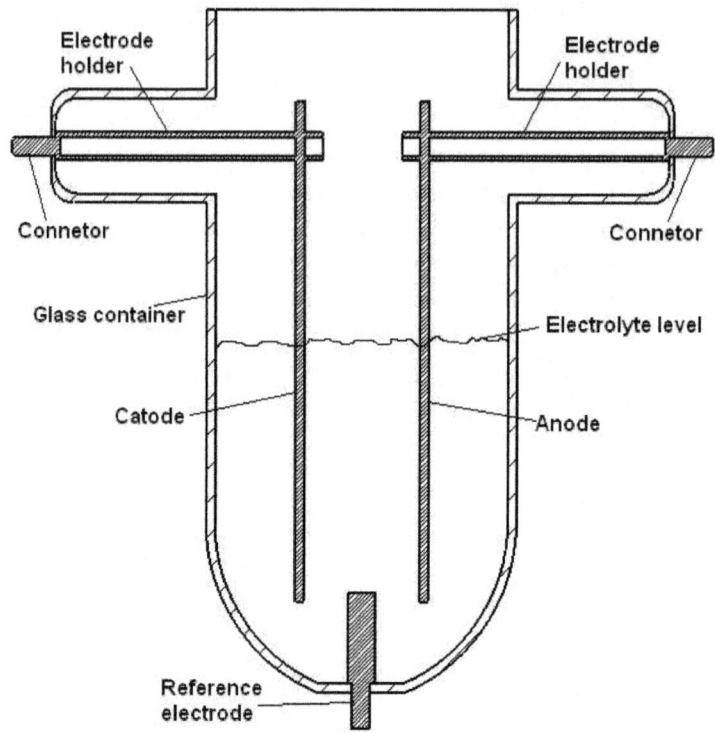

Figure 5. Water electrolysis cell to measure the kinetics of pulse electrolysis.

RESULTS AND ANALYSIS

Current and Energy Efficiencies

Average values of voltage and current, as well as flow of generated hydrogen gas depending on KOH concentrations are shown in table 3. Theoretically maximal current is calculated knowing the hydrogen gas flow by assumption that 2 electrons generate one molecule of hydrogen. Using data from Table 3, current and energy efficiency coefficients are calculated for pulse electrolysis process (see Table 4) on the assumption that the pulse transformer primary side and secondary side are two separate systems, which are only bind by average value of current flowing in the cell.

Table 3. Parameters of registered voltage and current pulses on an electrolysis cell

KOH Concentration [mol/kg]	Average value for current pulse [mA]	Average voltage value [V]	Current value calculated from mass of generated hydrogen [mA]	Hydrogen flow [cm³/min]
0.1	6.5	2.1	3.2	0.043
1	8	2.1	3.7	0.054
2	8.3	2.1	4	0.057
3	8.6	2.1	4.2	0.059

This assumption is not entirely correct, but acceptable. When viewed from the primary circuit side, the pulse generator is with a reactive element included in its scheme - an induction coil (the primary winding of pulse transformer).

Table 4. Current and energy efficiency coefficients

KOH concentration [mol/kg]	Current efficiency coefficient[%]	Energy efficiency coefficient[%]
0.1	49	66
1	46	64
2	48	68
3	49	68

By disconnecting secondary side, primary side does not consume anything(except the power that is distributed on elements with active resistances included in the primary circuit). When connecting the secondary side, the active 1 V amplitude of the voltage pulse in the primary side is unable to consume more, because it is necessary to exceed electrolysis overvoltage – at least 1.23 V (ratio of windings in coil is 1:1).

Therefore, the average current values in Table 3 are replaced with the current consumed in the power supply system. Voltage value is read from the oscilloscope by measuring the voltage pulse on primary coil. Thus, equipment errors associated with variations in voltage values are excluded. Then, the resulting pulse is averaged over time and the resulting voltage values are shown in second column of Table 5.

It should be mentioned that adjusted energy efficiency coefficients were calculated without any reference to the circuit elements and the quantity of generated gas flow. As it is seen from Table 4.3., it is necessary to determine current and voltage values with oscilloscope within scheme of this experiment, which eliminates the pulse schemes for analogue measuring errors.

Table 5. Adjusted parameters of voltage, current and efficiency

KOH concentration	Power supply voltage [V]	Average current value on the cell [mA]	Hydrogen flow [cm^3/min]	Energy efficiency coefficient [%]
0.1	1.43	6.5	0.043	97-
1	1.48	8	0.054	96
2	1.53	8.3	0.057	94
3	1.49	8.6	0.059	97

Pulse Kinetics at Different Concentration Solutions and Electrode Distances

In Figure 6 the voltage and current pulse oscillograms are shown for steel electrode plates in 0.1 M KOH solution, where the maximum voltage pulse value is approximately 5.5 V when distance between the electrodes is 5 mm and it drops to about 3 V when the distance between the electrodes is 1 mm. In 0.3 M KOH solution (curves similar to 0.1 M solution) the maximum voltage pulse value is 3.5 V, when distance between the electrodes is 5 mm and it drops to 2.6 V when distance between the electrodes is reduced to 1 mm. In even more concentrated solution, ie., 0.5 M KOH, the maximum voltage pulse value at the electrode distance of 5 mm is approximately 2.9 V and when the electrode distance is 1 mm, it drops to 2.4 V. Current peak value does not change significantly depending on the electrode distance, or the concentration, but there are observed changes in the discharge tail length, suggesting that higher charge in electrolysis cell flows at more concentrated electrolyte solution.

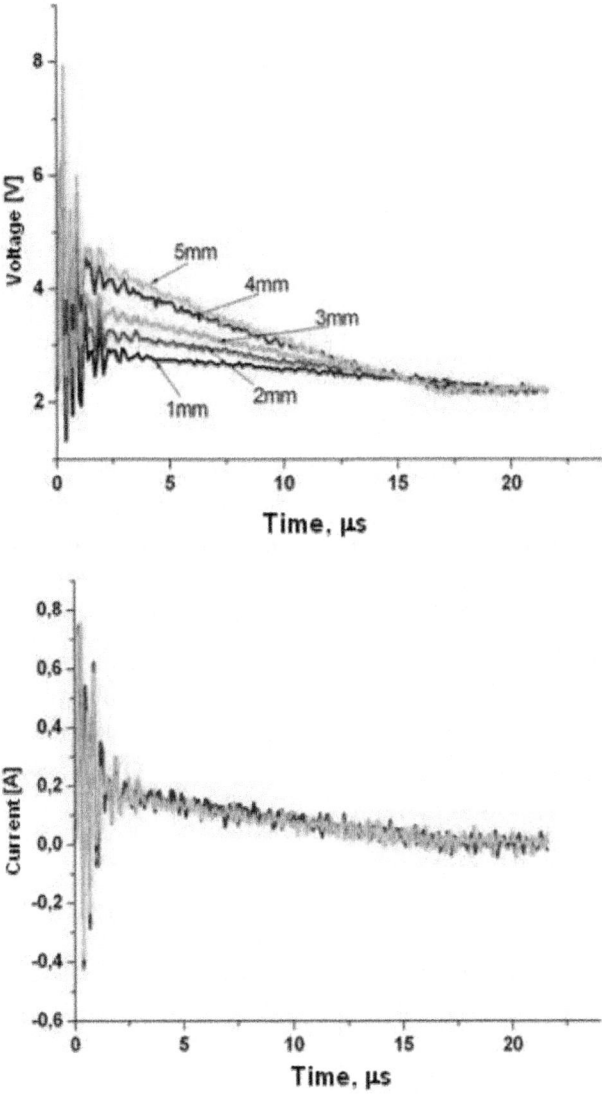

Figure 6. Current and voltage pulses registered with oscilloscope in 0.1 M KOH.

When looking at pulse generation scheme in the experimental method section (Figure 1), it is clear that high-voltage pulse generated in the transformer is reactive in nature. Reactive pulse amplitude will depend on the quality factor of capacitive element. Capacitor with a large leak (concentrated electrolyte solution) will not be able to hold the reactive pulse with large amplitude, though in the previous figures it is shown that the amplitude of those achieved in the secondary circuit on the electrolysis cell is greater than the direct pulse amplitude. This means that at the first moment when short inductive pulse is applied, the water electrolysis cell behaves as good capacitor, also at the voltage region, in which water electrolysis can occur. But after starting the discharge tail, the energy stored in the capacity transforms into the chemical energy in the process of water electrolysis.

The Concentration of Dissolved Hydrogen at Cathode

Current and voltage pulses registered with oscilloscope (Fig. 7) show that changing the electrode material, the rising front and relaxation of voltage pulses does not change, while voltage pulse amplitude decreases with increasing current pulse amplitude when solution concentration increases. Current pulses also are not different on the platinum and tungsten electrodes with identical concentration of KOH solutions (Fig. 7). To evaluate pulse energy supplied to the cell, pulse voltage and current values were multiplied and resulting curves was integrated with time (Table 6). Each row in the Table 6 shows electrode material and the concentration of the solution, and in the next column – calculated supplied energy to the system during the pulse.

Figure 8 presents each electrode's voltage and current oscillograms in the same time scale in order to better evaluate the phase shift angle between current and voltage. There are not noticeably significant differences observable between the phase shift angles depending on the electrode material.

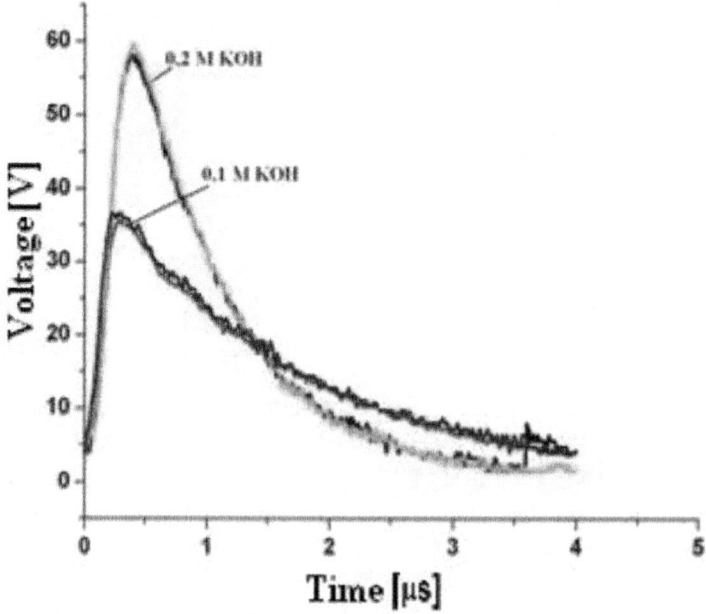

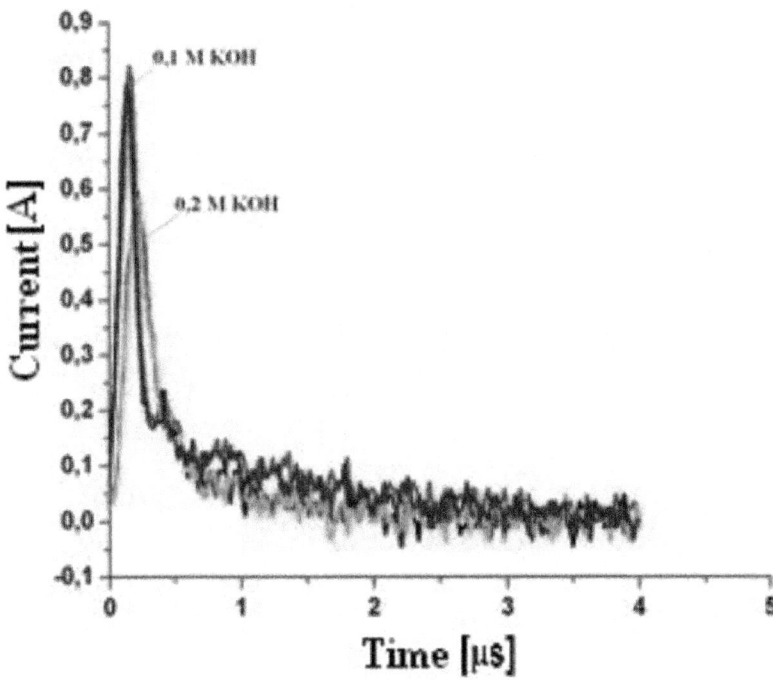

Figure 7. Current and voltage pulse oscillograms of Pt and W electrodes (Pt – black and blue, W – green and red accordingly).

Table 6. Energy supplied to the cell during the pulse, calculated from voltage and current oscillograms

Electrode material and solution concentration	Energy, mJ
Pt in 0.1M KOH solution	8.5
Pt in 0.2M KOH solution	7.7
W in 0.1M KOH solution	8.2
W in 0.2M KOH solution	7.6

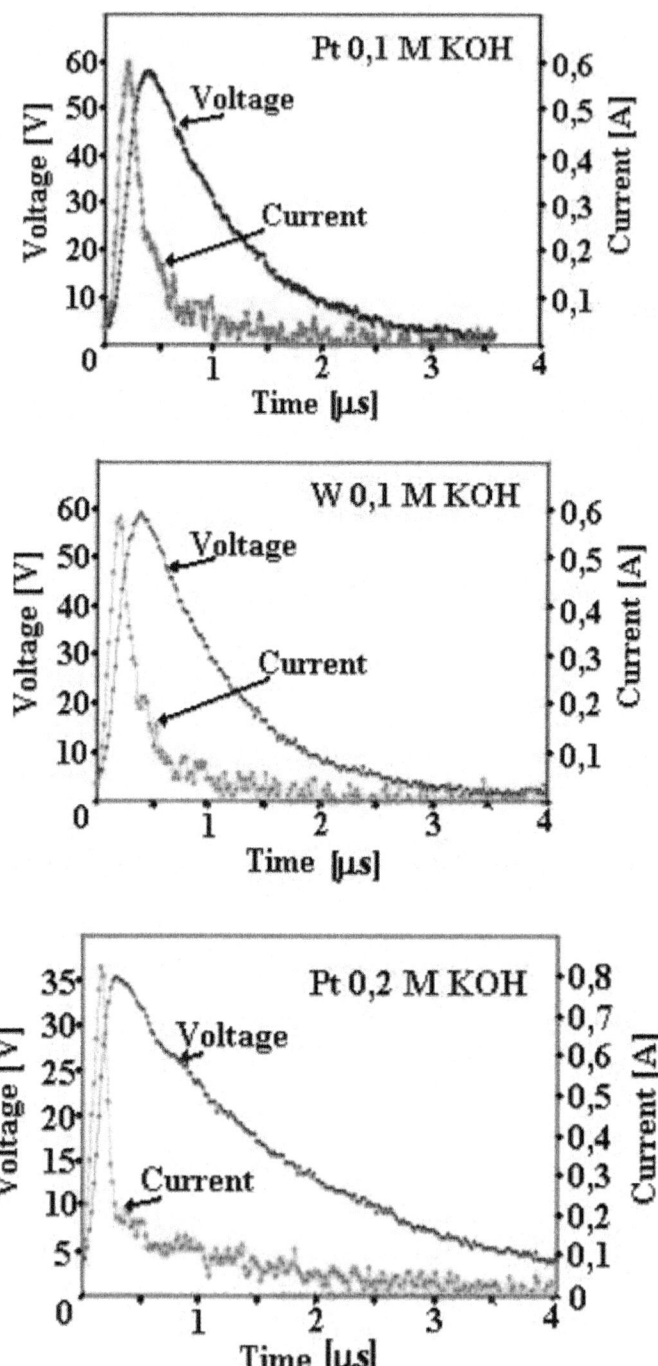

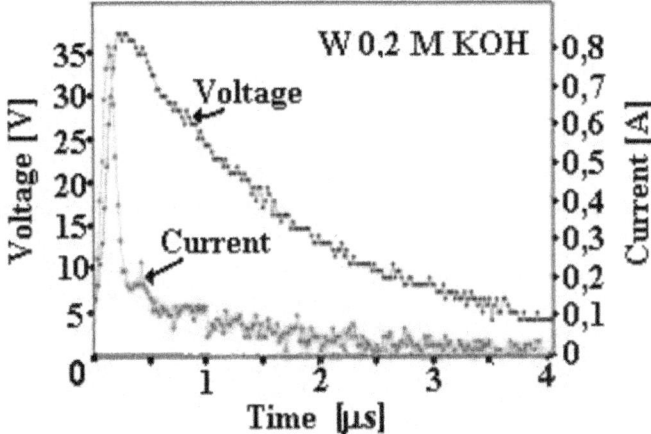

Figure 8. Current and voltage pulse oscillograms of Pt and W electrodes in 0.1 M and 0.2 M KOH.

In each experiment with microsensor to measure the concentration of dissolved hydrogen, the measuring time lasted 100 s (curves at Fig. 9). As it is seen, the curves with largest slope are an electrolyte with a higher concentration, and the tungsten electrode, rather than platinum.

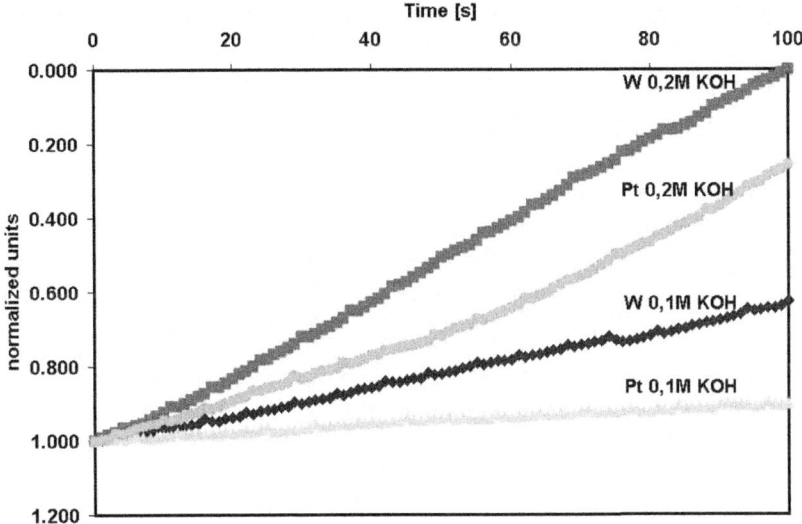

Figure 9. Changes in concentration of dissolved hydrogen gas during pulse electrolysis.

It means that on the tungsten electrodes the concentration of dissolved hydrogen increases faster than on the platinum electrodes. As it is seen from cathodic region of voltamperic curves (Fig. 10), for platinum electrode the characteristic hydrogen adsorption/absorption peak at negative currents appears at potential -0.5 V, but not for tungsten electrode.

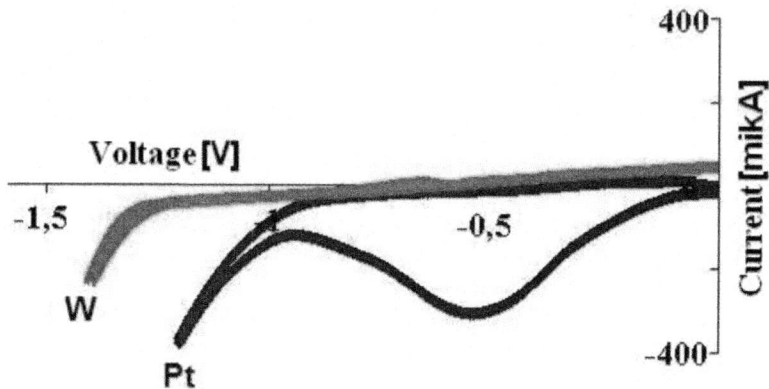

Figure 10. Voltamperic characteristics for platinum and tungsten electrodes in 0.1 M KOH solution measured in two electrode configuration at scan rate 10 mV/s.

The pulse energy of inductive reverse voltage pulses is limited. Voltage and current during the pulse reacts in such way that their multiplication and next integration in time would be equal in the same concentration of electrolyte without reference to the material of electrodes that are used. Pulse energy decreases whilst solution concentration increases, suggesting that reactive energy component has decreased. Therefore it is observable that phase shift angle between current and voltage is smaller in a more concentrated solution. Since the inductive voltage pulse energy is limited, on platinum electrode it is consumed in the adsorption area, thus structuring hydrogen adsorption monolayer on the platinum electrode. There is no hydrogen adsorption/absorption peak for tungsten electrode and during a very short voltage pulse, electrons from the metal discharge directly on hydrogen ions at interface electrode/electrolyte and hydrogen molecules are formed intensively which are detected with dissolved hydrogen microsensor.

Pulse Kinetic Measurements in Highly Dilute Solutions

Voltage pulse growth at various concentrations of KOH solution (Figure 11) is equal in all concentrations, while the discharge tile after voltage pulse at various concentrations is different. Amplitude of voltage pulse is maximal in deionized water, but pulse dynamics in cell with slightly diluted electrolyte is exactly what is in the case of the open circuit, only the amplitude is smaller. Continuing to increase the concentration of electrolyte in cell, the value of voltage pulse amplitude continues decrease, while the discharge tail will increase.

Current changes the direction from negative to positive with increasing concentration of electrolyte passing through the point where the current pulse has not descending a long tail (Fig. 12). Current pulse in deionized water most of the momentum is negative. By increasing the concentration of solution up to 1 mM, current pulse appears in both positive and immediately following a negative pulse, while discharge tail almost disappears. Continuing to increase the concentration of electrolyte, the negative values of current pulse disappear and the discharge tail remains positive and increasing, which indicates that the charge injected in the cell during pulse increases. More increase of concentration does not change the view of current pulse and it remains like from the previous concentrations (Fig. 12). When electrolyte concentration increases, voltage peak drop is observed (Fig. 13). The peak value of voltage pulse is decreasing exponentially, and it stabilizes around the value of 9 V for solutions, while in deionized water that value is over 600 V. These curves almost coincide in different alkali solutions, while in the sulfuric acid the peak values are falling faster.

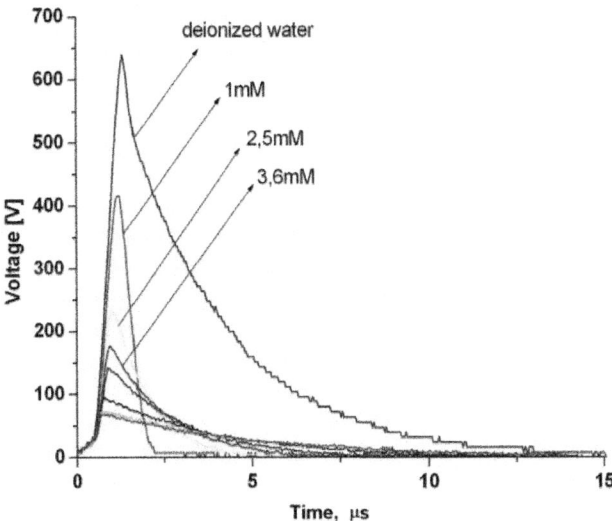

Figure 11. Inductive discharge voltage pulses with different concentrations of KOH.

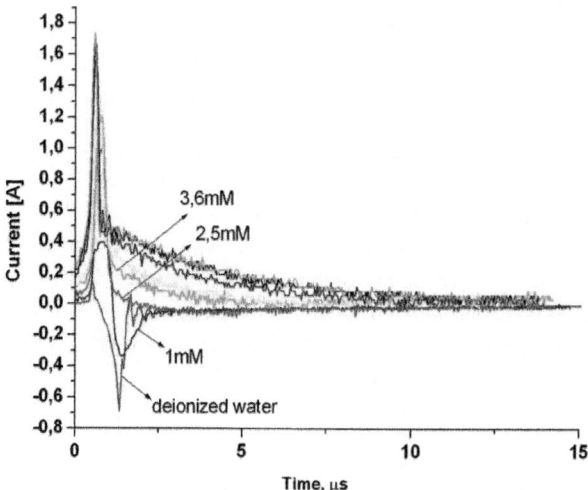

Figure 12. Current pulses initiated by inductive voltage pulses on cell with different concentrations of KOH solution

The pulse charge (integral of the current pulse) increases with increase of electrolyte concentration and tend saturate at some value (Figure 14).

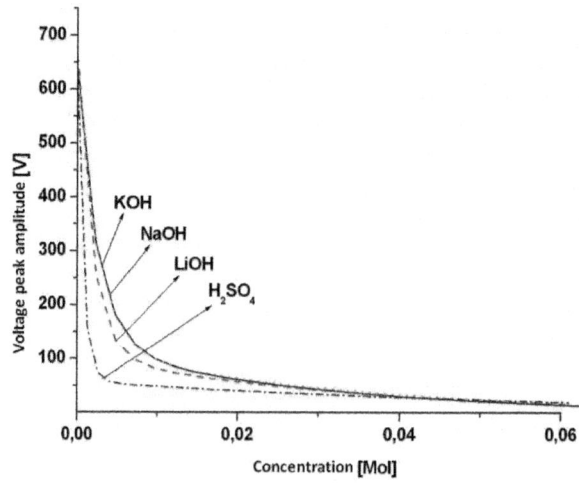

Figure 13. Decrease of voltage pulse amplitude with increasing concentration of electrolyte.

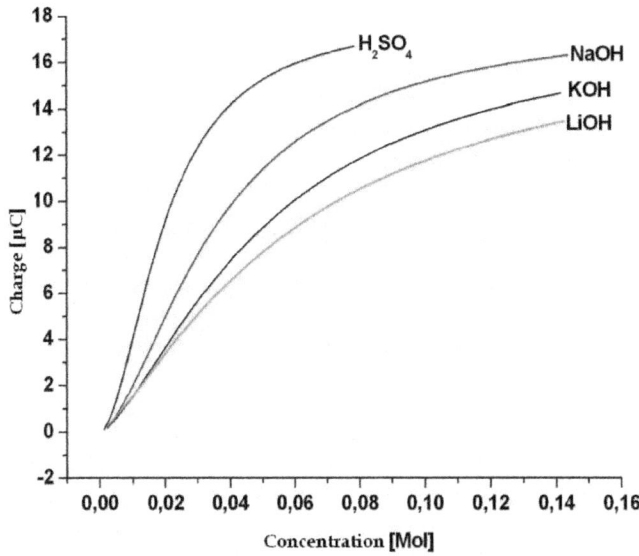

Figure 14. Changes of pulse charge (integrated current pulse) in various solutions with increasing concentration of electrolyte.

In alkali solutions, the charge behavior is nearly identical for alkali tested, while in the cell with sulfuric acid solution, the increase of charge is more rapid. Concerning occurrence of the negative currents following hypothesis is proposed. Voltage pulse kinetics demonstrates that around the electrode spatial charge density appears, i.e., when voltage rapidly grows in two-electrode system, electrons are emitted from the cathode environment. Since water ion concentrations in deionized water is low (H_3O^+ molar concentration is in the order of 10^{-7} M), then, most likely, the emitted electrons are solvated between polar water molecules and than will attach a neutral water molecule, which is described by following hydration reaction:

$$H_2O + e^- \leftrightarrow H + OH^-$$ (19)

If OH^- ions and solvated electrons don't manage to discharge at the cathode, then around the cathode a spatial charge appears. In case of arising the spatial charge around electrode, it is more likely of electrons to move back into the metal. If the electron donor is the OH^- ion, then oxygen evolution should appear at the cathode. In principle, according to the experimental circuit, such electron returning back in metal in large amounts what results from the negative current pulse value presented in Fig. 12, is not possible since this current component is blocked by the diode incorporated in the circuit. Therefore behind the diode, parasitic element with the inductive nature must exist in the measurement circuit (Fig. 1) which becomes comparatively small and solvated electrons are discharged by ions in electrolyte, therefore decreasing negative current. To confirm this hypothesis, it is necessary to determine if oxygen does not appear near the cathode (the solvated electrons OH⁻ form allows a reverse reaction (28) on the cathode).

Measurement of Dissolved Oxygen near the Cathode

The concentration of dissolved oxygen in a solution near the cathode during pulse electrolysis in dependence of time is presented in Fig. 15. During the first 60 seconds current pulses has an explicit negative peak in the cell. After 60 seconds, the generator is set to manage the negative current peak disappeared. From Fig. 15 it is seen that when a negative current peak occurs, the oxygen evolves at the cathode, but when the negative current during voltage pulses is prevented, the oxygen at cathode is no longer released and concentration decreases.

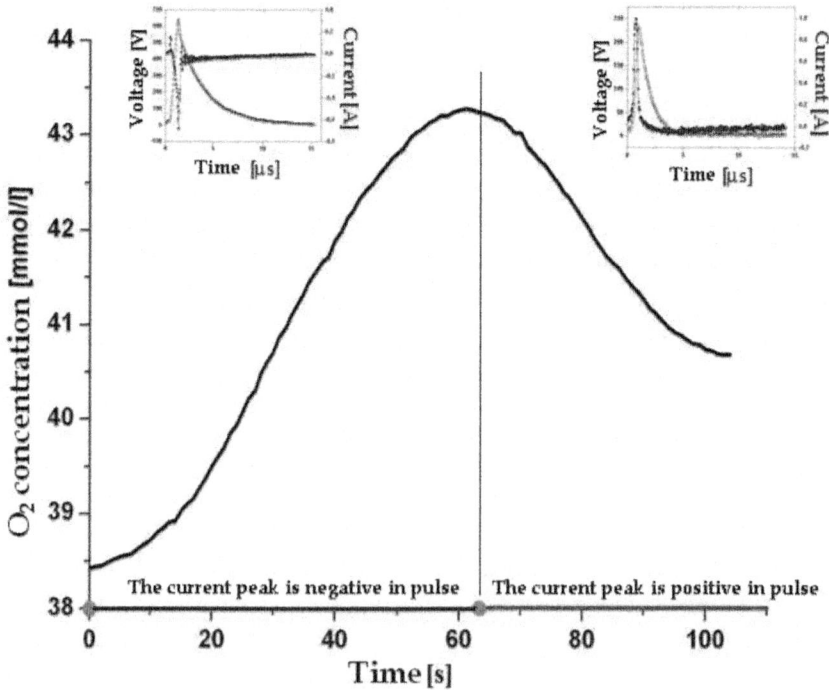

Figure 15. Dissolved oxygen concentration near the cathode in two regimes of pulse generator – when current peak is negative (left, dissolved oxygen concentration is increasing) and positive (right, dissolved oxygen concentration is decreasing).

CONCLUSIONS

Reactive short voltage pulse generator is designed to power water electrolysis cells of different constructions, both with spatially separated and with variable distance electrodes. Required value of electrolysis voltage in the primary circuit of power supply can be reduced by inserting the electrolysis cell in secondary circuit of power supply together with inductive element and reverse diode. For example, in this work electrolysis is provided with direct pulse amplitude 1 V, which induces a short high voltage pulse (tens, hundreds of volts, depending on the conductivity of electrolyte) in the secondary circuit. For studying the process of electrolysis the microelectrode sensors are used to measure concentration of dissolved hydrogen and oxygen gas in the direct vicinity of cathode for the first time.

By changing the distance between the electrodes and concentration of electrolyte, it is experimentally proved that the electrolysis cell is capacitor with high Q factor when short voltage pulse (width below 1 µs) is applied. During this short time the capacitor (electrolysis cell) is charged, which can be interpreted as charging of double-layer on interface cathode/electrolyte. After the interruption of short voltage pulse the energy accumulated in double-layer capacitor slowly discharges (pulse discharge tail), thus activating the process of electrolysis. Consequently, it is shown that with short voltage pulse electrolysis the charging of cell can be separated from the electrochemical reactions in electrolysis process. Kinetics of charging the electrolysis cell with reactive high voltage short pulses does not depend on concentration of the electrolyte, whereas kinetics of the subsequent long-discharge process depends on the electrolyte concentration (faster in dilute solutions, slower in more concentrated solutions). If concentration of electrolyte in the electrolysis cell is above 3 mM, reactive voltage pulse energy does not depend on concentration of the solution. The current polarity switch from cathodic to anodic and back is observed in oscillograms, when de-ionized water electrolysis is performed with short reactive voltage pulses. Polarity changes only during short time when reactive voltage pulse is applied; at the beginning of the discharge tail the current is cathodic again. Measurements of dissolved oxygen concentration by microsensor in the immediate vicinity of the cathode show that oxygen concentration increases (reverse process to hydrogen evolution) in the presence of anodic current. The hypothesis is proposed that high-voltage pulse causes the emission of electrons from the cathode metal into electrolyte, where at first the electrons are solvated, then dissociating the water molecules forms H atoms and OH⁻ ions; next generated OH⁻ ions can discharge on cathode at the moment when applied voltage pulse reduces, giving the release of oxygen detected by microsensor. Current efficiency of 50% is registered in high-voltage reactive short pulse electrolysis, while the energetic efficiency is in range 70-100%.

Platinum and tungsten electrodes are studied to find the impact of electrode material on the process of short reactive voltage pulse electrolysis. Experimental results show that the voltage and current characteristics of inductive voltage short pulse electrolysis are the same for both metals, but the concentration of dissolved hydrogen grows faster at the tungsten electrode. Delay in the release of hydrogen on a platinum electrode is explained by the tendency of platinum to adsorb hydrogen on the surface.

ACKNOWLEDGEMENTS

The authors acknowledge laboratory colleagues Liga Grinberga and Andrejs Lusis for stimulating discussions, and Vladimirs Nemcevs for technical assistance. Authors thank Professor Robert Salem for helpful guidance and advices, and deep compassion to his family on the death of Professor in 2009. Financial support from the European Social Fund project "Support for doctoral studies at the University of Latvia" is acknowledged by M. Vanags. All authors thank the National Research Program in Energy supporting the development of hydrogen infrastructure in Latvia.

REFERENCES

1.   Bockris O`M and Potter E.C.1952Water Electrolysis with Inductive Voltage PulsesJournal of the Electrochem. Society, 99169 EOF

2.   J. Bockris, O'M, I. A. Ammar, A. K. S. Huq, 1957The Mechanism of the Hydrogen Evolution Reaction on Platinum, Silver and Tungsten surfaces in Acid Solutions. J Chem Phys, 61879886

3.   J. Bockris, M. O`, B. Dandapani, D. Cocke, J. Ghoroghchian, 1985Water Electrolysis with Inductive Voltage PulsesInternational J. Hydrogen Energy, 10179 EOF201 EOF

4.   Bott A.W.2000Controlled Current Techniques. Current Separations 18/4125127

5.   S. B. Chambers, 2002Method for producing orthohydrogen and/or parahydrogen. US Patent 6419815 (2002).

6.   M. S. Chandrasekar, M. Pushpavanam, 2008Water Electrolysis with Inductive Voltage PulsesElectrochimica Acta5333133322

7.   P. Delahay, 1966Electrode Processes without a Priori Separation of Double-Layer Charging, The Journal of Physical Chemistry, 7023732379

8.   R. De Levie, 1999The electrolyses of water. Journal of Electroanalytical

Chemistry 4769293

9.   J. Dimants, B. Sloka, J. Kleperis, I. Klepere, 2011Renewable Hydrogen Market Development: Forecasts and Opportunities for Latvia. Paper 319GOVProceedings of International Conference on Hydrogen Production ICH211June 19-22, 2011, Thessaloniki, Greece, p.1-6

10. A. A. El -Meligi, N. Ismail, 2009Hydrogen evolution reaction of low carbon steel electrode in hydrochloric acid as a sourse for hydrogen production, International Journal of Hydrogen Energy 349197

11. J. Ghoroghchian, J. Bockris, M. O`, 1985Water Electrolysis with Inductive Voltage PulsesInternational Journal of Hydrogen energy, 10101 EOF112 EOF

12. F. Gutmann, O. J. Murphy, 1983Water Electrolysis with Inductive Voltage PulsesIn boock: R.E. White, J.O'M. Bockris, B.E. Conway (Eds.), Modern aspects of electrochemistry, 15p. 1, Plenum, New York (1983) 513

13. M. Heyrovsky, 2006Research Topic- Catalysis of Hydrogen Evolution on Mercury Electrodes, Croatica Chemica Acta 79 (1) (2006) 1-4

14. T. Hirato, Y. Yamamoto, Y. Awakura, 2003Water Electrolysis with Inductive Voltage PulsesSurface and Coatings Technology 169/170135138

15. G. Horányi, G. G. Láng, 2006Water Electrolysis with Inductive Voltage PulsesJournal of Colloid and Interface Science29618

16. St. Horvath, (1976) Electrolysis apparatus. US Patent 3954592 (1976).

17. Hydrogen Pathway2011Welcome to the Roads2HyCom Hydrogen and Fuel Cell Wiki: Cost Analysis; 18.11.2011; Available from: www.ika. rwth-aachen.de/r2h/index

18. N. Ibl, J. Puippe, Cl, H. Angerer, 1978Water Electrolysis with Inductive Voltage PulsesSurface Technology6287300

19. G. Kisis, M. Zeps, M. Vanags, 2009Parameters of an efficient electrolysis cell, Latvian Journal of Physics and Technical Sciences. Riga, 2009, N3. 6 p.

20. W. Kreuter, H. Hofmann, 1998Water Electrolysis with Inductive Voltage PulsesInt. J. Hydrogen Energy 238661666

21. Kristalik L.I.1965Barrierless Electrode Process, Russian Chemical Reviews, 34

22. K. Kuroda, H. Shidu, R. Ichino, M. Okido, 2007Water Electrolysis with Inductive Voltage PulsesMaterials Transactions48328331

23. S. A. Meyer, 1986Electric pulse generator. US Patent 4613779 (1986).

24. S. A. Meyer, 1989Gas generator voltage control circuit. US Patent 4798661 (1989).

25. S. A. Meyer, 1992Process and apparatus for the production of fuel gas and the enhanced release of thermal energy from such gas. US Patent 5149407 (1992).

26. J. R. Miller, P. Simon, 2008Water Electrolysis with Inductive Voltage PulsesThe Electrochemical Society Journal Interface, Spring:3132

27. Murphy G.F.1983In: White RE, Bockris JO'M, Conway BE, editors. Modern aspects of electrochemistry, 15NewYork: Plenum Press; 1983. 513

28. K. Nisancioglu, J. Newman, 2012Water Electrolysis with Inductive Voltage PulsesJournal of The Electrochemical SocietyE59E61.

29. M. Noel, K. I. Vasu, 1990Water Electrolysis with Inductive Voltage PulsesOxford and IBH Publishing Co. Pvt. Ltd., New Delhi, 1990., 695 p., 8-12040-478-5

30. K. B. Oldham, J. C. Myland, 1993Water Electrolysis with Inductive Voltage PulsesUnited Kingdom: Academic Press Limited, 1993, 474 p.

31. Our Common Future1987Book, Oxford: Oxford University Press. 019282080

32. H. K. Puharich, 1983Method & Apparatus for Splitting Water Molecules. US Patent 4,394,230 (1983).

33. R. R. Salem, 2004Chemical Thermodynamics, Fizmatlit, 2004., 352 p, 5-92210-078-5

34. R. M. Santilli, 2001Durable and efficient equipment for the production of a combustible and non-pollutant gas from underwater arcs and method therefore. US Patent 6183604 (2001).

35. T. Sasaki, A. Matsuda, 1981Water Electrolysis with Inductive Voltage PulsesJ. Res. Inst. Catalysis, Hokkaido Univ., 29

36. Articles. Selected, Hydrogen. of, The. Phenomena, Editors. T. Book, Nejat Vezġroğlu, M. Oktay Alniak, Ġenay Yalçin; 1.Basım: Ekim 2009978-6-05593-623-53945

37. Shaaban A.H.1994Pulsed DC and Anode Depolarisation in Water Electrolysis for Hydrogen Generation, HQ Air Force Civil Engineering Support Agency Final Report, August, 1994. 21.11.2011. Available from: http://www.free-energy-info.co.uk/1 pdf

38. N. Shimizu, S. Hotta, T. Sekiya, O. Oda, 2006Water Electrolysis with Inductive Voltage PulsesJournal of Applied Electrochemistry36419423

39. E. Spirig, 1978Water decomposing apparatus. US Patent 4113601 (1978).

40. The Hydrogen Economy2004Opportunities, Costs, Barriers, and R&D Needs. 21.11.2011; Available from: http://www.nap.edu/openbook. php?record_id=10922&page=R1

41. C. D. Themu, 1980High voltage electrolytic cell. US Patent 4316787 (1980).

42. Unisense   Microsensors2011Tool   Guide   15.11.2011.   From: UNISCIENCE   Science: Availablefrom: http://www.unisense.com/ Default.aspx?ID=458);http://www.unisense.com/Default.aspx?ID=443

43. M. Vanags, P. Shipkovs, J. Kleperis, G. Bajars, A. Lusis, 2009Water Electrolyses- Unconventional Aspects" In Book: Selected Articles Of Hydrogen Phenomena. Editors: T. Nejat Vezġroğlu, M. Oktay Alniak, Ġenay Yalçin; 1.Basım: Ekim 2009, 978-6-05593-623-53945

44. M. Vanags, J. Kleperis, G. Bajars, 2011aWater Electrolysis with Inductive Voltage PulsesInternational Journal of Hydrogen Energy361316 EOF1320 EOF

45. M. Vanags, J. Kleperis, G. Bajars, 2011bWater Electrolysis with Inductive Voltage PulsesLatvian Journal of Physics and Technical Sciences33440

46. E. Zoulias, E. Varkaraki, N. Lymberopoulos, C. N. Christodoulou, G. N. Karagiorgis, (2002) A Review On Water Electrolysis. Centre for Renewable Energy Sources (CRES), Pikermi, Greece. 21.11.2011. Available   from:http://www.cres.gr/kape/publications/papers/ dimosieyseis/ydrogen/A%20REVIEW%20ON%20WATER%20 ELECTROLYSIS.pdf

Chapter 3

ELECTRICAL IMPEDANCE TOMOGRAPHY OF ELECTROLYSIS

Arie Meir[1], Boris Rubinsky[2]

[1] Biophysics Graduate Program, University of California, Berkeley, California, United States of America

[2] Department of Mechanical Engineering, University of California, Berkeley, California, United States of America

ABSTRACT

The primary goal of this study is to explore the hypothesis that changes in pH during electrolysis can be detected with Electrical Impedance Tomography (EIT). The study has relevance to real time control of minimally invasive surgery with electrolytic ablation. To investigate the hypothesis, we compare EIT reconstructed images to optical images acquired using pH-sensitive dyes embedded in a physiological saline agar gel phantom treated with electrolysis. We further demonstrate the biological relevance of our work using a bacterial E.Coli model, grown on the phantom. The results demonstrate the ability of EIT to image pH changes in a physiological saline phantom and show that these changes correlate with cell death in the E.coli model. The results are promising, and invite further experimental

INTRODUCTION

Tissue ablation with minimally invasive surgery is important for treatment of many diseases and has an increasing role in treatment of solid neoplasms. A variety of biophysical and biochemical processes are used for this purpose. They include thermal ablation with heating, cooling or freezing, electroporation, injection of chemical agents, photodynamic effects, sonoporation effects and many others. Electrolysis, the passage of a low magnitude direct ionic current through the tissue, between two electrodes, is a biochemical/biophysical process that has been considered for tissue ablation since the 19th century [1]. Electrolysis affects the ionic species in tissue, which change into compounds that can ablate cells. The advantage of electrolysis in comparison to other ablation techniques can be attributed to its simplicity and low cost of instrumentation, which might make it a suitable treatment modality for resource constrained communities where more expensive medical treatment is often not available [2].

The work of Nordenstrom and colleagues [3,4], is among the early modern work on electrolysis. Important recent work was published on understanding of the effects of electrolysis on tissue through histology, mathematical modeling of involved electrochemical processes, and clinical work, e.g. [5–15] and [16–19]. Several findings were made and several research techniques were developed that have inspired this paper: it was shown that the electrolysis induced pH changes can be used to reliably monitor the extent of tissue ablation [20]. These findings have led to several basic studies on quantifying the process of electrolysis through the use of transparent gels with pH dyes [12,21,22].

While minimally invasive thermal ablation is now commonly used in surgery "it has become common since the advent of modern imaging" [23]. Electrolysis is currently limited by the lack of an effective means to monitor the extent of tissue ablation deep in the body. The finding that pH changes are indicative of electrolytic tissue ablation [20], and the interesting results obtained with pH dyes marked gels [24–26], have suggested to us a way to monitor and image electrolysis. The pH fronts developing during the electrolysis process [6,24] are caused by evolution of protons (H^+) and hydroxide (OH^-) ions. While the relationship between pH level and electric conductivity is not straight-forward and depends on relative concentrations of other ions, the increased concentrations of protons and hydroxide ions would affect the local electrical conductivity of the tissue being electro-treated. The smaller relative sizes of the proton and the hydroxide ions could lead to a higher mobility compared to that of other ions typically found in biological solutions—we address this question further in the later discussion section. Historically, one of the foci of our lab's work has been in the field of electroporation, i.e. the effect of applying brief pulses of high-magnitude electric field to a tissue [27,28]. The central hypothesis of this work is based on numerous empirical observations by our group of conductivity changes during electro-stimulation occurring even when the stimulation voltage was low and did not cause electroporation. We hypothesize in this work that the pH change induced changes in electric conductivity, could be used to detect, monitor or image the process of electrolysis deep in the body, in real time.

It was further hypothesized that one possible technique to image electrolysis induced pH changes in real time, deep in the body, is Electrical Impedance Tomography (EIT). Electrical impedance tomography is used in a variety of scientific fields, from geology, to semiconductor characterization, to medical imaging. EIT produces an image of the electrical properties of the examined media. In a typical EIT application, electrodes are placed around the volume of interest, and small, sinusoidal currents are injected into the tissue, while voltages are measured on its boundary. Using the finite element method, the complex impedance of the analyzed domain is modeled, and a solution for the approximate impedance configuration that fits the measurements is obtained [29–31]. During the last four decades, substantial basic and applied research was done in the field of EIT. Our group has focused on applications for EIT in the context of monitoring minimally invasive surgery procedures such as cryosurgery [32], tissue viability [28,33] and electroporation [27],[34].

The primary goal of this study is to explore the hypothesis that changes in pH during electrolysis can be detected with EIT, for possible applications in monitoring tissue ablation with electrolysis. To explore the hypothesis we conduct an experimental study using a pH dye stained physiological saline agar-gel based phantom as a model for a living tissue, from an electrochemical standpoint. To investigate the hypothesis, we compare EIT reconstructed images to optical images acquired using pH-sensitive dyes embedded in the agar phantom that is exposed to electrolysis. In addition to validating the EIT-based approach using pH-sensitive dyes, we demonstrate a biological application of our EIT work by comparing a spatial map of bacterial viability exposed to electrolysis with the EIT image of the phantom during electrolytic treatment. Our results are promising, and invite further experimental explorations.

METHODS AND MATERIALS

Tissue Model

Our tissue model consists of a physiological saline based agar gel phantom with electrical conductivity designed to simulate that of a tissue. To construct the phantom, 0.5% Bacto-Agar (Fisher Scientific) was mixed with 0.9 g/l Sodium Chloride (Fisher Scientific) in distilled water. The solution was then brought to a boil and poured into the Petri dishes. The conductivity of the agar phantom was measured to be approximately 0.14 S/m which is close to the range of hepatic tumor conductivity [35]. During the experiments, the EIT electrode holder was placed in the Petri dish with the electrodes galvanically coupled to the gel phantom (Fig 1).

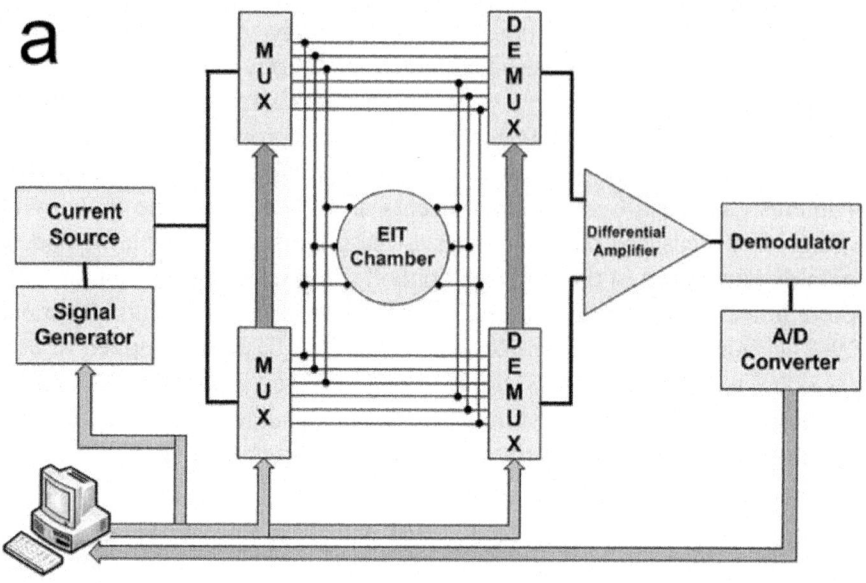

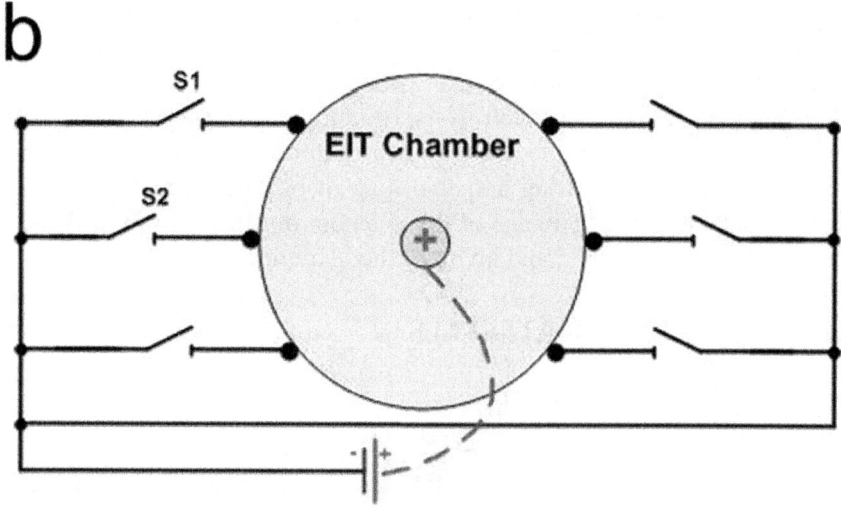

C

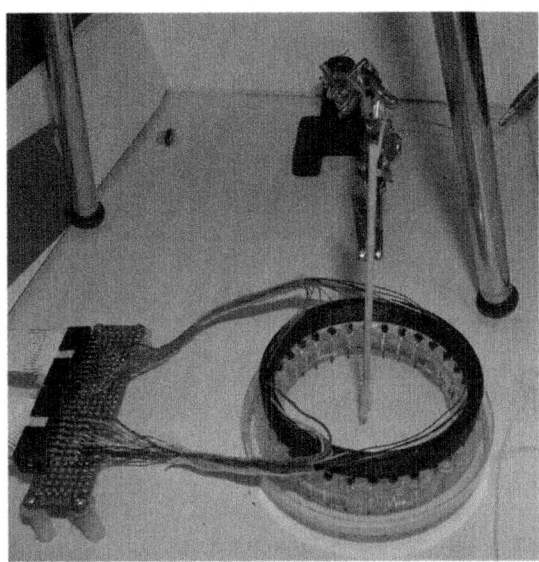

Figure 1. Electrical Impedance Tomography System (a) EIT System Schematic (reproduced from [29], MUX = multiplexer, DEMUX = demultiplexer) (b) Schematic of Experimental EIT Chamber for Electrolysis Experiments (c) EIT Chamber With Central Electrolysis Electrode in Our Experimental Setup.

Experimental Model

To test the feasibility of EIT as a means to monitor the onset and extent of electrolysis in tissue, we have devised the following experiment: 1) A reference EIT image of the tissue phantom is taken, 2) Electrolytic stimulation is applied, 3) Another EIT reference image of the tissue phantom is taken. We leverage the differential nature of EIT images to represent the changes in conductivity, which are used as surrogates to regions of altered pH level. As a control study, we use pH sensitive dyes in order to estimate the boundary of the region where the pH has changed due to electrolysis. We use a digital camera (Casio Exilim EX-ZR100) to acquire optical images of the experimental chamber and correlate these images with the EIT reconstruction images. The results of several representative studies are presented in the following section: in each study we have repeated experimental steps 2) and 3), above, multiple times, in order to observe the evolution of the pH front over time.

Bacterial Model

Lyophilized *E.Coli* of HB101 strain (BioRad) were grown in LB broth overnight and plated on LB broth based agar gel filled petri dishes. The LB broth for the overnight growth consisted of 1% BactoTryptone (BD), 0.5% Yeast Extract (BD), 1% NaCl (Sigma Aldrich) and 1.5% Agarose (Sigma Aldrich). For pouring the plates, we have held the sodium salt from the broth, in order to control the conductivity of the resulting gel. 6mm glass beads (Sigma Aldrich) were used for plating to ensure uniform coverage. After plating, the beads were removed and the plates were incubated for 15 minutes at 37°. The conductivity of the gel was measured around 0.2 *S/m*. At the experimental stage, the petri dish was separated from its lid and the EIT electrode array was lowered into the gel. On top of the EIT chamber, a 2 electrode holder with auxiliary electrodes was introduced into the gel. For the bacteria-focused experiments, only the auxiliary electrodes were used for stimulation, as opposed to the pH-sensitive dye experiments where we have also used the EIT electrodes for electrolytic stimulation. The stimulation sequence was applied using specified current and time parameters, with EIT snapshots being taken in the process as a monitoring step. The results section includes the exact current and time parameters used for each study. After the stimulation the petri dishes were covered and incubated for 24 hours. To evaluate viability we have visually inspected the petri dishes for areas where bacterial growth was inhibited.

EIT Instrumentation

An EIT data acquisition system consists of a collection of electrodes, which are used to inject known sinusoidal AC current into the observed sample. Due to the sample's conductivity, a potential develops on the sample. Due to the low frequency of the stimulation current, we ignore phase information and approximate impedance by Ohmic conductivity (resistivity). This potential is measured on the boundary using the electrodes not used for current injection. A schematic of a typical EIT system is presented in Fig 1A. In this work, we have used the EIT system described in [36], with $N = 32$ electrode surrounded circular chamber. We used an adjacent stimulation scheme [37], leading to each data set containing $\frac{N \cdot (N-3)}{2} = 464$ independent measurements. While recent literature recommends against this stimulation scheme [38], the EIT system available to us in the lab has this stimulation scheme hardwired. While this clearly poses a limitation, and most likely leads to reduced image quality, we intend to address it in future work. After the data has been acquired, the data processing module of an EIT system attempts to reconstruct a conductivity map of the domain of interest from a set of known injected currents and measured resulting voltages, typically at the boundary of the geometric domain. In a typical EIT reconstruction algorithm, a map of impedance is guessed and the voltages resulting from injected currents calculated by solving Laplace equation in the domain. These voltages are compared to the measured voltage and the difference is then used as feedback for an iterative scheme. The guessed map of impedance is then updated, until the calculated and measured voltages agree within a certain tolerance. Here, we have used the EIDORS framework (v3.7.1) [39] with the complete electrode model (CEM). We modeled our electrodes as 2D needle electrodes surrounding a circular phantom area. The reconstruction was done by using a Gauss-Newton solver (EIDORS function inv_solve_diff_GN_one_step) with total variation (TV) regularization [40,41]. This approach works by attempting to minimize a cost function representing the overall voltage measurement discrepancy between the input (measured) voltages and the reconstruction algorithm's internal model. The total variation functional is an attempt to regularize the reconstruction by making sure that the jumps in the final conductivity are bounded. The value of the regularization hyper-parameter was selected experimentally by taking a value that locally minimizes the residual error. This approach can be seen as a simplified version of the L-Curve method described in [42]. We used the value 0.003.

Experimental Setup

The system is composed of 32 stainless steel electrodes mounted on a holder (Diameter = 75mm) lowered into a circular Petri dish (diameter = 85mm) chamber (Fig 1C). Following the work of [43] we have used internal electrodes in the experimental model to test the sensitivity limits of the system. The chamber contains the pH dye infused agar gel phantom which is imaged using EIT and optical digital camera. All the EIT stimulation currents had amplitude of 350μA and a frequency of 5kHz.

RESULTS AND DISCUSSION

Anode Centered Experiment

In this experiment, we have placed a thin, stainless steel rod (diameter 0.6mm) in the center of the agar gel filled chamber. The central rod was connected using a copper wire to the positive terminal of the power supply and acted as the anode during this part of the experiment. For the cathode, all the 32 electrodes of the EIT were connected to each other by closing the switches $S1$. $S32$ presented schematically in the diagram on Fig 1B. The negative power supply terminal was then connected to the unified EIT electrodes. The EIT electrodes acted a distributed cathode in this case. As a control study we have employed two pH sensitive dyes: 1% phenolphthalein (Sigma-Aldrich) which turns pink/purple above pH 8.8 and acts as a basic indicator, and 2.4% pH indicator (Fresh water test-kit, API) which turns yellow at pH 6.0. Both pH indicators were added to the agar gel phantom before its solidification.

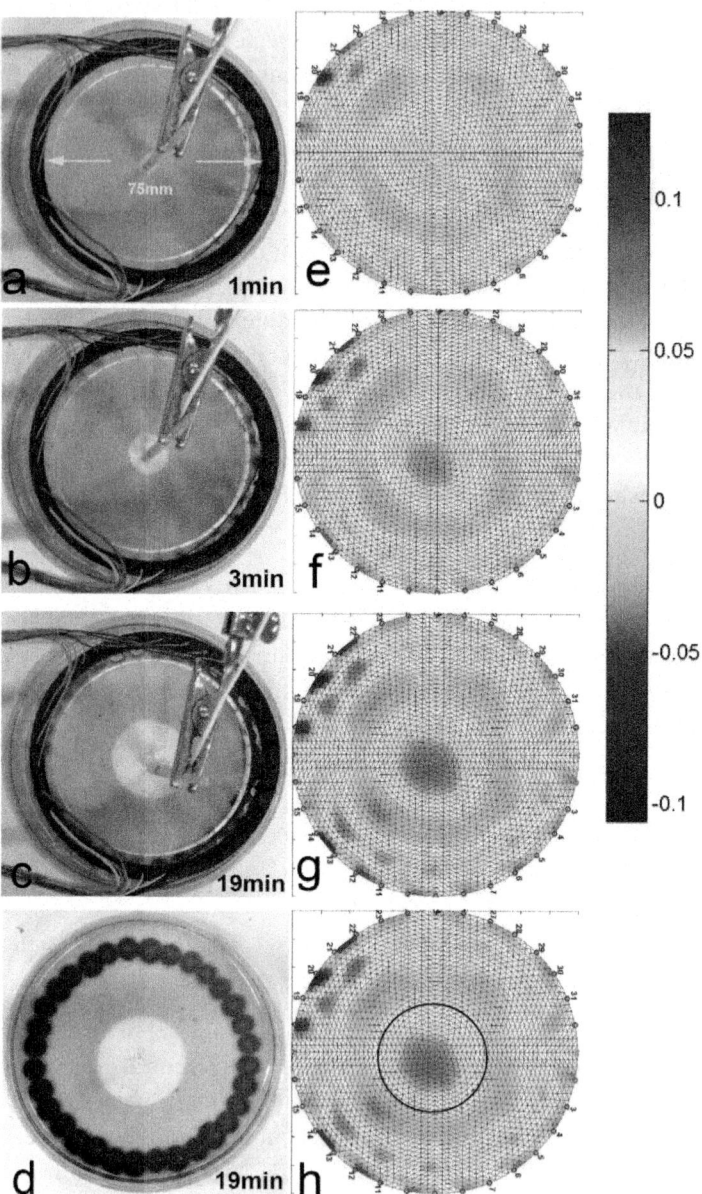

Figure 2. Anode Centered Electrolysis Experiment.

A photographic picture of our experimental chamber is presented in Fig 1C. The protocol of our experiment involved taking a control set of images: EIT and optical, before every electro-stimulation step. The electro-stimulation parameters including time and stimulation current are specified in Table 1. These parameters are typical to tissue ablation electrolytic processes, at the lower range of the parameters [4,44]. Fig 2 summarizes the results of our experiment by showing a sequence of image pairs: each EIT image is accompanied by its matching optical image which we used as a validation method. We have also excluded the possibility that the changes in impedance are caused by the pH-sensitive dye by running a control study where EIT data was acquired from agar gels without pH-sensitive dyes. The units of the color bars next to EIT images are specified in relative impedance changes i.e. 0.1 stands for 10% impedance change (increase for warmer colors and decrease for cooler ones).

EIT images: (a) After 1 minute, (b) After 3 minutes, (c) After 19 minutes, (d) After 19 minutes with overlaid outline of the pH altered region. Optical images: (1) After 1 minute, (2) After 3 minutes, (3) After 19 minutes (4) After 19 minutes with increased contrast

Table 1. Summarizes the experimental parameters used in this work

Experiment	Current	Total Charge	Time
Anode centered	1ma	1.14C	19min
Cathode centered	1ma	2.16C	36min
Two electrodes	2ma	1.44C	12 min
Bacterial viability	2ma	5.4C	45min

doi:10.1371/journal.pone.0126332.t001

The current was delivered at $1mA$, and the delivered charge dosage was $1.14C$, which falls within a range of a typical electro-chemo therapy stimulation charge dosage [4,44]. Fig 2E— 2G show the EIT images at selected time points whereas Fig 2A— 2C show the corresponding optical images. Fig 2H shows the final result of the gel model after the EIT electrodes have been removed. It can be seen that the EIT images of the pH from near the central electrodes are in good correspondence with the pH indicator dye: the central spot around the anode grows over time in both the optical and the EIT images. The data shows a good qualitative correspondence between the EIT reconstructed images and their optical counterparts. We have chosen to include representative images corresponding to times t = 1 minutes, t = 3 minutes and t = 19 minutes. The contrast of the image in Fig 2H was increased to show the altered pH indicator at the perimeter, close to the distributed anode.

The color-bar presented to the right of the figure helps interpret the EIT results: the EIT images are taken in differential mode which means that the images show differences relative to a reference image taken before any electrolytic stimulation was applied. Warmer colors correspond to increased conductivity while colder colors correspond to decreased conductivity in the sample.

The change in conductivity in the center of the gel phantom captured by the EIT system can be explained by noting the relatively small radii of protons (H^+, $0.88fm$) compared to the radii of other physiological ions such as chlorine ions (Cl^- 167 pm) and Na^+(116 pm). The smaller ions produced in the electrolytic reactions led to increased local mobility which in turn led to increased conductivity around the anode. While the relationship between pH and change in conductivity depends on multiple factors including the concentrations of various ions and their respective mobilities, the results indicate an observable correlation between changes in pH and changes in conductivity. To facilitate the comparison of results, we superimposed the margin of the altered region in the pH dye infused gel onto the EIT image in Fig 2H. The margin of the marked region indicates the boundary of the highest pH of 6. For our stimulation conditions, the area with altered pH takes a circular shape. The region within the marked boundary corresponds to pH range of 1–6. Our future work is envisioned to explore the relationship between pH and the electric conductivity. To control for confounding impedance changes we have measured changes in temperature during the electrolytic stimulation, and no observable temperature change was detected. In our previous work with higher current densities we have observed electro-osmotic effects [45], which could further contribute to the changes in impedance, but under the current stimulation regimes, these effects were not observable in this experiment. Characterizing the impact of electro-osmosis on changes in impedance is another research direction which we chose to focus on in future explorations.

Cathode Centered Experiment

In this part of the experiment, we have reversed the roles of the anode and the cathode. The same pH indicators were used as before: 1% phenolphthalein which turns pink/purple above pH 8.8 and acts as a basic indicator, and 2.4% pH indicator (Fresh water test-kit, API) which turns yellow at pH 6.0. As in the previous section, both pH indicators were added to the agar gel phantom before its solidification. The protocol of this experiment involved taking another control set of images: EIT and optical, before every electro-stimulation step. The electro-stimulation parameters including the current and the stimulation duration are specified in Table 1. Fig 3presents the results of the experiment by showing a sequence of image pairs: each EIT image is accompanied by its matching optical image which we used as a validation method. The units of the color bars next to EIT images are specified in relative impedance changes i.e. 0.1 stands for 10% impedance change (increase for warmer colors and decrease for cooler ones). The overall charge dosage was charge dosage was $2.16C$. While this dosage falls within a range of a typical electro-chemo therapy procedure, it is a larger charge dosage compared to the anode centered experiment. We have administered more charge in the cathode-centric experiment because the altered pH front indicated by the pH-sensitive dye (phenolphthalein) was growing slower in the cathode-centered case. A possible explanation to this difference is the relative size of the H^+ and the OH^- ions, and we discuss this discrepancy in more details in a later section (*Bacterial Sterilization Model*). Fig 3E –3G show the EIT images at selected time points whereas Fig 3A— 3C show the corresponding optical images. Fig 3D shows the final result of the gel model after the EIT electrodes have been removed. It can be seen that the EIT images are in good correspondence with the pH indicator dye: the central spot around the cathode grows over time in both the optical and the EIT images. Moreover, while it is too subtle to see in the optical images, the EIT imaging clearly shows a circular feature at the periphery of the EIT chamber. This peripheral region with reduced pH level can be clearly distinguished in Fig 3D by its distinguished yellowish color. It can only be seen after the EIT electrodes have been removed. The data shows a good qualitative correspondence between the EIT reconstructed images and their optical counterparts. We have chosen to include representative images corresponding to times t = 2 minutes, t = 6 minutes and t = 36 minutes. An accumulation of liquid, presumed to be water can be observed around the cathode in the form of a growing bubble. We have attributed it to the osmotic effects of electrolysis reported by other researchers. Fig 3D shows the optical image after the EIT electrodes have been lifted at the end of the experiment. The contrast of the image in Fig 3D was increased to show the altered pH indicator at the perimeter, close to the distributed anode.

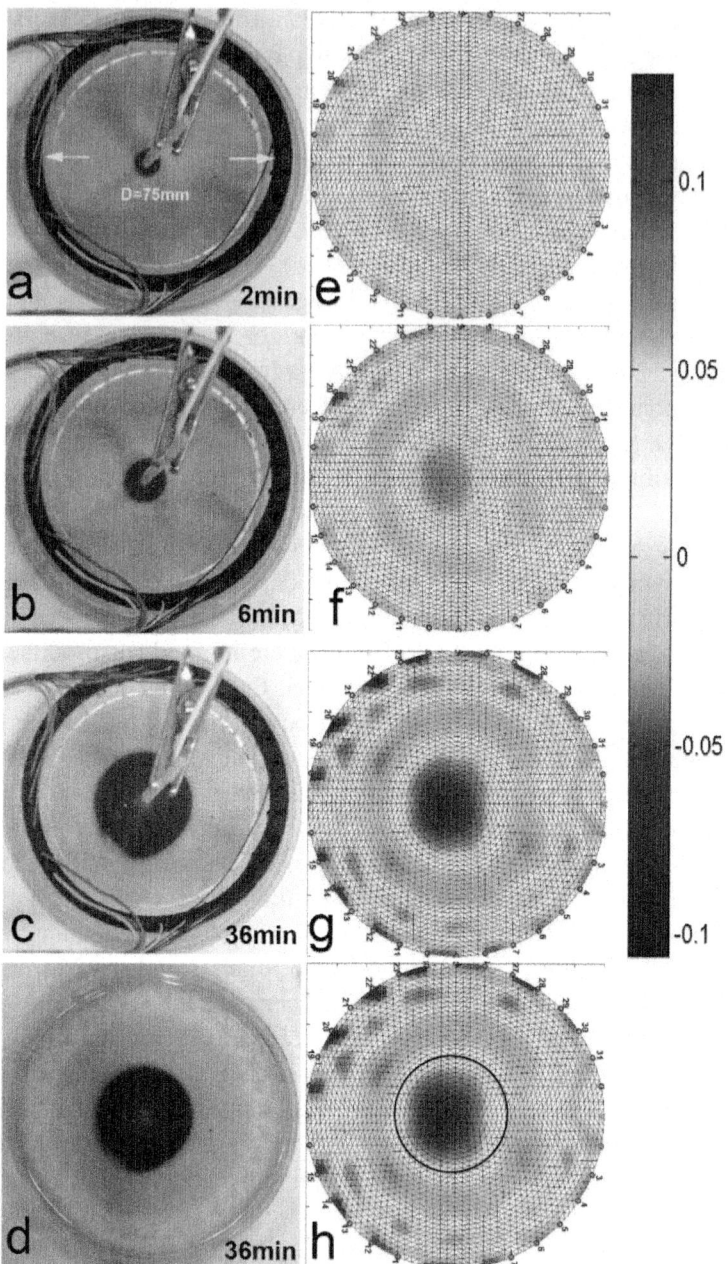

Figure 3. Cathode Centered Electrolysis Experiment. EIT images: (a) After 2 minute, (b) After 6 minutes, (c) After 36 minutes (d) After 36 minutes with overlaid outline of the pH altered region. Optical images: (1) After 2 minute, (2) After 6 minutes, (3) After 36 minutes (4) After 36 minutes with increased contrast

The change in conductivity in the center of the gel phantom captured by the EIT system can be explained by noting the relatively small radii of hydroxyl ions (OH^-, $110pm$) compared to the radii of other physiological ions such as chlorine ions (Cl^- $167\,pm$) and Na^+($116\,pm$). The smaller ions produced in the electrolytic reactions led to increased local mobility which in turn led to increased conductivity around the cathode. This behavior is similar to the anodic case, although the degree of change in conductivity is different, possibly due to the difference in relative sizes of protons and hydroxyl ions. To facilitate the comparison of results, we superimposed the margin of the altered region in the pH dye infused gel onto the EIT image inFig 3H. The margin of the marked region indicates the minimal pH of 8.8. For our stimulation conditions, the area with altered pH takes a circular shape. The region within the marked boundary corresponds to pH range of 8.8–14. Like in the previous experiment, to control for confounding impedance changes we have measured changes in temperature during the electrolytic stimulation, and no observable temperature change was detected. In our previous work with higher current densities we have observed electro-osmotic effects which could further contribute to the changes in impedance, but under the current stimulation regimes, these effects were not observable in this experiment.

Two Internal Electrodes Experiment

In this part of the experiment, instead of using the EIT electrodes as a distributed electrode, we have utilized two graphite electrodes made of pencil lead (Pentel super HB 0.7mm). The electrodes, mounted in a horizontal holder were placed perpendicularly to the gel phantom. The electrodes were inserted 5mm deep into the gel. We have used a 5% pH indicator (RC Hagen wide range). As in the previous experiments, the pH indicator was added to the agar gel phantom before its solidification.

The protocol of this experiment involved taking another control set of images: EIT and optical, before every electro-stimulation step. The electro-stimulation included a sequence of direct current injections at 2mA of the following durations: [1min, 1min, 1min, 1min, 1min, 1min, 1min, 5min]. Fig 4 presents the results of the experiment by showing a sequence of image pairs: each EIT image is accompanied by its matching optical image which we used as a validation method. The units of the color bars next to EIT images are specified in relative impedance changes i.e. 0.1 stands for 10% impedance change (increase for warmer colors and decrease for cooler ones). The delivered charge dosage was $1.44C$ which falls within a range of a typical electro-chemo therapy stimulation charge dosage [4,44]. Fig 4E— 4G show the EIT images at selected time points whereas Fig 4A— 4C show the corresponding optical images. It can be seen that the EIT images are in good correspondence with the pH indicator dye: the central spot around the anode (red) grows over time in both the optical and the EIT images, and the same is observed for the spot around the cathode (blue). We have chosen to include representative images corresponding to times t = 1 minutes, t = 3 minutes, t = 6 minutes and t = 12 minutes. It is notable that both the optical and the EIT approaches are able to image the collision of the basic and the acidic fronts (Fig 4H and 4D).

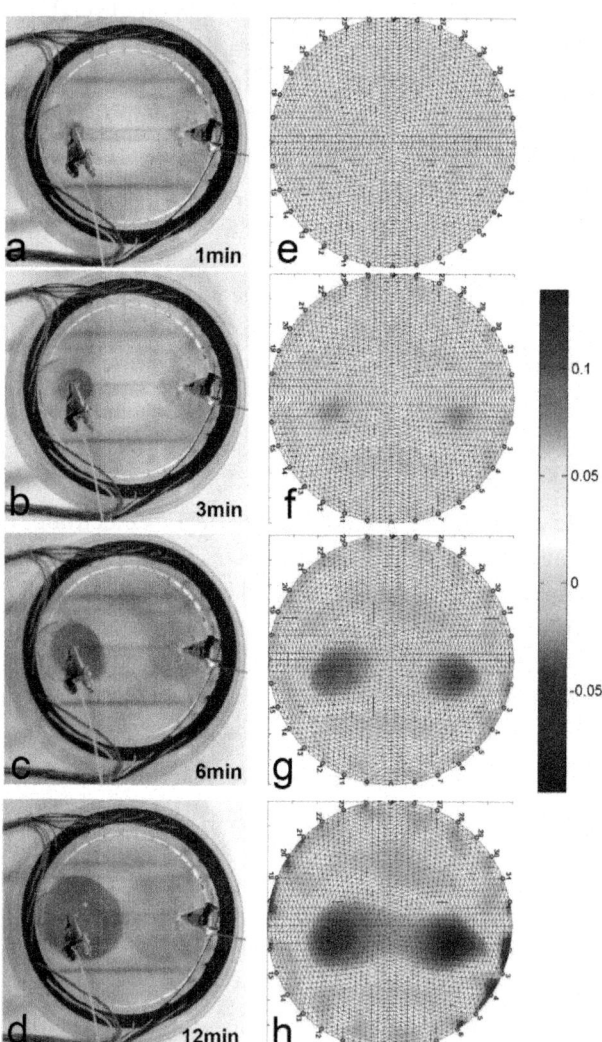

Figure 4. Two Electrodes Electrolysis Experiment. EIT images: (a) After 1 minute, (b) After 3 minutes, (c) After 6 minutes, (d) After 12 minutes. Optical images: (1) After 1 minute, (2) After 3 minutes, (3) After 6 minutes (4) After 12 minutes.

As in the previous two experiments, to control for confounding impedance changes we have measured changes in temperature during the electrolytic stimulation, and no observable temperature change was detected. In our previous work with higher current densities we have observed electro-osmotic effects which could further contribute to the changes in impedance, but under the current stimulation regimes, these effects were not observable in this experiment, even though the currents were higher than in the case with one central electrode.

Bacterial Sterilization Model

To confirm the efficacy of our method in a biological model, we have used EIT for imaging electrolysis in an agar dish plated with *E.Coli* bacteria. The liquid bacterial culture was first plated as described in our methods, and then a current of 2mA was administered using the auxiliary electrodes. The total administered charge dosage was $5.4C$ which falls within a range of a typically delivered charge during an electro-chemo therapy stimulation[4,44]. Fig 5 shows a comparison between the EIT imaging data and a bacterial viability pattern captured using an optical, digital camera after 24 hour growth period. The units of the color bars next to EIT images are specified in relative impedance changes i.e. 0.1 stands for 10% impedance change (increase for warmer colors and decrease for cooler ones). Fig 5A–5C indicate two growing regions of increased conductivity, around the auxiliary electrodes through which electrolytic stimulation was applied. We have chosen to include representative images corresponding to times t = 15 minutes, t = 30 minutes and t = 45 minutes. Fig 5D shows the optical image of the viability pattern taken 24 hours post-stimulation. It is interesting to note that both the EIT images as well as the optical image exhibit asymmetry with regards to the anodic and the cathodic regions under our experimental conditions. The brighter upper spots in the EIT images, in particular in the one shown in Fig 5C indicates that the conductivity of the anodic region has changed to a larger degree than the conductivity of the cathodic region. This discrepancy can be attributed to the relative radii of protons (H^+, $0.88fm$) and hydroxide ions (OH^-, $110pm$). Due to their relative smaller size, the protons are more mobile hence contributing to a larger extent to the conductivity increase around the anode. The increased mobility causes the bactericidal pH region around the anode to be larger than around the cathode. This is supported by the viability observations presented in Fig 5D. To clarify, the circular pattern of dots around the bacterial culture dish corresponds to the EIT electrodes imprinted in the gel when the EIT chamber was lowered.

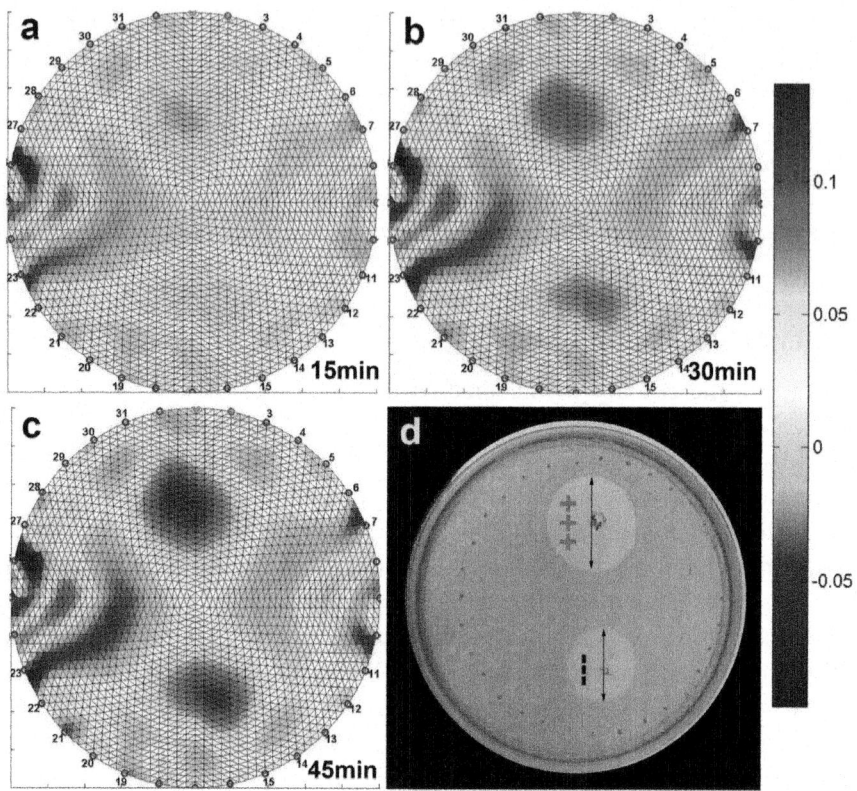

Figure 5. Bacterial Viability Experiment. EIT images: (a) After 15 minutes, (b) After 30 minutes, (c) After 45 minutes. (d) Optical image of growth patterns after 24 hour incubation

LIMITATIONS AND FUTURE WORK

While our results show a degree of correlation between EIT measurements and optical measurements obtained using pH-sensitive dyes, several key questions arise which need to be answered before EIT can be successfully used in clinic as an imaging technique for electrolysis. These include inquiries into the nature of the relationship between pH and conductivity, or the differential changes around the cathodic and the anodic regions. Since other imaging techniques for electrolysis were not reported in the literature, our current benchmark is optical pH-sensitive dye images, but other imaging modalities such as MRI might potentially result in a better comparison standard. There are several limitations inherent to our work: first, our tissue model is limited to an agar gel based phantom which is a simplistic way to represent a biological tissue. Further experiments are envisioned with living tissues replacing the gel phantom. In addition to the inherent limitation of the model, the EIT system we have employed in this work is outdated and was used as a first order, crude approximation in order to establish if further experimental effort can be justified. Clearly, optimizing EIT acquisition and reconstruction parameters would result in better image quality which might increase the likelihood of developing a clinically viable imaging modality. Assuming that the method works on a living tissue, it could potentially be applied in a clinical setting: in this scenario EIT might become a monitoring method for electrolytic ablation in the treatment of breast cancer, see e.g. [46]. This could allow a more controlled real-time monitored treatment at a low cost, which could provide a new set of medical tools to resource constrained communities. The key limitations are foreseen to be the tissue types where the therapy could be applied. For example, when the tissue exhibits high baseline conductivity, the changes in pH might not affect the conductivity to such an extent that can be detected using an off-the-shelf EIT system. In a highly conductive tissue, the pH changes can fall under the noise margin of EIT measurements. We have limited this work to conductivity corresponding to a hepatic tumor, in which detection feasibility was demonstrated. We intend to explore the relationship between pH and conductivity, which will allow us to characterize the conductivity regions in which our approach could be potentially useful.

Demonstrating the proof of concept for a new application of the EIT technology was the main goal of this study, i.e. our purpose was to show that EIT is sensitive enough to image electrolysis and we do not claim any improvements in the technology itself. We believe that using EIT for this new application, i.e. monitoring electrolysis could become an enabling technology for planning and monitoring low-cost minimally invasive surgery procedures involving electrolysis. The configuration employed in this study and the *E.Coli* experiments have an immediate application in relation to the use of electrolysis and electrolytic products to sterilize wounds and destroy microorganisms on the surface of the body, the skin [1,47]. Placing an EIT array around a surface wound, treated with electrolysis, could provide a means to monitor the treatment. While EIT has numerous technical limitations, the application in the configuration tested which focuses on the outer surface of the body is less restricted by the typical constraints of EIT. The microorganism study employed here is directly relevant and illustrative of this application. Another field of application is in relation to the combination between reversible electroporation and electrolysis[25,26,48,49]. We have recently shown that generating electrolytic products, prior, during or after reversible electroporation can be seen as a new method for tissue ablation[48,49]. The ability to monitor the extent of electrolysis could provide greater control over tissue ablation with this combination, with possible applications to treatment of melanomas.

Reversible electroporation combination with bleomycin is a tissue ablation technology known as electrochemotherapy, also used primarily for treatment of melanoma [50]. Reversible electroporation is also an important technology for gene vaccination[51]. However, it was recently found that electrolytic products can be generated during reversible electroporation for insertion of genes into cells. These products may cause cell death and be detrimental to gene therapy protocols[25]. Using EIT to detect the production of electrolysis during reversible electroporation may benefit gene vaccination; in particular in configurations in which the electroporation is near the surface of the body, the skin.

Given the scope of our work, we see how it suffers from the generic limitations of EIT i.e. nonlinear current spread and the uncertainty associated with the position of the electrodes. Having said that, these are challenges that are not unique to the application we suggest and we do not claim to solve them in this paper. Furthermore, as discussed earlier, the configuration analyzed here may be less susceptible. Our wish in this work was to demonstrate that EIT can reliably detect pH changes. We see this work as a first step towards developing a novel imaging approach and this paper was focused on validating the fundamental feasibility of the method.

CONCLUSION

In summary, we report experimental findings that support the hypothesis that electrolysis induced pH-changes lead to local conductivity changes in a physiological gel tissue model. It is these changes in conductivity that can be captured in real time by EIT. Our results indicate the feasibility of using EIT as a means to monitor dynamic changes in local pH level of a biological sample during an electrolysis process. Our work uses agar-based gel model with conductivity in the range of a biological tissue, and is validated vs. optical images utilizing pH indicator dyes. In addition we demonstrate the relevance of our work in the biological context by correlating bacterial viability data with EIT measurements. Our results are promising, and invite further experimental explorations.

AUTHOR CONTRIBUTIONS

Conceived and designed the experiments: AM BR. Performed the experiments: AM. Analyzed the data: AM BR. Contributed reagents/materials/analysis tools: BR. Wrote the paper: AM BR.

REFERENCES

1.   Amory R (1886) A treatise on electrolysis and its therapeutical and surgical treatement in disease. New York: William Woof &Co.

2.   Van Lerberghe W (2008) The world health report 2008: primary health care: now more than ever: World Health Organization.

3.   Nordenstrom BEW (1978) Preliminary clinical trials of electrophoretic ionization in the treatement of malignant tumors. IRCS Medical Sc 6: 537.

4.   Nordenstrom BE (1989) Electrochemical treatment of cancer. I: Variable response to anodic and cathodic fields. American journal of clinical oncology 12: 530–536. pmid:2556014 doi: 10.1097/00000421-198912000-00015

5.   Nilsson E, Berendson J, Fontes E (1998) Development of a dosage method for electrochemical treatment of tumours: a simplified mathematical model. Bioelectrochemistry and Bioenergetics 47: 11–18. doi: 10.1016/s0302-4598(98)00157-3

6.   Nilsson E, Berendson J, Fontes E (1999) Electrochemical treatment of tumours: a simplified mathematical model. Journal of Electroanalytical Chemistry 460: 88–99. doi: 10.1016/s0022-0728(98)00352-0

7.   Nilsson E, von Euler H, Berendson J, Thorne A, Wersall P, et al. (2000) Electrochemical treatment of tumours. Bioelectrochemistry 51: 1–11. pmid:10790774 doi: 10.1016/s0302-4598(99)00073-2

8.   von Euler H, Nilsson E, Olsson JM, Lagerstedt AS (2001) Electrochemical treatment (EchT) effects in rat mammary and liver tissue. In vivo optimizing of a dose-planning model for EChT of tumours. Bioelectrochemistry 54: 117–124. pmid:11694391 doi: 10.1016/s1567-5394(01)00118-9

9.   von Euler H, Nilsson E, Lagerstedt AS, Olsson JM (1999) Development of a dose-planning method for electrochemical treatment of tumors: A study of mammary tissue in healthy female CD rats. Electro- and Magnetobiology 18: 93-+. doi: 10.3109/15368379909012903

10.   Bergues Pupo AE, Bory Reyes J, Bergues Cabrales LE, Bergues Cabrales JM (2011) Analytical and numerical solutions of the potential and electric field generated by different electrode arrays in a tumor tissue under electrotherapy. Biomedical Engineering Online 10. doi: 10.1186/1475-925x-10-85

11.   Placeres Jimenez R, Bergues Pupo AE, Bergues Cabrales JM, Gonzalez Joa JA, Bergues Cabrales LE, et al. (2011) 3D Stationary Electric Current Density in a Spherical Tumor Treated With Low Direct Current: An

Analytical Solution. Bioelectromagnetics 32: 120–130. doi: 10.1002/bem.20611. pmid:21225889

12. Turjanski P, Olaiz N, Abou-Adal R, Suarez C, Risk M, et al. (2009) pH front tracking in the electrochemical treatment (EChT) of tumors: Experiments and simulations. Electrochimica Acta 54: 6199–6206. doi: 10.1016/j.electacta.2009.05.062

13. Camue Ciria HM, Morales Gonzalez M, Ortiz Zamora L, Bergues Cabrales LE, Sierra Gonzalez GV, et al. (2013) Antitumor effects of electrochemical treatment. Chinese Journal of Cancer Research 25: 223–234. doi: 10.3978/j.issn.1000-9604.2013.03.03. pmid:23592904

14. Yoon D-S, Ra Y-M, Ko D-g, Kim Y-M, Kim K-W, et al. (2007) Introduction of electrochemical therapy (EChT) and application of EChT to the breast tumor. Journal of Breast Cancer 10: 162–168. doi: 10.4048/jbc.2007.10.2.162

15. Czymek R, Dinter D, Loeffler S, Gebhard M, Laubert T, et al. (2011) Electrochemical Treatment: An Investigation of Dose-Response Relationships Using an Isolated Liver Perfusion Model. Saudi Journal of Gastroenterology 17: 335–342. doi: 10.4103/1319-3767.84491. pmid:21912061

16. Griffin D, Dodd N, Zhao S, Pullan B, Moore JV (1995) Low-level direct electrical current therapy for hepatic metastases. I. Preclinical studies on normal liver. British journal of cancer 72: 31–34. pmid:7599063 doi: 10.1038/bjc.1995.272

17. Griffin D, Dodd NJ, Moore JV, Pullan B, Taylor T (1994) The effects of low-level direct current therapy on a preclinical mammary carcinoma: tumour regression and systemic biochemical sequelae. British journal of cancer 69: 875–878. pmid:8180017 doi: 10.1038/bjc.1994.169

18. Miklavčič D, Serša G, Kryžanowski M, Novakovič S, Bobanović F, et al. (1993) Tumor treatment by direct electric current-tumor temperature and pH, electrode material and configuration. Bioelectrochemistry and bioenergetics 30: 209–220. doi: 10.1016/0302-4598(93)80080-e

19. Miklavčič D, Fajgelj A, Serša G (1994) Tumor treatment by direct electric current: electrode material deposition. Bioelectrochemistry and Bioenergetics 35: 93–97. doi: 10.1016/0302-4598(94)87017-9

20. Finch JG, Fosh B, Anthony A, Slimani E, Texler M, et al. (2002) Liver electrolysis: pH can reliably monitor the extent of hepatic ablation in pigs. Clinical Science 102: 389–395. pmid:11914100 doi: 10.1042/cs20010222

21. Olaiz N, Suarez C, Risk M, Molina F, Marshall G (2010) Tracking

protein electrodenaturation fronts in the electrochemical treatment of tumors. Electrochemistry Communications 12: 94–97. doi: 10.1016/j.elecom.2009.10.044

22. Ivic MLA, Perovic SD, Zivkovic PM, Nikolic ND, Popov KI (2003) An electrochemical illustration of the mathematical modelling of chlorine impact and acidification in electrochemical tumour treatment and its application on an agar-agar gel system. Journal of Electroanalytical Chemistry 549: 129–135. doi: 10.1016/s0022-0728(03)00251-1

23. Chu KF, Dupuy DE (2014) Thermal ablation of tumours: biological mechanisms and avdances in therapy. Nature Reviews/Cancer 14: 199–208. doi: 10.1038/nrc3672. pmid:24561446

24. Turjanski P, Olaiz N, Abou-Adal P, Suarez C, Risk M, et al. (2009) pH front tracking in the electrochemical treatment (EChT) of tumors: Experiments and simulations. Electrochimica Acta 54: 6199–6206. doi: 10.1016/j.electacta.2009.05.062

25. Turjanski P, Olaiz N, Maglietti F, Michinski S, Suarez C, et al. (2011) The role of pH fronts in reversible electroporation. PloS one 6: e17303. doi: 10.1371/journal.pone.0017303. pmid:21559079

26. Maglietti F, Michinski S, Olaiz N, Castro M, Suárez C, et al. (2013) The Role of Ph Fronts in Tissue Electroporation Based Treatments. PloS one 8: e80167. doi: 10.1371/journal.pone.0080167. pmid:24278257

27. Davalos RV, Otten DM, Mir LM, Rubinsky B (2004) Electrical impedance tomography for imaging tissue electroporation. Biomedical Engineering, IEEE Transactions on 51: 761–767. pmid:15132502 doi: 10.1109/tbme.2004.824148

28. Davalos R, Rubinsky B (2004) Electrical impedance tomography of cell viability in tissue with application to cryosurgery. Transactions of the ASME Journal of Biomechanical Engineering 126: 305–309. pmid:15179863 doi: 10.1115/1.1695577

29. Yorkey JT, Webster JG (1987) A comparison of impedance tomographic reconstruction algorithms. Clin Phys Physiol Meas 8: 55–62. pmid:3568572 doi: 10.1088/0143-0815/8/4a/007

30. Boone K, Barber D, Brown B (1997) Imaging with electricity: Report of the European Concerted Action on Impedance Tomography. Journal of Medical Engineering & Technology 21: 201–232. doi: 10.1016/j.cmpb.2015.04.003. pmid:25985887

31. Barber D, Brown B (1984) Applied potential tomography. Journal of physics E Scientific instruments 17: 723–733. doi: 10.1088/0022-3735/17/9/002

32. Davalos RV, Rubinsky B, Otten DM (2002) A feasibility study for electrical impedance tomography as a means to monitor tissue electroporation for molecular medicine. Ieee Transactions on Biomedical Engineering 49: 400–403. pmid:11942732 doi: 10.1109/10.991168

33. Edd JF, Rubinsky B (2006) Assessment of the viability of transplant organs with 3D electrical impedance tomography. 2005 27th Annual International Conference of the IEEE Engineering in Medicine and Biology Society (IEEE Cat No05CH37611C): 4 pp.|CD-ROM.

34. Granot Y, Ivorra A, Maor E, Rubinsky B (2009) In vivo imaging of irreversible electroporation by means of electrical impedance tomography. Physics in Medicine and Biology 54: 4927–4943. doi: 10.1088/0031-9155/54/16/006. pmid:19641242

35. Laufer S, Ivorra A, Reuter VE, Rubinsky B, Solomon SB (2010) Electrical impedance characterization of normal and cancerous human hepatic tissue. Physiological measurement 31: 995. doi: 10.1088/0967-3334/31/7/009. pmid:20577035

36. Granot Y, Ivorra A, Maor E, Rubinsky B (2009) In vivo imaging of irreversible electroporation by means of electrical impedance tomography. Physics in medicine and biology 54: 4927. doi: 10.1088/0031-9155/54/16/006. pmid:19641242

37. Brown B (2003) Electrical impedance tomography (EIT): a review. Journal of medical engineering & technology 27: 97–108. doi: 10.1016/j.cmpb.2015.04.003. pmid:25985887

38. Adler A, Gaggero PO, Maimaitijiang Y (2011) Adjacent stimulation and measurement patterns considered harmful. Physiological measurement 32: 731. doi: 10.1088/0967-3334/32/7/S01. pmid:21646709

39. Vauhkonen M, Lionheart WR, Heikkinen LM, Vauhkonen PJ, Kaipio JP (2001) A MATLAB package for the EIDORS project to reconstruct two-dimensional EIT images. Physiological Measurement 22: 107. pmid:11236871 doi: 10.1088/0967-3334/22/1/314

40. Borsic A, Graham BM, Adler A, Lionheart W (2010) In vivo impedance imaging with total variation regularization. Medical Imaging, IEEE Transactions on 29: 44–54. doi: 10.1109/TMI.2009.2022540. pmid:20051330

41. Vogel CR, Oman ME (1996) Iterative methods for total variation denoising. SIAM Journal on Scientific Computing 17: 227–238. doi: 10.1137/0917016

42. Hansen PC (1992) Analysis of discrete ill-posed problems by means of the L-curve. SIAM review 34: 561–580. doi: 10.1137/1034115

43. Farooq A, Tehrani JN, McEwan AL, Woo EJ, Oh TI (2014) Improvements and artifact analysis in conductivity images using multiple internal electrodes. Physiological measurement 35: 1125. doi: 10.1088/0967-3334/35/6/1125. pmid:24845453

44. Nordenström B (1983) Biologically closed electric circuits: Clinical, experimental and theoretical evidence for an additional circulatory system: Ursus Medical AB.

45. Meir A, Hjouj M, Rubinsky L, Rubinsky B (2015) Magnetic resonance imaging of electrolysis. Scientific reports 5. doi: 10.1038/srep08095

46. Cherepenin V, Karpov A, Korjenevsky A, Kornienko V, Mazaletskaya A, et al. (2001) A 3D electrical impedance tomography (EIT) system for breast cancer detection. Physiological measurement 22: 9. pmid:11236894 doi: 10.1088/0967-3334/22/1/302

47. Tamai M, Matsushita S, Miyanohara H, Imuta N, Ikeda R, et al. (2013) Antimicrobial effect of an ultrasonic levitation washer disinfector with silver electrolysis and ozone oxidation on methicillin-resistant Staphylococcus aureus. The Journal of dermatology 40: 1020–1026. doi: 10.1111/1346-8138.12327. pmid:24304000

48. Phillips M, Rubinsky L, Meir A, Raju N, Rubinsky B (2014) Combining Electrolysis and Electroporation for Tissue Ablation. Technology in cancer research & treatment: 1533034614560102.

49. Phillips M, Raju N, Rubinsky L, Rubinsky B Modulating electrolytic tissue ablation with reversible electroporation pulses. TECHNOLOGY 0: 1–9.

50. Marty M, Sersa G, Garbay JR, Gehl J, Collins CG, et al. (2006) Electrochemotherapy–An easy, highly effective and safe treatment of cutaneous and subcutaneous metastases: Results of ESOPE (European Standard Operating Procedures of Electrochemotherapy) study. European Journal of Cancer Supplements 4: 3–13. doi: 10.1016/j. ejcsup.2006.08.002

51. Gehl J, Skovsgaard T, Mir LM (2002) Vascular reactions to in vivo electroporation: characterization and consequences for drug and gene delivery. Biochimica et Biophysica Acta (BBA)-General Subjects 1569: 51–58. pmid:11853957 doi: 10.1016/s0304-4165(01)00233-1

Chapter 4

SYNERGISTIC COMBINATION OF ELECTROLYSIS AND ELECTROPORATION FOR TISSUE ABLATION

Michael K. Stehling[1], Enric Guenther[1], Paul Mikus[1], Nina Klein[2], Liel Rubinsky[1], Boris Rubinsky[1]

[1] Inter Science GmbH, Biophysics, Luzern, Switzerland

[2] Institut fuer Bildgebende Diagnostik - Tumortherapy Center, R&D, Offenbach, Germany

ABSTRACT

Electrolysis, electrochemotherapy with reversible electroporation, nanosecond pulsed electric fields and irreversible electroporation are valuable non-thermal electricity based tissue ablation technologies. This paper reports results from the first large animal study of a new non-thermal tissue ablation technology that employs "Synergistic electrolysis and electroporation" (SEE). The goal of this pre-clinical study is to expand on earlier studies with small animals and use the pig liver to establish SEE treatment parameters of clinical utility. We examined two SEE methods. One of the methods employs multiple electrochemotherapy-type reversible electroporation magnitude pulses, designed in such a way that the charge delivered during the electroporation pulses generates the electrolytic products. The second SEE method combines the delivery of a small number of electrochemotherapy magnitude electroporation pulses with a low voltage electrolysis generating DC current in three different ways. We show that both methods can produce lesion with dimensions of clinical utility, without the need to inject drugs as in electrochemotherapy, faster than with conventional electrolysis and with lower electric fields than irreversible electroporation and nanosecond pulsed ablation.

INTRODUCTION

Minimally invasive surgery employs various tissue ablation technologies, each with their own advantages, disadvantages and specific use. Electric currents passing through a biological medium produce a number of biophysical and biochemical effects which are used in tissue ablation. This study deals with the use of a combination of two different electricity driven phenomena, electrolysis and electroporation.

The electrochemical reactions known as electrolysis occur at the surface of electrodes submerged in an ionic conducting media, during the passage of an electric current [1]. New chemical species are generated at the interface of the electrodes as a result of electron transfer between the electrodes and the ions in solution. The new chemical species diffuse away from the electrodes, into tissue, in a process driven by electrochemical potentials. In physiological solutions, electrolytic reactions yield changes in pH, resulting in an acidic region near the anode and a basic region near the cathode. The cytotoxic environment developing due to local changes in pH, as well as the presence of some of the new chemical species formed during electrolysis of the solution, cause cell death. Electrolysis is harnessed for tissue ablation in medicine, since the early 1800's [2]. The field has experienced a revival in the mid 1970's, with the work of Nordenstrom [3,4]. During the last two decades, substantial research was done on tissue ablation by electrolysis [5–24]. The cited studies include cell and animal experiments, mathematical modeling and clinical work. From an operational standpoint, electrolysis requires very low voltages and currents, providing advantages relative to other ablation techniques, e.g. reduced instrumentation complexity. It is, however, a lengthy procedure, controlled by the process of diffusion and the need for high concentrations of electrolytically-produced ablative chemical species.

Permeabilization of the cell membrane through the application of very brief, high-magnitude electric field pulses characterizes the bioelectric phenomenon of electroporation [25–34]. The effect on the cell membrane is a function of the electric field strength and pulse time length [35–37]. Lower electric fields produce reversible pores in the lipid bilayer, allowing the introduction of molecules such as genes and drugs into cells [32,38]. Higher electric fields produce irreversible defects (pores), resulting in a cell membrane that does not reseal after the field is removed [35]. Reversible and irreversible electroporation have numerous medical applications [39]. Reversible electroporation techniques have been combined with anticancer drugs such as bleomycin to target cancerous tissues for successful clinical use in the field of electrochemotherapy [40]. Reversible electroporation for electrochemotherapy employs voltage over distance between electrodes in the range of between 300 kV/cm and 1–1.5 kV/cm and usually eight pulses [40]. The use of non-thermal irreversible electroporation (NTIRE) for tissue ablation is a more recent addition to the armamentarium of tissue ablation techniques available to surgeons [41–44]. Its use results in direct cell death without the need to introduce drugs or other molecules to facilitate the treatment process [45]. Irreversible electroporation usually employs up to one hundred pulses of microsecond length and electric fields in the single kV/cm range. Another recent non-thermal and non-chemical approach, known as nanosecond pulsed electric fields, uses much shorter pulses in the nanosecond domain, and increased electric field strengths in the range of tens or hundreds of kV/cm [46–48].

A major advantage of tissue ablation by electroporation is the relative speed of the procedure in comparison to any other ablation technique. Furthermore, because the procedure primarily affects the cell membrane, critical features of the extracellular matrix are spared, and this ablation modality can be used to treat tissues, such as the pancreas, without concern for collateral damage [45]. Current pulsed electric field tissue ablation approaches, however, have their respective disadvantages. Electrochemotherapy requires the injection of drugs into the tissue, while the high electric fields used in irreversible electroporation were found to cause muscle contractions, electric discharge associated with high pressure waves and may affect the electrical function of the heart.

Two recent studies, performed in a small animal model, have shown that combining electroporation with electrolysis has a synergistic effect and yields a tissue ablation technique that has certain advantages over tissue ablation with electroporation or electrolysis delivered separately [49,50] We will refer to the combination of electroporation and electrolysis "synergistic electroporation and electrolysis" (SEE). We have developed several methods of SEE. One of them employs electrolysis as the central tissue ablation modality and adds reversible electroporation electrochemotherapy-type pulses to permeabilize the cell membrane, making the cell more susceptible to lower amounts of products of electrolysis [50]. Several different combinations of electrolysis and electroporation are possible [50]. Another method of SEE is based on the fact that electric currents produce a variety of different effects, simultaneously [49] Electric currents which accompany the electroporation pulses also generate products of electrolysis. Therefore, we applied multiple reversible electroporation electrochemotherapy-type pulses to also generate products of electrolysis that are toxic and thus cause cell death to the reversible permeabilized cell membrane. This method of SEE ablation causes cell death with lower electric field pulses than the pulses used in irreversible electroporation or nanopulse ablation alone [49].

The study reported in was acute and done on a small animal model. While the principles of SEE were demonstrated in [49,50] the tissue ablation parameters and geometrical configurations used in those papers are not clinically applicable. Therefore, the primary goal of this pre-clinical study is to expand on the concepts introduced in [49,50] and explore the use of SEE in settings that are more relevant to clinical use. To this end we have tested a variety of potentially clinical SEE protocols. in an *in vivo* pig liver model. Our results confirmed the findings of [49,50] in a large animal model, and provide further clinical insight into the use of the combination of electrolysis and electroporation for tissue ablation with relevance to clinical procedures.

MATERIALS AND METHODS

The study was conducted on *in vivo* pig liver and approved by the PMI's Institutional Animal Care and Use Committee IACUC (PMI—Pre Medical Inovation, San Carlos, CA, USDA number: 93-R-0506, Study number: ANS 2094). We used three female pigs, weight 30 kg to 40 kg, treated in accordance with Good Laboratory Practice regulations as set forth by the 21 Code of Federal Regulations (CFR) Part 58. Each procedure started with anesthetization of the animal under general anesthesia per SOP #33156. Preanesthetic medication was Telazol 4.0 mg/kg (2.0 ml) IM and Atropine 0.02 mg/kg (1.8 ml) IM. Anesthetic induction was done by Isoflurane with Oxygen, 2%/2L/minute via mask. Possible postoperative pain was ameliorated by Buprenorphine 0.01 mg/kg IM Pre-med at recovery and Carprofen 4 mg/kg at extubation/ recovery. Antibiotics administered during surgery was Cefazolin 25 mg/kg IV every 2 hours. In addition, pancuronium (0.1 mg/kg, at a dose of 1 mg/ ml) was administered through an IV prior to the procedure, to reduce muscle contractions during the application of the electrical pulses. Pancuronium (0.05 mg/ml at 1 mg/ml) was administered throughout the procedure as needed. The liver was exposed via a midline incision. The treatment was delivered using two 18 gauge Titanium needles (Inter Science GmbH, Ch) with a sliding insulating sheath inserted in the liver. We have used Ti needles to eliminate possible electrolytic products involving the electrode materials. The 18 gauge variable length electrodes were custom designed for the delivery of both electroporation and electrolytic pulse sequences.

Two electrodes were inserted in the liver in a roughly axial parallel configuration, normal to the liver surface, under ultrasound monitoring (Fig 1). Ultrasound images were also taken throughout the procedure. Square DC electroporation pulses were applied to the liver through the electrodes from a DC pulse generator (BTX, Harvard Instruments, USA). The electrodes were also connected to an Arbitrary Function Generator (AFG 3102, Tektronix, Beaverton, OR) to produce a constant current for a fixed period of time for electrolysis. The parameters varied in this study were: the voltage and the current of the electroporation and electrolysis pulses, the number of pulses, the distance between electrodes as well as the sequence of electrolysis and electroporation pulses. Animals were sacrificed at 24 hours. The pigs were euthanized using Euthasol 1 ml/lb IV.

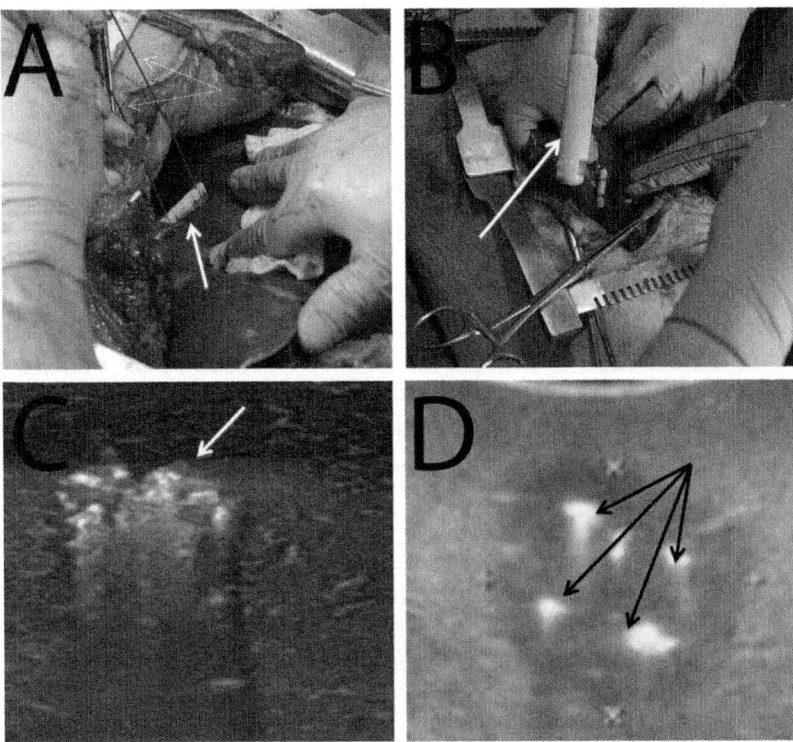

Figure 1. Photographs of the surgery and the ultrasound images taken during the procedure. (A) Placement of two electrodes (dashed arrows) through a grid (solid line arrow) into the liver. (B) Use of an intraoperative ultrasound (solid arrow) to monitor the placement of the probes and the tissue ablation. (C) Ultrasound images of an electrolytic treatment with two electrodes (note the bright hyperechoic area—marked by arrow). (D) Ultrasound images of an electroporation treatment with four electrodes (note the bright hyperechoic area—marked by arrows). (Bottom right figure used with permission of the publisher [44]).

To fix the liver in its current state for microscopic viewing, a Foley catheter was placed into the descending aorta and the hepatic vein was snipped off for drainage of the affluent. The liver was flushed with physiological saline for ten minutes at a hydrostatic pressure of 80 mmHg from a pressurized IV drip. Immediately following saline perfusion, a 10% formalin fixative was perfused in the same way for ten minutes. The liver lobe in which the SEE lesion was made was removed and stored in the same formalin solution. For microscopic analysis, the tissue was bread loafed perpendicular to the capsule surface and parallel to the needle tracts. All cassettes were processed routinely from 10% phosphate buffered formalin to wax blocks. Five micrometer sections were made from each block and stained with hematoxylin and eosin for histologic examination. The stained samples were scanned with a digital microscope D-Sight Fluo 2.0 (A. Menarini, Diagnostics, S.r.l, Firenze, Italy) in preparation for histological examination. The digitized histological images were examined blindly by an independent histology service company and reports were prepared (Narayan Raju, Inc). The focus of the histology was to verify the extent and nature of tissue ablation with SEE; in particular in relation to locations relative to the electrodes and the continuity of the lesion between the electrodes.

To compare lesion sizes, the relative ablation extent was calculated. For the axial histological cuts, a line was drawn between the centers of the insertion sites of the electrodes. Then, both the ablation extent and the total length was measured and put in relation. For the coronary cut slides, the measurement was taken at 0.5 cm of the exposure length to be able to compare to the axial cuts, which were cut roughly at the same length.

RESULTS

Twelve lesions were produced in the liver of three pigs with a variety of SEE parameters and electrode placement configurations. This was a first large animal study of SEE and therefore we pursued several different goals, such as: testing the effects of SEE parameters, methods and combinations, pre-clinical practice and safety. All the animals survived the procedure without any complication.

The experimental parameters were chosen in such a way as to gain insight into the effects of the various combinations of electrolysis and electroporation on the extent of tissue ablation. Similarly to the evaluation technique in [49,50] we used the extent of the ablated tissue between the electrodes as a measure of the treatment effectiveness. Specifically, we evaluated to what extent the region between electrodes was completely ablated. Table 1 provides details on the treatment parameters in the lesions analyzed in this study and the extent of tissue ablation between the electrodes.

Table 1. Details of the treatment parameters

Experimental parameters				1. Electroporation sequence				Electrolysis		2. Electroporation sequence				Relative Ablation
Fig #	Lesion #	d /cm	e /cm	U /V	p/μs	N	f/Hz	I /mA	t/min	U /V	p /μs	N	f/Hz	ablation/total length
2A, 2B	1	1.5	1	n/a	n/a	n/a	n/a	50	5	n/a	n/a	n/a	n/a	0.40
2C, 2D	2	1.5	1	500	100	16	1	n/a	n/a	n/a	n/a	n/a	n/a	0.00
2E, 2F	3	1.5	1	500	100	8	1	50	5	n/a	n/a	n/a	n/a	0.61
2G, 2H	4	1.5	1	500	100	8	1	50	5	500	100	8	1	0.71
3A	5	2.5	1	n/a	n/a	n/a	n/a	60	10	n/a	n/a	n/a	n/a	0.30
3B	6	2.5	1	3000	100	16	1	n/a	n/a	n/a	n/a	n/a	n/a	0.41
3C	7	2.5	1	3000	100	8	1	60	10	3000	100	8	1	0.77
4A, 4B	8	1.5	1	1000	100	8	1	n/a	n/a	n/a	n/a	n/a	n/a	0.25
4C, 4D, 5, 6	9	1.5	1	1000	100	297	1	n/a	n/a	n/a	n/a	n/a	n/a	0.66
7A–7D, 8, 9	10	1.5	1	1000	100	8	1	75	10	1000	100	8	1	1
7E	11	1.5	1	1000	100	8	1	100	10	n/a	n/a	n/a	n/a	0.97
7F	12	1.5	1	n/a	n/a	n/a	n/a	100	10	1000	100	8	1	0.81

doi:10.1371/journal.pone.0148317.t001

The columns of the table give figure number, lesion number, distance between the electrodes (d; cm) and exposed electrode length (e; cm) for each lesion shown in the figures. This is followed by data for the first electroporation sequence with the parameters voltage (U; V), pulse length (p; μs), amount of pulses applied (N) and frequency (f; Hz). The next column shows the parameters of the electrolysis sequence with the parameters current (I; mA) and time length of delivery (t; min). The next column shows the data for the second electroporation sequence (same parameters as the first sequence). The last column gives the ratio of the extent of tissue ablation between the electrodes and the distance between the electrodes. The line was drawn at about 0.5 cm of exposed length to be able to compare the coronary histological cuts with the axial ones, as the latter was cut at roughly 0.5 cm.

Fig 1 shows photographs from the surgical procedure and ultrasound images taken during the procedure. Ultrasound monitoring was utilized in all the experiments, including electrolysis and multiple pulse electroporation. Fig 1C shows an ultrasound image generated around two Ti needles during a treatment with electrolysis. For comparison, we show in Fig 1D an ultrasound image of a clinically relevant irreversible electroporation procedure performed using four stainless steel needles in a square configuration [44]. The electroporation study in [44] was done by applying a sequence of electroporation pulses, between each adjacent electrodes on each face of the square-like arrangement of electrodes. The sequence consisted of eight 2.5 kV pulses delivered for 100 microseconds at 10 Hz. It is interesting to notice the hyperechoic areas around the electrodes in both electroporation and electrolysis.

Fig 2 shows results from a set of experiments, particularly designed to use below-conventional irreversible electroporation voltages, to produce a non-contiguous lesion between the electrodes. This was done to verify conclusions from [49,50] in regards to the synergistic effects of electroporation and electrolysis on tissue ablation. The figure is a comparison between extent of tissue ablation from electrolysis, electroporation and a combination of electroporation and electrolysis. The left column (Fig 2A, 2C, 2E and 2G) shows results from macroscopic sections obtained after flushing the liver with saline and formalin. The dark areas are regions where the blood circulation has ceased and are indicative of regions of cell death [43,44]. The right column (Fig 2B, 2D, 2F and 2H) is for tissues stained with Mason trichromatic stain. The area of cell death from the central core of the lesions (the site of the electrodes) extends to and is demarcated by an outer boundary of pale hepatic cells considered non-viable. The top row (Fig 2A and 2B) shows the appearance of tissue treated by electrolysis with 50 mA for 5 minutes. The lesions have formed around the electrodes, with a gap of intact tissue between them. The minimal width of the unaffected tissue is greater than 1 cm. The second row (Fig 2C and 2D) shows the appearance of tissue treated with 16 electroporation pulses (parameters (a), as described in the text below). No tissue ablation is observed. The ablated tissue on the top of the liver, indicated with an arrow, is a thermal mark made with a bowie to identify the location in which the electrodes were inserted in the tissue. In the samples seen in the second to last row (Fig 2E and 2F) a combination of electroporation and electrolysis was delivered as follows: (a) electroporation; 500 V, 100 microsecond pulse, 1 Hz, 8 pulses, followed by (b) electrolysis; 50 mA for 5 minutes. Here, the ablation ratio is with 0.61 significantly bigger than with electrolysis (0.4) or electroporation (0.0) alone. The samples in the bottom row were treated with a combination of electroporation (a), electrolysis (b) and a second sequence of electroporation with identical electroporation parameters. It is evident that the ablated regions near the electrodes in Fig 2G and 2H are much larger than for the case (a) of electrolysis or case (b) of electroporation alone. While the ablated zone between the electrodes is not continuous, the minimal width of the gap of untreated tissue between the ablated zone is less than 5 mm and much smaller than for electrolysis alone. The relative ablation is 0.71, which is larger than the lesions in Fig 2E and 2F.

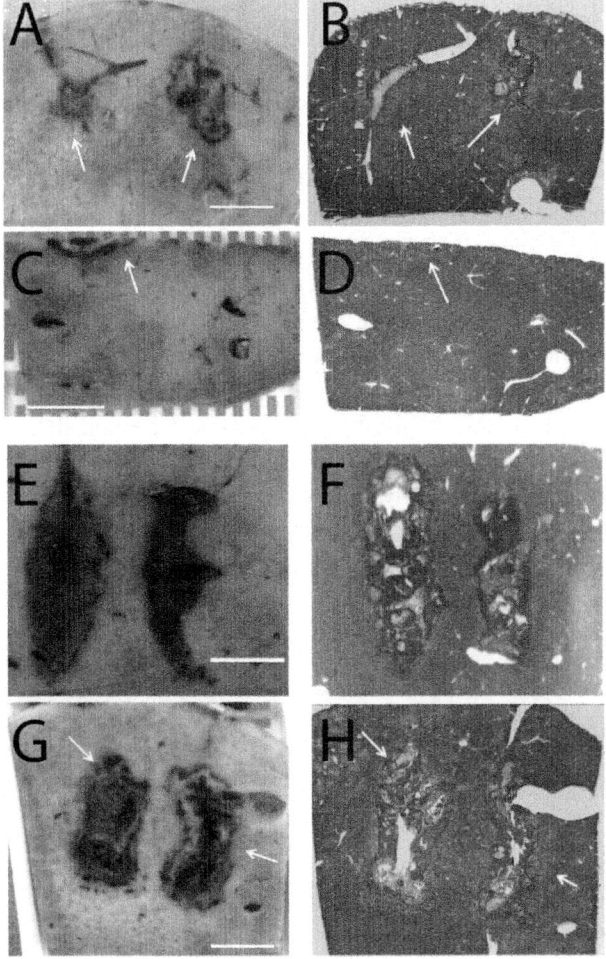

Figure 2. Comparison between extent of tissue ablation from electrolysis, electroporation and combination electroporation and electrolysis. Applied parameters are given in Table 1. (A),(C),(E),(G) Gross macroscopic sections. (B),(D),(F),(H) Trichromatic stained slides. (A) and (B) Electrolysis; 50 mA for 5 minutes. (C) and (D) Electroporation; 500 V, 100 microsecond pulse, 1 Hz, 16 pulses. (E) and (F) Electroporation; 500 V, 100 microsecond pulse, 1 Hz, 8 pulses, followed by electrolysis; 50 mA for 5 minutes. (G) and (H) Electroporation; 500 V, 100 microsecond pulse, 1 Hz, 8 pulses, followed by electrolysis; 50 mA for 5 minutes, followed by electroporation; 500 V, 100 microsecond pulse, 1 Hz, 8 pulses. Scale bar shows 1 cm. Arrows point to lesions.

Fig 3 shows results from another set of experiments, particularly designed to use a larger than conventional irreversible electroporation distance between electrodes to produce a non-contiguous lesion between the electrodes. This was done to verify conclusions from [49,50] in regards to the synergistic effects of electroporation and electrolysis on tissue ablation. Again, the figure compares the extent of tissue ablation from electrolysis, electroporation and a combination of electroporation and electrolysis with two Ti electrodes nominally separated by 2.5 cm (indicated by the arrows). The figure shows results from macroscopic sections obtained after flushing the liver with saline and formalin. The dark areas indicate regions of cell death, as they are the site of ceased blood circulation [44]. The figure compares three types of treatments: A) treated with electrolysis only; B) treated with electroporation only; C) treated with: (a) electroporation, followed by (b) electrolysis, followed by (c) electroporation. The samples were oriented in such a way as to facilitate comparison of the tissues between the electrodes. It is evident that the damage from electrolysis only is negligible and centered around the electrodes. The damage from these high electric field electroporation pulses is more substantial, but the unaffected region between electrodes is of significant size, well over 1 cm. The damage from the combination is greater than from electrolysis or electroporation alone and extends in parts across the entire region between the electrodes.

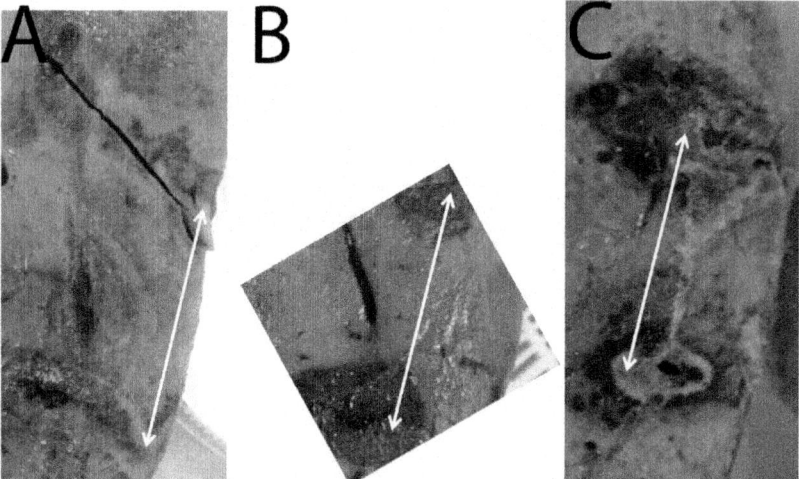

Figure 3. Comparison between extent of tissue ablation from electrolysis, electroporation and the combination electroporation and electrolysis. In this set of experiments, the two Ti electrodes were separated by 2.5 cm. Parameters are given in Table 1. Gross macroscopic section. (A) Electrolysis; 60 mA for 10 minutes. (B) Electroporation; 3000 V, 100 microsecond pulse, 1 Hz, 16 pulses. (C) Electroporation; 3000 V, 100 microsecond pulse, 1 Hz, 8 pulses, followed by electrolysis; 60 mA for 10 minutes, followed by electroporation; 3000 V, 100 microsecond pulse, 1 Hz, 8 pulses.

Fig 4 is relevant to the clinical use of the combination electrolysis and electroporation. It is an implementation of the concept introduced in [49] concerning the use of multiple low voltage electric pulses for tissue ablation. The figure compares the ablation caused from 8 typical electrochemotherapy magnitude type pulses (Fig 4A and 4B) to the ablation caused by 297 such pulses (Fig 4C and 4D. The extent of tissue ablation is clearly much larger in the 297 pulse study than in the 8 pulse study. Comsol Multiphysics calculated isoelectric fields produced by the voltages used in this study are superimposed on the images in the bottom panels. It is evident that in the 8 pulses case the ablation extends to electric fields higher than 500 V/cm, which is typical to irreversible electroporation cell damage. In the case of 297 pulses the damage extends to electric fields of 200 V/cm, which are typical to the combination electrolysis electroporation cell damage. Microscopic details from the 297 pulse treatment are shown in Figs 5 and 6.

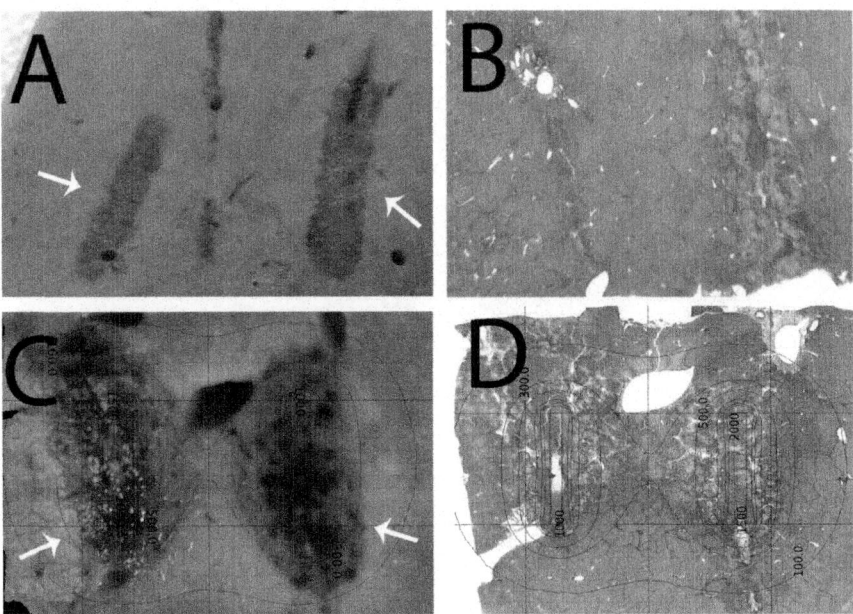

Figure 4. Comparison between extent of tissue ablation by different numbers of electroporation pulses. A detailed list of the parameters can be read in Table 1. (A) and (C) Gross macroscopic sections. (B) and (D) Trichromatic stained slides. Electroporation pulse parameters—1000 V, 100 microseconds, 1 Hz. (A) and (B) 8 pulses. (C) and (D) 297 pulses. Calculated isoelectric field lines are superimposed on the panels in the bottom row. They are also applicable to the top row. In all panels, the anode is the right electrode, while the cathode is on the left.

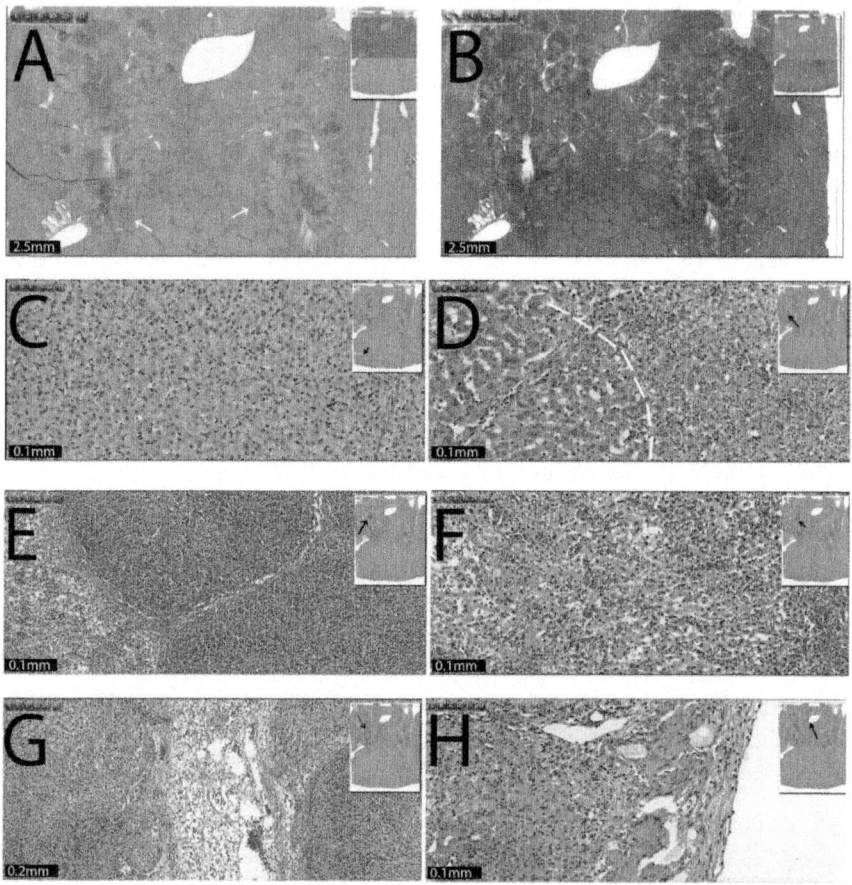

Figure 5. Electroporation with 1000 V, 100 microseconds, 1 Hz, 297 pulses. Applied parameters are given in Table 1. The left electrode is always the anode, while the cathode is on the right side. (A) Magnified H&E and (B) Mason trichromatic stained tissues, focusing on the region between the electrodes. (C) Control. (D) Interface between affected and normal area. (E) The core of the treated region on the left hand side (x10). (F) Magnification (x20) from the core on the left hand side. (G) Left hand side electrode core near a dehydrated region (x10). (H) The left hand side lesion near a large blood vessel. Rectangles and black arrows show the site from which the magnified sample was taken. Scale bars and magnification are given in the figures.

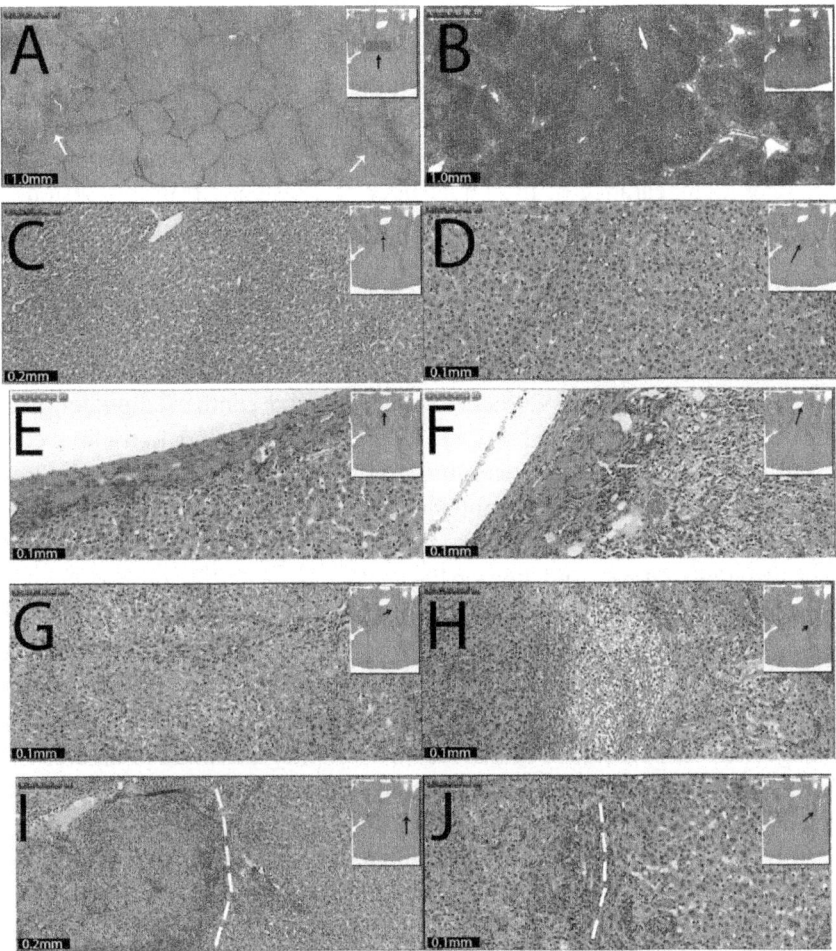

Figure 6. Microscopic details from the 297 pulse treatment. (A) and (B) Low magnification micrographs from different sites of the middle region between electrodes. (C) and (D) 10x and 20x magnification of the tissue in the area midway between the electrodes. (E) and (F) 20x magnification of tissue adjacent to the large blood vessel, with sites in the area midway between the electrodes (E) and tissue near the large blood vessel on the side of the right electrode (F). (G) and (H) Micrographs from sites in the core of the region affected by the right hand side electrode. (I) and (J) Micrographs from sites at the right hand side, outer edge of the right electrode. The sites from which the magnified micrographs were taken are marked with a square in the insert. Scale bars and magnification are given in the figures.

The top row two panels in Fig 5 are higher magnification of the H&E stain (Fig 5A) and Mason trichromatic stain (Fig 5B) images from Fig 4. Both kinds of staining show a section of the liver with parallel vertical lesions traversing the parenchyma, following the tract of the electrodes. The lesions are characterized as acute tissue necrosis with edema and hemorrhage. The outer boundaries of the tissue injury between the electrodes are pale and swollen. There is a wedge of liver tissue between the lesions (electrode) that appears to be normal and intact (viable). Fig 5C functions as a control and shows the microscopic appearance of the intact liver at a distance from the lesion. The location of the site from which the magnification was taken is shown in the rectangle and marked with an arrow in the insert. This section of the liver has normal and intact cellular details. Fig 5D shows the left margin of the treated zone. There is clear demarcation between the normal hepatocytes on the left compared to the ablated cells on the right (dashed line). The affected cells are swollen and/or contracted with pale cytoplasm and condensed nuclei. Edema and congestion are also noted. Fig 5E shows a 10x magnification of the core of the left lesion. The H&E stained micrograph shows a section of the liver with acute cellular necrosis throughout the field. A more severe effect is present in the left side of the panel, where the cellular architecture is completely destroyed with edema and hemorrhage. The hepatocytes on the right side of the panel are swollen and/or condensed with loss of cytoplasmic detail. There is mild edema and congestion. Fig 5F shows a higher magnification (x20) of Fig 5E. The image is showing necrotic hepatocytes with disrupted sinusoidal pattern. Cells are swollen and/or contracted with dark nuclei. There is congestion/hemorrhage. Fig 5G shows a 10x magnification of an area of the core near a dehydrated region. The section shows an area of extreme treatment related hepato-cellular destruction surrounding the core of the electrode pathway. Notice marked loss of hepatocytes in the core, which is filled with fibrous substance and red blood cells. Fig 5H illustrates an ablated region on the left hand side of a large blood vessel. The section shows a wide field of pale ablated cells adjacent to the vessel. There are small and large pockets containing proteinaceous pink substance. The liver cells on the left side of the panel are compacted. No cellular outline is discernible. Some endothelial cells of the blood vessel may be preserved.

Fig 6 also shows the results of the 297 electroporation pulses case (lesion 9) at other sites in the multiple electroporation treated tissue. Fig 6A and 6B show lower magnification images of the tissue mid-way between electrodes. A large band (or plate) of intact hepatic cells (lobules) are seen between the outer margins of the left and right lesions. Fig 6C and 6D show 10x and 20x magnification micrographs of the tissue in the treated region midway between the electrodes. Well preserved, normal hepatocytes can be seen. Due to the massive damage of the surrounding area, some very mild sinusoidal congestion is visible. Fig 6E shows a magnification of tissue adjacent to a large blood vessel in the area midway between the electrodes. Histologically, the hepatocytes and the endothelial cells of the blood vessel appear intact. In Fig 6F can be seen a section of the ablated region on the right hand side near a large blood vessel. The section shows a wide field of pale ablated cells adjacent to a large blood vessel. The liver cells on the right side of the panel are compacted. No cellular outline is discernible. Some endothelial cell of the blood vessel may be preserved.

The fourth row (Fig 6G and 6H) shows micrographs from the core of the area affected by the right hand side electrode. Fig 6G displays a section of the liver with acute cellular necrosis throughout the field, while Fig 6H shows an area in which the cellular structure is completely disintegrated with marked loss of hepatocytes in the core, filled with fibrous substance and red blood cells. The fifth row (Fig 6I and 6J) illustrates micrographs from sites at the right hand side outer edge of the right side electrode. There is clear demarcation between the normal hepatocytes on the right compared to the ablated cells on the left (dashed lines). The affected cells are swollen and/or contracted with pale cytoplasm and condensed nuclei. Edema and congestion is also visible.

Fig 7A–7D shows results from a possible clinical protocol, in which low voltage electroporation pulses are combined with electrolysis. The protocol consists of: typical electrochemical magnitude type electroporation (1000 V, 100 microsecond pulse, 1 Hz, 8 pulses), followed by electrolysis (75 mA for 10 minutes), followed by another typical electrochemical magnitude type electroporation. The treatment was delivered with two Ti needles at a nominal distance of 1.5 cm and an exposed length of 1 cm. The anode is the top electrode and the cathode is the bottom electrode. Fig 7A shows a macroscopic section of tissue, while stained slides can be observed in Fig 7B (H&E staining) and 7C (Mason trichromatic stain). Fig 7D is a magnified micrograph of the region between the electrodes. Here, unlike the case for 297 pulses, the lesion is continuous between the electrodes. The figures of the stained tissues show two distinct adjacent lesions arranged somewhat is a "dumb-bell" shape. The upper lesion is relatively more pronounced and contains severe necrotic acellular debris with charring and edema. The necrotic center is surrounded by a thick circular zone of coagulation necrosis, followed by an outermost zone marked by hemorrhagic liver cell necrosis. The lower lesion is slightly less severe and is characterized by a central region of coagulation necrosis, with the outermost circular thick zone hemorrhagic necrosis merging with the hemorrhagic zone of the upper lesions, without discernible interface or normal liver tissue between the two electrodes induced lesions. Fig 7E and 7F show the results of an experiment with the same combination, but with the permutations electroporation/electrolysis and electrolysis/electroporation. Comparing the lesions shows that all three permutations produced a similar extent of continuous ablation between electrodes, which illustrates that the exact sequence of the permutations of electrolysis and electroporation does not affect the results. On the contrary, the delivery of 297 electroporation pulses of the kind did not produce an uninterupted lesion.

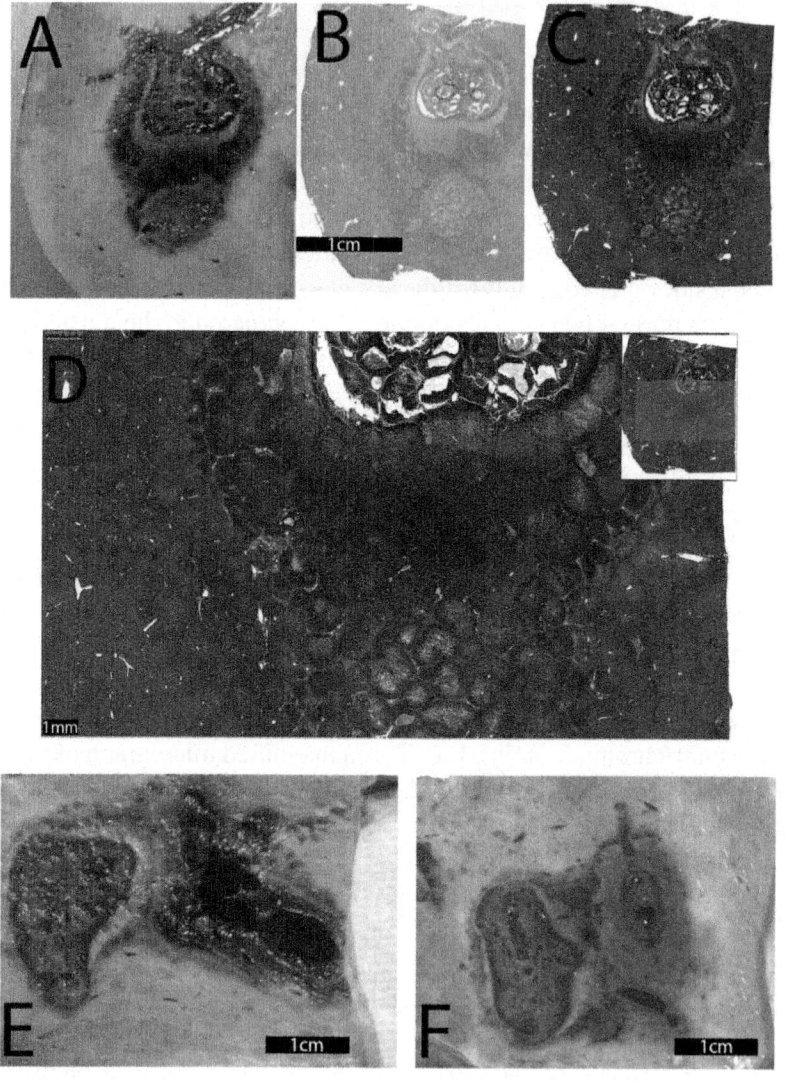

Figure 7. Electroporation followed by electrolysis followed by electroporation. The parameters were: Electroporation—1000 V, 100 microseconds, 1 Hz, 8 pulses followed by electrolysis—10 minutes, 75 mA, followed by electroporation—1000 V, 100 microseconds, 1 Hz, 8 pulses. Used were two Ti electrodes separated by nominal 1.5 cm and exposed length of 1 cm. In this experiment, the top electrode was the anode,

while the cathode was placed at the bottom. (A) Macroscopic image. (B) H&E staining. (C) trichromatic stain. (D) Magnified Mason trichromatic stained tissue, focusing on the region between the electrodes. The insert shows a highlighted rectangle representing the region from where the magnification was taken. (E) Electroporation (1000 V, 100 microseconds, 1 Hz, 8 pulses), followed by electrolysis (10 minutes, 100 mA). (F) Electrolysis (10 minutes, 100 mA) followed by electroporation (1000 V, 100 microseconds, 1 Hz, 8 pulses)

The lesion in Fig 8 is in higher magnification than illustrated in Fig 7, but shows the same treatment taken at different sites. The sites from which the images are taken are marked in the insert with a highlighted rectangle and a black arrow. The magnification and scale bar are given in the figures. The control image in Fig 8A shows a section of the liver stained with H&E, which contains normal liver cells with no evidence of treatment related injury. Fig 8B shows the edge of the lesion at the anode side. This section of the liver stained with H&E shows acutely necrotic liver cells at the margin of the lesion. A more profound necrosis can be seen on the right side where the cellular architecture is affected and has hemorrhage (dashed line). A relatively narrow region of delineation occurs between the affected tissue on the right and the normal tissue on the left. It is seen as a pale zone that is less affected and is mixed with cells that appear normal. The cells on the very left appear normal. Fig 8C displays an area towards the core of the anode treated region. This section of the liver is stained with H&E and shows severe coagulation with congestion/hemorrhage. This is followed by another site towards the core of the treated region near the anode (Fig 8D). This section of the liver, stained with H&E, has massive tissue necrosis with complete architectural disruption. The most extreme right edge is fragmented and probably dehydrated (shrinkage), followed by lobules that have thin strips of liver cells separated by clear spaces (edema). There is also coagulation necrosis and hemorrhage. The third row (Fig 8E and 8F) shows the appearance of the tissue at the core, near the anode at 20x (Fig 8E) and 10x (Fig 8F) magnification. The H&E stained section shows a markedly necrotic center containing acellular tissue debris bordered by a dark band of coagulated tissue with loss of cellular details. This is followed by a zone of dehydrated or condensed hepatocytes with edema. All cells in this region are necrotic. Fourth and fifth row (Fig 8G–8J) show micrographs of the edge of the anode affected region towards the cathode. The sections are stained with H&E and trichromatic stains and captured at 10x (Fig 8G and 8H) and 20x (Fig 8I and 8J) magnification. Sections have severe necrosis of the liver cells in each of the lobules. The cells are completely effaced. Sinusoidal edema is also present.

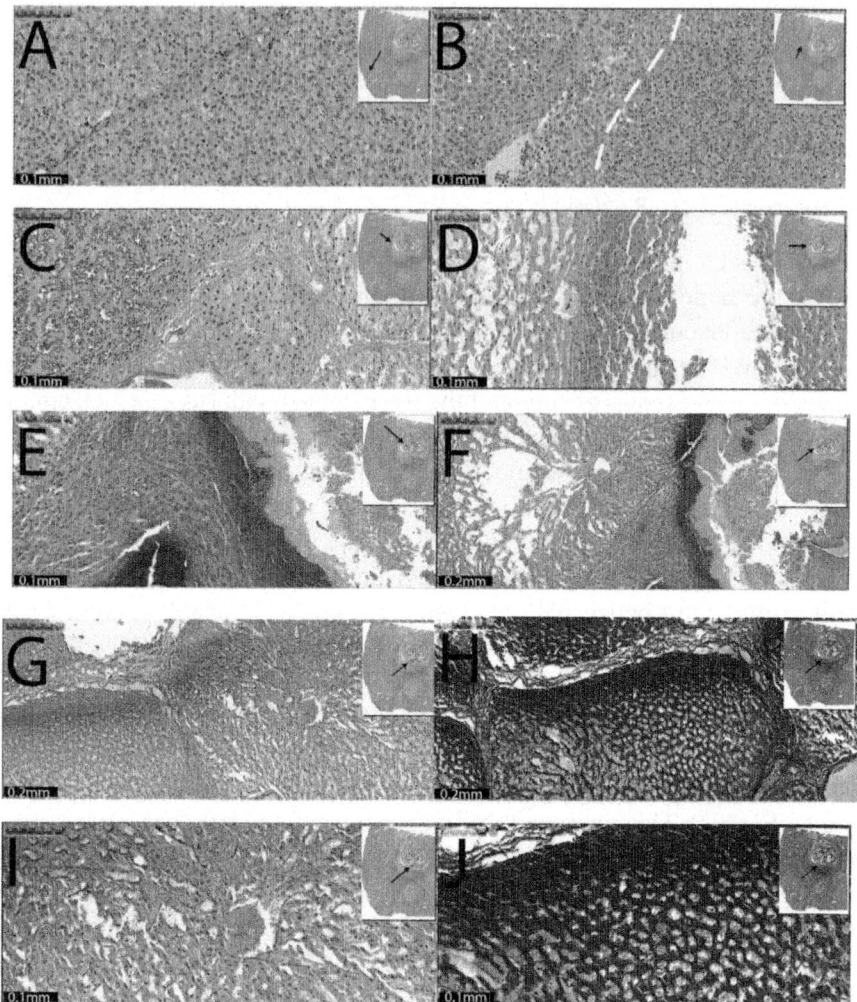

Figure 8. Electroporation followed by electrolysis followed by electroporation. The lesions shown are in higher magnification than illustrated in Fig 7, but are from the same treatment taken at different sites. (A) Control. (B) This section of the liver stained with H&E contains normal liver cells with no evidence of treatment related injury; outer edge of lesion at the anode side. (C) and (D) Micrographs towards the core of the anode treated region. (E) and (F) Micrographs at the anode core with 20x (E) and 10x (F) magnification. (G)—(J) Sites at the margin of the anode affected region towards the cathode, with (G) and (H) in 10x and (I) and (J) in 20x magnification. Tissues were stained in H&E ((G), (I)) and Mason trichromatic stain ((H), (J)).

Fig 9 shows higher magnification micrographs of the treatment in Fig 7, but samples were taken at different sites which are marked in the insert with a highlighted rectangle and an arrow. The magnification and scale bar are given in the panels. Fig 9A shows a site midway between the anode and cathode, towards the anode in H&E staining. The liver cells in the lobules are affected by acute necrosis and have either swollen and/or condensed cytoplasm with dark attenuated nuclei. Some sinusoidal edema is also present. Fig 9B shows a site midway between the anode and cathode towards the cathode. This H&E section shows the necrotic hepatic region gradually transiting into a zone of hemorrhagic necrosis below, towards the cathode. The liver cells have inconspicuous cell boundaries and their nuclei are attenuated. The transition between the tissue ablation modality near the anode to that near the cathode is continued in the panels of the second row (Fig 9C and 9D). Fig 9C shows an H&E stained section which has hemorrhagic necrosis on the top and gradually transitions into more compact and necrotic lobules at the bottom. Further towards the cathode (Fig 9D) is a H&E stained section that has hemorrhagic necrosis on the top with abrupt transiting into cellular necrosis and edema at the bottom. Third and fourth rows (Fig 9E–9H) show the core of the ablated region near the cathode at 10x and 20x magnification, with H&E staining (Fig 9E and 9G) and with Mason trichromatic stain (Fig 9F and 9H). The sections expose acute necrosis of the liver cells in individual lobules. Notice the liver cells are becoming desiccated and individualized with sinusoidal edema. There is a substantial difference between the appearance of the tissue at the core of the lesion near the anode and near the cathode. The anode inflicted massive necrosis that spread to an extensive area of the liver surrounding the core. The core was reduced into mangled acellular tissue debris and edema. The immediately surrounding zone had a thick plate of coagulated liver cells followed by a wide zone of hemorrhagic necrosis. The outermost zone was delineated by acute necrosis and cell swelling. In comparison, the cathode at the bottom inflicted a less severe but complete necrosis of the liver cells confined to narrow proximity. The affected fields somewhat maintained the lobular pattern but the hepatocytes sustained cell death. Fifth row (Fig 9I and 9J) shows micrographs at the outer margin of the cathodic lesion. Fig 9I (10x magnification) shows the lesion with trichromatic stain, and Fig 9J (20x magnification) was stained with H&E. Both stained slides show a field of hemorrhagic necrosis (top left) interfacing with a field of acutely necrotic liver cells in the bottom half of the section.

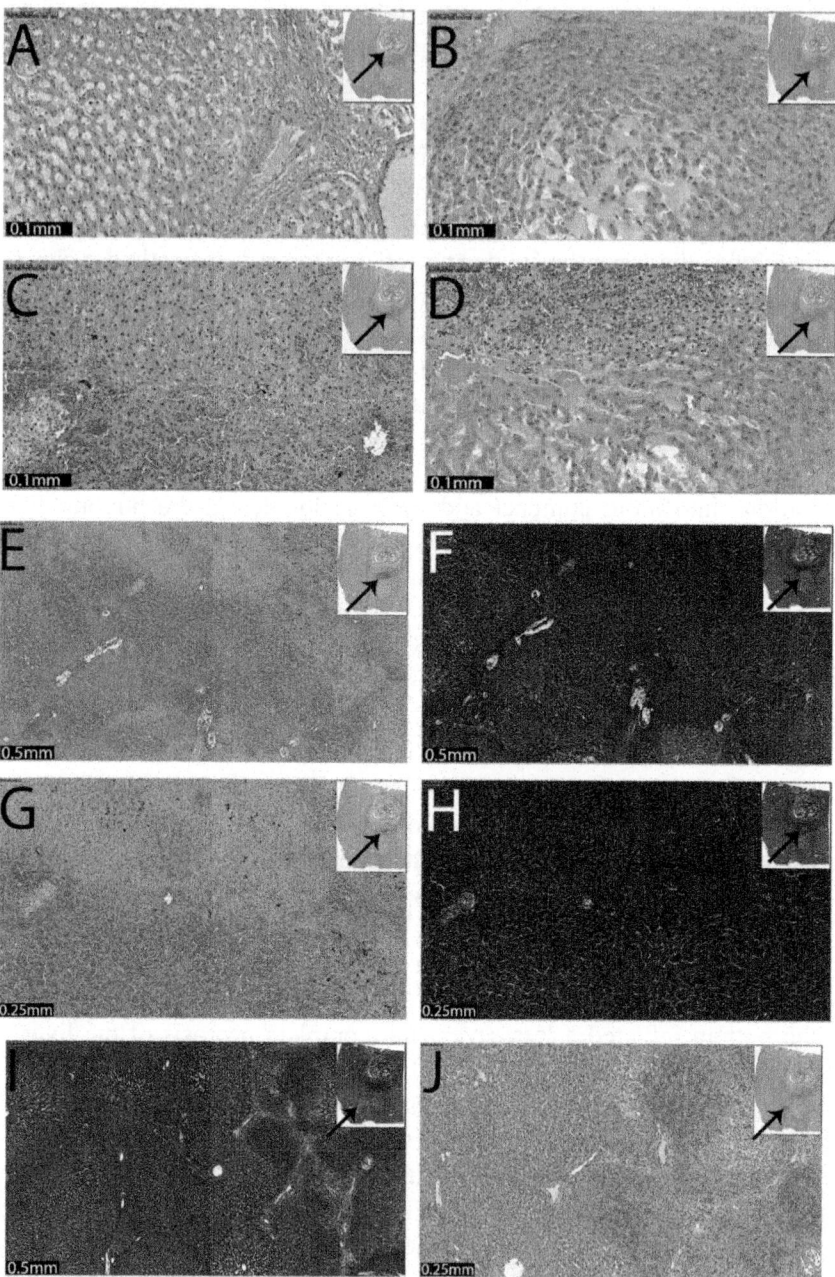

Figure 9. Higher magnification micrographs of the treatment in Fig 7. (A) and (B) Midway between the anode and cathode. (C) and (D) Midway between anode and cathode towards the cathode (x10). (E) The core of the ablated region near the cathode in 10x magnification with H&E (F) and Mason trichromatic

stain. The core of the ablated region near the cathode with 20x magnification and H&E staining (G) and with Mason trichromatic stain (H). (I) and (J) Micrographs at the outer margin of the cathodic lesion: (I) 10x magnification and trichromatic stain, (J) 20x magnification and H&E stain.

DISCUSSION

The primary goal of this study was to expand on our earlier small animal study [49,50] and examine tissue ablation protocols that combine electrolysis and electroporation in a configuration and with parameters that are relevant to clinical applications. Since our earlier work on tissue ablation with non-thermal irreversible electroporation (NTIRE) [44, 51], medical imaging has become standard in clinical use of NTIRE. We therefore considered using ultrasound imaging in our study as well. While performing the electrolysis studies under ultrasound, we have observed in all the experiments the bright hyperechoic appearance in the region adjacent and between the electrodes. An example is shown in Fig 1C. Since electrolysis is known to produce gases near the electrodes, the most likely explanation is that these bright areas are the reflection of ultrasound waves from the gas tissue interface. This led us to re-examine ultrasound images from our earlier irreversible electroporation studies [44,51]. Fig 1Dillustrates a typical site of NTIRE treated tissue immediately after the procedure. Comparing the panels in the bottom row (Fig 1C and 1D), it is now obvious that the bright spots near the electrodes in the NTIRE procedure are due to gases produced by electrolysis during the electroporation pulses. This has important clinical value. It is common to experience loud and sudden explosion-like sounds during NTIRE protocols. The observed bright areas now explain the mechanism: These sounds are most likely the result of an electric discharge across the electrolysis generated layer of gas around the electrodes. Electroporation employs voltages as high as 3000 V. When the field across the layer is higher than about 3000 V/cm, the gases ionize and an electric discharge akin to lightning occurs. For a voltage of 3000 V, a 100 micron thick layer of gas will be sufficient to cause a discharge. This discharge generates high pressure waves that could be detrimental to the treated organ, in particular if it is encapsulated such as the brain, bone or the prostate. Monitoring the formation of bright spots at the electrodes during electroporation could be used to avoid the electric discharge.

Fig 2 shows a set of four experiments that confirm the synergistic effect of the combination of electrolysis and electroporation. The top row (lesion 1) shows the extent of tissue ablation in a pure electrolytic procedure. The current delivered and the time of delivery are substantially lower and shorter, respectively, than conventional protocols for tissue ablation by electrolysis.

Indeed the ablation occurs primarily near the electrodes, with an ablation ratio of 0.4. The second row (lesion 2) shows the effect of electroporation with parameters which are typical to reversible electroporation. Indeed, the panels show that the applied voltage has no effect on the tissue. The third row shows the results of the combination of electroporation and electrolysis with the same parameters used for lesion 1 and 2. The ablated area has significantly increased, as can be seen in Fig 2E and 2F. The last row results (lesion 4) are consistent with the synergistic effect of the combination of two sequences of electroporation and electrolysis, also reported in [50]. The extent of tissue ablation is substantially larger than in the other tissue treatment panels, with an ablation ratio of 0.71. As discussed in the introduction, the mechanism probably involves using conventional electrolysis as the central tissue ablation modality and additionally using electroporation pulses to permeabilize the cell membrane to make the cell more susceptible to lower amounts of products of electrolysis [50]. This figure suggests some possible clinical implications. It seems that 500 V, which is usually used in reversible electroporation, may be sufficient to produce an approximately 2 cm ablation lesion, when delivered in combination with electrolysis, from two Ti electrodes separated by 1 cm.

Another set of three experiments that confirm the synergistic effect of electrolysis and electroporation are shown in Fig 3. In this case we increased the distance between the electrodes to 2.5 cm. Fig 3A shows the effect of delivering a low non-clinical dose of electrolysis, which shows that little damage was produced, just adjacent to the electrodes. The majority of the tissue in the 2.5 cm gap between the electrodes is unaffected. Fig 3B displays tissue treated with electroporation with a voltage that is typical for non-thermal irreversible electroporation (NTIRE) protocols. However, the distance between the electrodes is larger than used in clinical practice. Indeed, the damage occurs primarily around the electrodes. The minimal width of the area unaffected by the treatment is about 1.5 cm. The survival of tissue between electrodes is of clinical concern in NTIRE. Fig 3C shows the outcome of a treatment that combines the electrolysis and the electroporation in the previous two panels. It is evident that although the tissue ablation has a dumb-bell shape, the ablation zone from the two electrodes touch. The appearance of the ablated tissue is different near the two electrodes. This phenomenon is consistent with the mechanisms of electrolytic ablation and was discussed in [49,50]. While much work remains to be done to optimize the protocols that combine electrolysis and electroporation, the two studies in Figs 2 and 3 illustrate the potential. The combination could generate tissue ablation with lower voltages or with larger distances between electrodes.

Fig 4 illustrates an attempt to implement the multiple low voltage electroporation pulse concept described in [49] for tissue ablation in a clinical relevant configuration. The study described in [49] shows that multiple low electric field pulses, of the kind used in reversible type electroporation, can cause tissue ablation. The mechanism involves the penetration of the electrolytic products generated during the delivery of the electroporation pulses into the reversible permeabilized cell membrane. Fig 4 evaluates the effect of the number of low electric field pulses on the extent of tissue ablation. The top row (lesion 8) shows the extent of tissue ablation when 8 pulses are applied. This is the type of protocol that would be used in reversible electroporation treatment of tissue. The results show that the damage is minimal, just adjacent to the electrode needles. Indeed, this is what is hoped for in reversible electroporation. The second row (lesion 9) shows the extent of tissue ablation when 297 pulses are applied. The frequency of 1 Hz was chosen because it is used in conventional NTIRE. However, voltages between 2 kV and 3 kV would be used for this geometrical configuration in conventional NTIRE. The number of pulses used in conventional NTIRE, between two electrodes, is substantially smaller. Therefore, the protocol tested here employs voltages that are a factor of 2 and 3 smaller than in NTIRE and employs numbers of pulses that are a factor of 3 to 6 larger than in conventional NTIRE. Fig 4C and 4D (lesion 10) point out that the extent of tissue ablation by the 297 pulse protocol is substantially larger than that for the 8 pulse protocol. This is consistent with the findings of [49] and demonstrates that the combination of low electric fields with pulse generated electrolytic products can be used for tissue ablation in a clinical relevant setting. The overlaid calculated isoelectric field lines show that while the extent of tissue ablation with 8 pulses is encompassed by isoelectric fields of about 1000 V/cm and higher, the extent of tissue ablation with 297 pulses seems to extend to isoelectric field lines of 300 V/cm and even 200 V/cm. The electric fields of 200 V/cm and 300 V/cm are in the domain of reversible electroporation, which cells survive in the absence of electrolysis. The ablation shown in Fig 4 has a dumb-bell shape, with what appears to be intact tissue in the middle area between the electrodes. The electric fields have a similar dumb-bell shape. Fig 5A and 5B are higher magnification of the H&E stained (Fig 4A) and Mason trichromatic stain (Fig 4B) images from Fig 4. They focus on the region between the electrodes. Both kinds of staining show a section of the liver with parallel vertical lesions traversing the parenchyma, following the tract of the electrodes. The lesions are characterized as acute tissue necrosis with edema and hemorrhage. The outer boundaries of the tissue injury between the electrodes are pale and swollen. There is a wedge of liver tissue between the lesions (electrode) that appear to be normal and intact (viable). Perhaps the most interesting observation from bottom row in Fig 4 (Fig 4C and 4D) and top row of Fig 5 (Fig 5A and 5B) is that the macroscopic margin of cell death, while convoluted, follows quite precisely the isoelectric field lines, showing that the mechanism is related to the process of electroporation. It is also interesting to notice that the isoelectric field lines that are followed correspond to parameters for reversible electroporation. This supports the idea that the mechanism of damage in this protocol involves the synergistic combination of electrolysis with electroporation. It should be noted that the length of this procedure was about 8 minutes with the BTX. It is also interesting to notice that the extent of tissue ablation with the multiple electroporation pulse protocol is somewhat similar to that obtained with a combination of 5 minutes electrolysis and 16 500 V electroporation pulses. This aspect will be discussed later in the context of Fig 7.

Next we will follow, with high magnification, a path from the outer edge of the left electrode (anode) to the outer edge of the right electrode (cathode). Fig 5C shows a control sample from the same liver for comparison with treated tissue. It will become evident, as we go through the micrographs that the patterns of cell death change from point to point in the treated region. This was also observed in [49]. Fig 5D illustrates the appearance of the treated tissue at the left hand side margin. There seems to be a relatively sharp transition between the normal hepatocytes on the left and the swollen and/or contracted cells with pale cytoplasm and condensed nuclei on the right. This sharp transition is a feature of cells affected by electroporation and was also observed in [43]. Edema and congestion are also noted in the treated region, which is a known feature of cell ablation by electroporation [43]. It is important to notice that the edema and congestion extend to the margin of the treated lesion. This is what gives the macroscopic gross sections the dark appearance and suggests that, indeed, the dark margin in the macroscopic gross sections can be taken to approximately represent the margin of the lesion.

Fig 5E shows a 10x magnification at a site in the core of the left hand side lesion. The H&E stained micrograph shows a section of the liver with acute cellular necrosis throughout the field. It should be noted that the lobular structure is nevertheless retained. Another interesting observation is the appearance of the ablated cells which is different in the right hand side from the left hand side of the panel. A higher magnification from the same site is shown in Fig 5F. It is evident that the damage to the cells is major. The image is showing necrotic hepatocytes with disrupted sinusoidal pattern. Cells are swollen and/or contracted with dark nuclei, and there is congestion/ hemorrhage. This region has experienced complete cell ablation.

The fourth row (Fig 5G and 5H) shows that adjacent to the electrodes, the core region seems to be completely devoid of any cellular structure with marked loss of hepatocytes in the core, which is filled with fibrous substance and red blood cells. This may be related to electro-osmotic water flow from the anode to the cathode, which is typical to processes of electrolysis. Fig 5G in particular is interesting because it focuses on the interface between ablated tissue and a large blood vessel. The section shows a wide field of pale ablated cells with small and large pockets containing proteinaceous pink substance. The liver cells on the left side of the panel are compacted. No cellular outline is discernible. However, it appears that the morphological structure of the large blood vessel is intact and it has remained patent, i.e. not occluded. The preservation of large blood vessels architecture by both electrolysis [23,24] and electroporation [45] separately, are considered a major advantage of these ablation modalities. This is what allows their use to treat tumors near large blood vessels. It now seems that the combination of electroporation and electrolysis also maintains the large blood vessels functionality. Obviously, this is only a 24 hours survival study and longer studies are required to further verify this important aspect of the combination of electroporation with electrolysis as an ablation procedure.

Fig 6A and 6B show a low magnification of the area midway between the electrodes. The liver in this area is normal and unaffected by the treatment, as evident from the higher magnification micrographs in Fig 6C, 6D and 6E. The third row shows the appearance of tissue near the large blood vessel midway between the electrodes (Fig 6E) and near the right electrode (Fig 6F). In contrast with the normal appearance of the tissue adjacent near the blood vessel, midway between the electrodes, the cells near the right electrodes are completely ablated. Nevertheless, as on the left side of the blood vessel, here also the structural integrity of the blood vessel has been apparently retained.

The forth row (Fig 6G and 6H) shows micrographs from the core of the area affected by the right hand side electrode. Fig 6G shows a section of the liver with acute cellular necrosis throughout the field. Fig 6H, on the other hand, illustrates an area in which the cellular structure is completely disintegrated with marked loss of hepatocytes in the core, which is filled with fibrous substance and red blood cells. The ablated tissue appearance near the left electrode is comparable to that near the right electrode. Similar to the left hand side margin of the lesion here also, in slides close to the edge of the lesion, there is clear demarcation between the normal hepatocytes on the right compared to the ablated cells on the left. Edema and congestion are also noted on the left hand side margin.

The study in Figs 3–6 show that multiple low voltage electroporation pulses create a much larger ablated zone. The ablated zone seems to be congruent with the isoelectric field lines of the parameters of reversible electroporation, which suggest that the mechanism of damage is the synergy of electroporation and electrolysis. The appearance of the ablated cells changes throughout the lesion. This seems to be a feature of the combination, because the mechanisms of damage are different throughout the treated zone. We believe that there are at least five mechanisms: dominant anodic ablation, dominant cathodic ablation, combination reversible electroporation and anodic compounds, combination reversible electroporation and cathodic compounds and a small region of irreversible electroporation. This suggests that using the combination electroporation and electrolysis requires a good understanding of the biophysics of the process and careful treatment planning. While the ablation is not continuous, it is obvious that these type of protocols combining electrolysis with electroporation through the delivery of multiple low voltage pulses could find clinical application. A major advantage over conventional NTIRE is the use of much lower voltages.

Figs 7A–7D, 8 and 9 show a different possible clinical protocol that employs the combination of electrolysis and electroporation in a mode explored in reference [49]. Specifically, we used conventional electrolysis as the central tissue ablation modality and added electroporation pulses to permeabilize the cell membrane, making the cell more susceptible to lower amounts of products of electrolysis. We tested the concept with a protocol that may become clinical. It combines low voltage electroporation pulses with short periods of electrolysis. The protocol employs voltages that are lower than in conventional NTIRE and the time of electrolysis is much shorter than in conventional electrolysis. Fig 7A–7D shows macroscopic images of the treated tissue. The lesion is made of two distinct adjacent lesions, arranged somewhat is a "dumb-bell" shape, similar to the shape obtained with the multiple pulse protocol. However, unlike the multiple pulse protocol, in this case the outer margins of the necrotic zones mingle with each other and the ablation is continuous, forming an oval shape. The macroscopic images show a variety of cell ablation modalities, which, as with the multiple electroporation pulses, can be attributed to the different mechanisms of cell death. For instance, the upper lesion around the anode is relatively more pronounced and contains severe necrotic acellular debris with charring and edema. The necrotic center is surrounded by a thick circular zone of coagulation necrosis followed by an outermost zone marked by hemorrhagic liver cell necrosis. The lower lesion is slightly less severe and is characterized by a central region of coagulation necrosis, with the outermost circular thick zone hemorrhagic necrosis merging with the hemorrhagic zone of the upper lesions with no discernible interface or normal liver tissue between the two electrodes induced lesions.

The lesions 11 and 12 shown in Fig 7E and 7F serve as a comparison of different permutations of the combination. Lesion 11 shows the effects of electroporation followed by electrolysis, while lesion 12 illustrates the affected tissue after electrolysis followed by electroporation. All three permutations produced a similar extent of ablation that was continuous between the electrodes. This shows that the exact sequence of the permutations of electrolysis and electroporation does not affect the results.

Figs 8 and 9 show higher magnification of the tissue ablation, in a progression from the outer edge of the ablated tissue near the anode to the outer edge of the ablated tissue near the cathode. This should facilitate a better understanding of the different mechanisms of action during the combination tissue ablation process.

In Fig 8A we see a section of the liver stained with H&E which contains normal liver cells with no evidence of treatment related injury. Fig 8B illustrates the edge of the lesion at the anode side. The appearance is rather similar to that in Fig 5D. Here there is also profound necrosis on the right hand side of the panel, with substantial hemorrhage, separated by a relatively narrow region of delineation between the affected tissue on the right and the normal tissue on the left. The cells on the very left appear normal. As in the multiple pulses case and in NTIRE, the interface between ablated and living cells is very narrow. The sharp delineation is considered an advantage of NTIRE and it apparently occurs also in this combination of electrolysis and electroporation. A possible explanation is that the opening of the cell membrane by electroporation is required for cell death. Electroporation is a relatively binary process, and this may explain the narrow range of transition. The hemorrhage is also typical to electroporation and, like in the previous study, supports the use of macroscopic gross sectioning of tissue to evaluate the extent of ablated tissue.

The second and third rows (Fig 8C–8F) show areas towards the core of the anode treated lesion. The damage here is much more massive than in the multiple electroporation pulses case. There is severe coagulation, massive tissue necrosis, as well as fragmented and dehydrated strips of liver tissue. In this case there was much more electrolysis than in the multiple electroporation pulse case, and this more massive mechanism of damage can be attributed to the electrolytic ablation. The dark band of coagulated tissue is probably due to the chlorine species that form at the anode in a chemical reaction. The entire core region seems to be severely dehydrated. This can be explained by the electro-osmotic flow of water from the anode to the cathode. In fact the low magnification images in Fig 7 support this model. It is possible to observe a rim of condensed red blood cells around the anode affected region, concentrating in the region between the anode and cathode. A possible explanation is that this could be due to the flow of water from the anode to the cathode has carried red blood cells, which were then deposited at the interface of the cathode affected region. The higher magnification micrographs in Fig 8 rows four and five (Fig 8G–8J) and Fig 9 rows one and two (Fig 9A–9C) show that indeed, the damage is much less severe than around the anode and there is a high conglomeration of red blood cells with hemorrhagic necrosis. The appearance of the necrotic core near the cathode is substantially different from that near the anode. In comparison to the anode, the cathode at the bottom inflicted a less severe but complete necrosis of the liver cells. It is interesting to notice that the lobular pattern is retained. The margin of the treated region near the cathode is shown in Fig 9I and 9J. Similarly to the margin of the lesion near the anode and the margins of the lesion near the anode and cathode with the multiple electroporation pulse protocol, the transition region between dead and live cells is narrow and characterized by hemorrhagic necrosis.

Evidently, in this protocol also, the mechanisms of damage vary as a function of the proximity to the anode or cathode. This phenomenon was also observed by others working in the field of electrolytic ablation [23,24]. and us There is a clinical significance to that. It means that in protocols that combine electrolysis and electroporation, it is important to place the cathode and anode in an optimal configuration in regards to the nature of the treated tissue. For example, it tentatively appears that it would be preferential to place the cathode near a sensitive tissue that is to be spared, such as a blood vessel.

CONCLUSION

Using a large animal model, we have confirmed the findings in [49,50] that a synergistic combination of electrolysis and electroporation can produce more effective ablation than either electrolysis or electroporation separately. Furthermore, this combination lends itself to the design of clinical protocols that employ lower voltages than NTIRE and shorter times than electrolysis. Obviously, this is only a first large animal study of this combination tissue ablation modality. Substantial research remains to be done to optimize the concept for clinical use.

ACKNOWLEDGMENTS

We would like to thank Dr. Narayan Raju for the histology.

AUTHOR CONTRIBUTIONS

Conceived and designed the experiments: MKS EG PM BR LR. Performed the experiments: MKS BR LR EG PM. Analyzed the data: NK BR EG PM. Contributed reagents/materials/analysis tools: EG NK BR. Wrote the paper: BR MKS NK EG.

REFERENCES

1.   Noad HM. Lectures on electricity; comprising galvanism, magnetism, electro-magnetism, magneto- and thermo- electricity, and electro-physiology. Third Edition ed. London: George Knight and Sons; 1849.

2.   Amory R. A treatise on electrolysis and its therapeutical and surgical treatement in disease. New York: William Woof &Co.; 1886.

3.   Nordenstrom BEW. Preliminary clinical trials of electrophoretic ionization in the treatement of malignant tumors. IRCS Medical Sc. 1978;6:537.

4.   Nordenstrom BEW. Survey of mechanisms in electrochemical treatment (ECT) of cancer. European Journal of Surgery. 1994;0(SUPPL. 574):93–109. BCI:BCI199598038108.

5.   Miklavcic D, Fajgelj A, Sersa G. Tumor treatement by direct electric-current- electrode material deposition Bioelectrochemistry and Bioenergetics. 1994;35(1–2):93–7. WOS:A1994PR99500017. doi: 10.1016/0302-4598(94)87017-9

6.   Miklavcic D, Jarm T, Cemazar M, Sersa G, An DJ, Belehradek J, et al. Tumor treatment by direct electric current. Tumor perfusion changes. Bioelectrochemistry and Bioenergetics. 1997;43(2):253–6. WOS:A1997YD72300010. doi: 10.1016/s0302-4598(96)05190-2

7.   Miklavcic D, Semrov D, Valencic V, Sersa G, Vodovnik L. Tumor treatment by direct electric current: Computation of electric current and power density distribution. Electro- and Magnetobiology. 1997;16(2):119–28. doi: 10.3109/15368379709009837 WOS:A1997XF46400004.

8.   Griffin DT, Dodd NJF, Moore JV, Pullan BR, Taylor TV. The effects of low-level direct-current therapy on a preclinical mammary-carcinoma tumor regretion and systemic biochemical sequelae.. British Journal of Cancer. 1994;69(5):875–8. doi: 10.1038/bjc.1994.169. WOS:A1994NJ83100013. pmid:8180017

9.   Griffin DT, Dodd NJF, Zhao S, Pullan BR, Moore JV. Low -level direct electrical-current therapy for hepatic metastases. 1. Preclinical studies on normal liver.. British Journal of Cancer. 1995;72(1):31–4. doi: 10.1038/bjc.1995.272. WOS:A1995RF10000005. pmid:7599063

10.   Haggendal E, Nilsson NJ, Norback B. Aspects of the autoregulation of cerebral blood flow. International anesthesiology clinics. 1969;7(2):353–67. MEDLINE:4911052. pmid:4911052 doi: 10.1097/00004311-196900720-00010

11.   Nilsson E, Berendson J, Fontes E. Development of a dosage method for

electrochemical treatment of tumours: a simplified mathematical model. Bioelectrochemistry and Bioenergetics. 1998;47(1):11–8. doi: 10.1016/s0302-4598(98)00157-3 WOS:000077969800003.

12. Nilsson E, Berendson J, Fontes E. Electrochemical treatment of tumours: a simplified mathematical model. Journal of Electroanalytical Chemistry. 1999;460(1–2):88–99. doi: 10.1016/s0022-0728(98)00352-0 WOS:000079018900009.

13. Nilsson E, von Euler H, Berendson J, Thorne A, Wersall P, Naslund I, et al. Electrochemical treatment of tumours. Bioelectrochemistry. 2000;51(1):1–11.   doi: 10.1016/s0302-4598(99)00073-2. WOS:000167099400001. pmid:10790774

14. von Euler H, Nilsson E, Lagerstedt AS, Olsson JM. Development of a dose-planning method for electrochemical treatment of tumors: A study of mammary tissue in healthy female CD rats. Electro- and Magnetobiology. 1999;18(1):93-+. doi: 10.3109/15368379909012903 WOS:000079639200011.

15. von Euler H, Nilsson E, Olsson JM, Lagerstedt AS. Electrochemical treatment (EchT) effects in rat mammary and liver tissue. In vivo optimizing of a dose-planning model for EChT of tumours. Bioelectrochemistry. 2001;54(2):117–24. WOS:000176715200003. pmid:11694391 doi: 10.1016/s1567-5394(01)00118-9

16. Camue Ciria HM, Morales Gonzalez M, Ortiz Zamora L, Bergues Cabrales LE, Sierra Gonzalez GV, de Oliveira LO, et al. Antitumor effects of electrochemical treatment. Chinese Journal of Cancer Research. 2013;25(2):223–34.   doi: 10.3978/j.issn.1000-9604.2013.03.03. WOS:000318377200015. pmid:23592904

17. Colombo L, Gonzalez G, Marshall G, Molina FV, Soba A, Suarez C, et al. Ion transport in tumors under electrochemical treatment: In vivo, in vitro and in silico modeling. Bioelectrochemistry. 2007;71(2):223–32. doi: 10.1016/j.bioelechem.2007.07.001. WOS:000251406800018. pmid:17689151

18. Turjanski P, Olaiz N, Abou-Adal P, Suarez C, Risk M, Marshall G. pH front tracking in the electrochemical treatment (EChT) of tumors: Experiments and simulations. Electrochimica Acta. 2009;54(26):6199–206. doi: 10.1016/j.electacta.2009.05.062

19. Olaiz N, Suarez C, Risk M, Molina F, Marshall G. Tracking protein electrodenaturation fronts in the electrochemical treatment of tumors. Electrochemistry Communications. 2010;12(1):94–7. doi: 10.1016/j. elecom.2009.10.044 WOS:000274233700023.

20. Ivic MLA, Perovic SD, Zivkovic PM, Nikolic ND, Popov KI. An electrochemical illustration of the mathematical modelling of chlorine impact and acidification in electrochemical tumour treatment and its application on an agar-agar gel system. Journal of Electroanalytical Chemistry. 2003;549:129–35. WOS:000183792100013. doi: 10.1016/s0022-0728(03)00251-1

21. Bergues Pupo AE, Bory Reyes J, Bergues Cabrales LE, Bergues Cabrales JM. Analytical and numerical solutions of the potential and electric field generated by different electrode arrays in a tumor tissue under electrotherapy. Biomedical Engineering Online. 2011;10. doi: 10.1186/1475-925x-10-85 WOS:000296095000001.

22. Placeres Jimenez R, Bergues Pupo AE, Bergues Cabrales JM, Gonzalez Joa JA, Bergues Cabrales LE, Godina Nava JJ, et al. 3D Stationary Electric Current Density in a Spherical Tumor Treated With Low Direct Current: An Analytical Solution. Bioelectromagnetics. 2011;32(2):120–30. doi: 10.1002/bem.20611. WOS:000286398900005. pmid:21225889

23. Czymek R, Dinter D, Loeffler S, Gebhard M, Laubert T, Lubienski A, et al. Electrochemical Treatment: An Investigation of Dose-Response Relationships Using an Isolated Liver Perfusion Model. Saudi Journal of Gastroenterology. 2011;17(5):335–42. doi: 10.4103/1319-3767.84491. WOS:000305196900008. pmid:21912061

24. Czymek R, Nassrallah J, Gebhard M, Schmidt A, Limmer S, Kleemann M, et al. Intrahepatic radiofrequency ablation versus electrochemical treatment in vivo. Surgical Oncology-Oxford. 2012;21(2):79–86. doi: 10.1016/j.suronc.2010.10.007 WOS:000304669500012.

25. Stampfli R. Reversible electrical breakdown of the excitable membrane of a Ranvier Node. An da Acad Brasileira de Ciencias. 1957;30(1):57–63.

26. Stampfli R. [Permeability of the membrane of Ranvier's node to potassium following excitation]. J Physiol (Paris). 1958;50(2):520–3. Epub 1958/03/01. pmid:13550223.

27. Stämpfli R, Willi M. Membrane potential of a ranvier node measured after electrical destruction of its membrane. Experimentia. 1957;13:297–8. doi: 10.1007/bf02158430

28. Hamilton WA, Sale AJH. Effects of high electric fields on microorganisms. 2. Mechanism of action of the lethal effect. Biochimica et Biophysica Acta. 1967;148:789–800. doi: 10.1016/0304-4165(67)90053-0

29. Sale AJH, Hamilton WA. Effects of high electric fields on microorganisms. 1. Killing of bacteria and yeasts. Biochimica et Biophysica Acta. 1967;148:781–8. doi: 10.1016/0304-4165(67)90052-9

30. Sale AJH, Hamilton WA. Effects of high electric fields on microorganisms. 3. Lysis of erythrocytes and protopasts. Biochimica et Biophysica Acta. 1968;163:37–43. pmid:4969954 doi: 10.1016/0005-2736(68)90030-8

31. Neumann E, Rosenheck K. Permeability changes induced by electric impulses in vesicular membranes. The Journal of Membrane Biology. 1972;29(10):279–90. doi: 10.1007/bf01867861

32. Neumann E, Schaeffer-Ridder M, Wang Y, Hofschneider PH. Gene transfer into mouse lymphoma cells by electroporation in high electric fields. EMBO J. 1982;1(7):841–5. pmid:6329708

33. Neumann E, Sowers AE, Jordan CA, editors. Electroporation and Electrofusion in Cell Biology. New York: Plenum Press; 1989.

34. Zimmermann U, Pilwat G, Riemann F. Dielectric breakdown of cell membranes. Biophys J. 1974;14(11):881–99. pmid:4611517 doi: 10.1016/s0006-3495(74)85956-4

35. Weaver JC, Chizmadzhev YA. Theory of electroporation: a review. Bioelectrochem Bioenerg. 1996;41:135–60. doi: 10.1016/s0302-4598(96)05062-3

36. Apollonio F, Liberti M, Marracino P, Mir L, editors. Electroporation mechanism: Review of molecular models based on computer simulation. Antennas and Propagation (EUCAP), 2012 6th European Conference on; 2012 26–30 March 2012.

37. Teissie J. Electropermeabilization of the Cell Membrane. Electroporation Protocols: Preclinical and Clinical Gene Medicine, 2nd Edition. 2014;1121:25–46. doi: 10.1007/978-1-4614-9632-8_2 WOS:000332321200003.

38. Titomirov AV, Sukharev S, Kistanova E. In vivo electroporation and stable transformation of skin cells of newborn mice by plasmid DNA. Biochimica et Biophysica Acta. 1991;1088(1):131–4. pmid:1703441 doi: 10.1016/0167-4781(91)90162-f

39. Miklavcic D, Mir LM, Vernier PT. Introduction to Third Special Electroporation-Based Technologies and Treatments Issue. Journal of Membrane Biology. 2013;246(10):723–4. doi: 10.1007/s00232-013-9595-y WOS:000325116100001.

40. Mir LM, Orlowski S, Belehradek JJ, Paoletti C. Electrochemotherapy potentiation of antitumour effect of bleomycin by local electric pulses. European Journal of Cancer. 1991;27(1):68–72. pmid:1707289 doi: 10.1016/0277-5379(91)90064-k

41. Davalos RV, Rubinsky B, inventors; The Regents of the University of

California, assignee. Tissue ablation with irreversible electroporation. USA2004.

42. Davalos RV, Mir LM, Rubinsky B. Tissue ablation with irreversible electroporation. Annals of Biomedical Engineering. 2005;33(2):223–31. doi: 10.1007/s10439-005-8981-8. WOS:000227162700011. pmid:15771276

43. Edd J, Horowitz L, Davalos RV, Mir LM, Rubinsky B. In-Vivo Results of a New Focal Tissue Ablation Technique: Irreversible Electroporation. IEEE Trans Biomed Eng 2006;53(5):1409–15. doi: 10.1109/tbme.2006.873745

44. Rubinsky B, Onik G, Mikus P. Irreversible electroporation: A new ablation modality—Clinical implications. Technology in Cancer Research & Treatment. 2007;6(1):37–48. WOS:000244732600006. doi: 10.1177/153303460700600106

45. .Martin RCG. Irreversible Electroporation of Locally Advanced Pancreatic Head Adenocarcinoma. Journal of Gastrointestinal Surgery. 2013;17(10):1850–6. doi: 10.1007/s11605-013-2309-z. WOS:000324870500018. pmid:23929188

46. .Schoenbach KH, Beebe SJ, Buescher ES. Intracellular effect of ultrashort electrical pulses. Bioelectromagnetics. 2001;22(6):440–8. pmid:11536285 doi: 10.1002/bem.71

47. Vernier PT, Sun Y, Chen M-T, Gundersen MA, Craviso GL. Nanosecond electric pulse-induced calcium entry into chromaffin cells. Bioelectrochemistry. 2008;73(1):1–4. doi: 10.1016/j.bioelechem.2008.02.003. pmid:18407807

48. Nuccitelli R, Pliquett U, Chen X, Ford W, James Swanson R, Beebe SJ, et al. Nanosecond pulsed electric fields cause melanomas to self-destruct. Biochemical and Biophysical Research Communications. 2006;343(2):351–60. pmid:16545779 doi: 10.1016/j.bbrc.2006.02.181

49. Phillips M, Rubinsky L, Meir A, Raju N, Rubinsky B. Combining Electrolysis and Electroporation for Tissue Ablation. Technology in cancer research & treatment. 2015;14(4):395–410. doi: 10.1177/1533034614560102 MEDLINE:pmid:25416745.

50. Phillips M, Raju N, Rubinsky L, Rubinsky B. Modulating electrolytic tissue ablation with reversible electroporation pulses. Technology. 2015;3(1):1–19. Epub March 2015. doi: 10.1142/s233954781550003x

51. Onik G, Mikus P, Rubinsky B. Irreversible electroporation: Implications for prostate ablation. Technology in Cancer Research & Treatment. 2007;6(4):295–300. WOS:000249073000005. doi: 10.1177/153303460700600405

Chapter 5

DIRECT ELECTROLYTIC AL-SI ALLOYS(DEASA) – AN UNDERCOOLED ALLOY SELF-MODIFIED STRUCTURE AND MECHANICAL PROPERTIES

Ruyao Wang and Weihua Lu

Institute of Material Science and Engineering, Donghua University, Shanghai, P.R.China

INTRODUCTION

Aluminum became attractive only after the invention of Hall-Heroult electrolysis process in 1886. In the earlier part of last century, the usage of aluminum products was restricted in decorative parts. After World War II, a dramatic expansion of the aluminum casting industry occurred. Many new alloys were developed to comply with the engineering requirements. Among the commercial aluminum alloy castings, Al-Si alloy is the most commonly used and constitutes 85-90% of the total aluminum cast parts produced. Al-Si alloys containing silicon as the major alloying element offer excellent castability, good corrosion resistance and machinability. Small amounts of Cu, Mg, Mn, Zn and Ni are being added to achieve strengthening of Al-Si alloys.

Al-Si alloys have been made for a long time by simply adding crushed silicon metal or a high-silicon aluminum base master alloy to molten aluminum in reduction cell or smelting furnace. In those processes pure silicon and aluminum are needed, and both metals are reduced from oxides in electrolytic cell. The idea of direct electrolytic reduction of silica dissolved in the cryolite bath in electrolytic cell has been developed at the end of nineteenth century. The idea to produce alloys in electrolytic process is not new. For several years before Hall-process the Cowles process, by which Cu-Al alloys in range of 30-40% Al were directly reduced from a mixture of Al_2O_3 and CuO or Cu by electric arc at high temperature, was used [1]. 1891 Menit firstly conducted the experiment to reduce the silica to silicon metal in Hall cell. In 1911 Frilley [2] achieved the production of Al-Si alloys containing less than 5% silicon by direct electrolytic reduction of alumina-silica and 5-96% silicon by aluminum-thermal reduction in laboratory. Frilley also obtained Mn-Si, Cr-Si, Fe-Si, Cu-Si and Si-Ni in electrolytic cells. Moreover, he found that the silicon appearance in Al-Si alloy with less than 10% Si was very fine and different from the existed alloy, but no attention had been paid on the change of structural characteristics of silicon due to limited usage of aluminum in industry at that time. Fridley's discovery revealed that electrolytic process is a powerful potential measure to improve the quality of alloy.

In the middle of last century a number of works had been reported to electrolyze Al-Si alloy in Hall cells, to which pure silica, quartzite containing more than 99% SiO_2 [3], sand stone with about 90% SiO_2 [4] glass scrap having 72% SiO_2 [5]. bauxite with 11% SiO_2 [6], sand and clay [7] were added. Recently the refractories from spent potlining were successfully introduced to alumina reduction cells to produce Al-Si alloys [8]. As well known, the purity of molten aluminum is of major concern in electrolytic reduction process. The impurity is considered as a negative factor, deteriorating operation conditions. Hence, the direct electrolytic reduction of silica in Hall cell is a difficult process. There are two severe problems related with silica added into molten cryolite, in which silica must be easily dissolved. One of them is how to compensate for alumina generated by the reaction of aluminum with the added silica for achieving a desired chemical composition of alloy. Other is that direct addition of silica or other silicates often results in the formation of the heavy ridges of silicate along the bottom of the cell, as a result the cell becomes inoperable, so limiting the size and placement of the ridge is a major concern in production. In 1970s C. J. McMinn and A.T. Tabereaux [9, 10] provided a procedure to strictly control the feed of alumina and silica into the cell, stabilizing the electrolytic process and successfully producing Al-Si alloys with up to 16%Si in Hall cell. However, they viewed this process to be economical when the price of silicon greatly increases. Production of Al-Si alloys in electrolytic reduction cell had not found industrial application.

Since 1970s many works have been carried out on direct electrolytic production of Al-Si alloys(DEASA) in China [11]. Most Chinese bauxites contain high content of silica, titania and small amount of rare earth oxides. It is very difficult to extract the pure alumina from bauxite by the Bayer process [12]. In electrolytic process the charge is composed of bauxite, from which the iron oxide is removed, and alumina, using which to regulate the proportion of bauxite added into salt bath in terms of the desired chemical composition of Al-Si alloy. Note that bauxite tested is easily to be dissolved into molten electrolyte compared to the commercial bauxites It would be an important factor to successfully produce Al-Si alloys in alumina reduction cells. At the end of last century several thousand tons of DEASA ingots containing Si content from 6% to 12% have been used in foundries to produce car parts such as engineering block and head, wheel and piston [12- 14]. Table 1 lists the chemical compositions of some DEASA ingots, which contain higher level of impurities such as Na, Sr, Ti and rare earth elements compared to commercial alloys. Undoubtedly it is related with bauxite composition.

Since 1980s author has focused attention on the microstructure of DEASA and its mechanical properties [15]. It has been found that the microstructure and fracture surface of DEASA ingots are very fine and similar to impurity-modified Al-Si alloy. Hence this phenomenon is characterized as self-modification due to no impurity- modifier added. The further research indicated that self-modification is attributed to the eutectic undercooling during solidification of DEASA. To answer the question why self-modified microstructure occurs and how it links with the electrolytic process, we must discuss some events related with electrolysis process. This chapter restricts the consideration into the structural characteristics of alloys and its original, which is related with electrolytic process. The details of electrolysis process can be referred to References [11, 12].

Table 1. Chemical analysis of DEASAs ingot wt%

Alloy*		Si	Cu	Mg	Mn	Ni	Zn	Fe	Cr	Ti	Na	RE	Sr
EZL101	top	7.9	<0.01	—	0.01	<0.01	0.01	0.25.	<0.01	0.33	—	0.002	0.001
No1	bottom	8.2	<0.01	—	0.01	0.01	0.01	0.24	<0.01	0.35	—	0.002	0.001
EZL101	No2	7.3	<0.01	0.36	0.01	0.02	<0.01	0.11	—	0.11	<0.0001	—	0.001
ESi 9**	No1	9.5	<0.02	0.010	0.060	0.15	0.03	0.65	0.02	0.48	0.0045	0.038	0.0034
	No2	9.2	<0.02	0.02	0.005	0.12	0.02	0.44	0.18	0.66	0.014	0.037	0.0026
EZL102		12.2	<0.01	0.15	0.005	—	<0.01	0.50	—	0.12	—	—	—
ZL108		11.60	1.95	0.65	0.62	0.30	—	0.25	—	0.20	0.0020	—	0.000
EZL109		12.1	<0.01	0.91	0.01	0.81	<0.01	0.25	—	0.09	0.0023	—	0.000
ZL101	(A356)	6.7	—	0.39	0.01	0.005	—	0.06	0.016	0.12	—	0.0005	0.002

BEHAVIOR OF ALLOY MELT IN ELECTROLYTIC PROCESS

The electrolysis cell runs at around 950°Cwith a voltage drop of 4.5-5.5 V across each cell[11]. The bauxite, from which iron oxide is removed, contains SiO_2, TiO_2, Fe_2O_3, Na_2O, CaO and rare earth oxides (RExOy),besides the Al_2O_3. During electrolysis process those compounds are reduced to Al, Si, Ti, Fe, Na, Ca and RE, respectively, which in atomic form continuously remove from electrolyte to the carbon bottom of the pot, forming a homogeneous Al-Si alloy melt with several impurities, as shown in Tab.1. Then the melt is siphoned out of the reduction cell at 24h intervals and held in a 10 ton insulated metal-mixer for homogenizing the composition, then poured into ingot mould with dimension of $100\times60\times600mm^3$ and weight of 10kg, without any impurity-addition or treatment. Hence, there are four factors i.e. homogeneous melt, superheating, impurity and electric field (current density and anode potential), influencing the structure of DEASA melt and its crystallographic characteristics and properties in solid state.

In many years a lot of studies have been done on the structure of liquid metals, including Al-Si alloy. The liquid metals can be considered as a system composed of ions and electrons, which are moving through the disordered liquid [16-18]. Below we discuss how superheating and electric field change the structure of Al-Si melt and its crystallization

Effect of Superheating on the Crystallographic Characteristics of Al-Si Alloys

As well-known melt superheating is a powerful factor influencing the microstructure and properties of commercial Al-Si alloys. The effect of superheating is associated with the temperature, holding time and cooling rate during solidification [19-24]. In 1990s many researchers [21,25] investigated the regularity of variety of viscosity and density of family of Al-Si alloy in liquid state with temperature, revealing that as temperature exceeding about 1000°C, these physical properties dramatically change. Therefore, they suggested that for the near eutectic Al-Si alloy containing 10-14% Si there is a critical temperature in range of 1050-1150°C, as shown in Fig.1, above which the silicon grains and other heterogeneous substances such as iron-rich particles are dissolved in melt, resulting in a homogeneous melt, which will change the crystallographic characteristics of alloy. This event has been proved by recent studies. At the beginning of this century X.F. Bian et al studied Al-13%Si alloy melt heated in the temperature range of 625-1250°C using high temperature X-ray diffractometer [22] and reported that when increasing the temperature to 875°C a sudden change of the atomic density and the coordination number of the Al-13%Si alloy melt occurs, demonstrating that the liquid structure has changed, which is caused by dissolving of Si-Si clusters into aluminum melt. In other study it has been found that at the temperature of about 1050°C the electrical resistivity of hypereutectic Al-16%Si alloy melt steeply changed and hereditary effect of different original structure can be eliminated after remelting, indicating that the change of liquid structure happened at temperature of 1050°C[26]. Hence Al-Si alloy melt at high temperature consists of two ion groups: Al-Si and Si-Si groups, which appear to consolidate the short-range order and the electrons are moving through the disordered melt [27-29]. Based on the experimental results P.J.Li [23] considered that in homogeneous Al-Si alloy melt the size of Si-Si and Al-Si micro- heterogeneous clusters range is from 10 to 100Å. M. Singh reported that in Al-Si alloys either hypoeutectic or hypereutectic silicon is present as silicon cluster essentially with the size of about 50-70 Å [27]. Moreover, as increasing the temperature, the size and number of ion groups simultaneously decrease.

P.C. Popel et al [21, 23] studied the influence of superheating on the crystallographic characteristics of alloys and revealed that superheating Al-Si alloy shifts its eutectic reaction toward higher level of silicon accompanying with the appearance of Al-dendrite. As temperature is higher than 780°C, eutectic silicon becomes finer with the fine α-Al dendrite. When heating temperature is in range of 900-1000°C, the size of silicon flake is less than 7μm. It is interesting that heating at temperature higher than 1000°C the modified silicon appears in eutectic alloy. The heating at 1000°C is capable of eliminating the occurrence of primary silicon and refining α-Al dendrite in Al-17%Si alloy. But when superheating hypereutectic Al-20%Si alloys at 950°C the primary silicon particles become finer [24, 30]. The higher the temperature, the finer the silicon grain. It would be expected that a higher superheating temperature is required for hypereutectic Al-Si alloys having higher silicon content to achieve a complete eutectic structure. It is worthy of note that if the holding time is insufficient to dissolve all the silicon particles present in original alloy, even the superheating at 1200°Cdoes not significantly change the crystallographic characteristics of alloy, and the modified structure does not appear [31].

For hypoeutectic alloys as temperature rises to 950°C the dendrite arm spacing (DAS) steeply decreased and the dislocation density in α-Al dendrite increased. Moreover, the eutectic silicon tended to a fine fibrous structure [32].

Overheating significantly increases the content of silicon, magnesium and iron in α-Al-dendrite in hypoeutectic alloy [33]. As overheating Al-8%Si alloy at temperature of 950°C for 10min silicon and magnesium content solved in α-Al-dendrite increases to 1.9% and 0.3%, respectively, much higher than their solubility in Al-matrix at room temperature. Undoubtedly, overheating is one of the factors strengthening the mechanical properties of alloys.

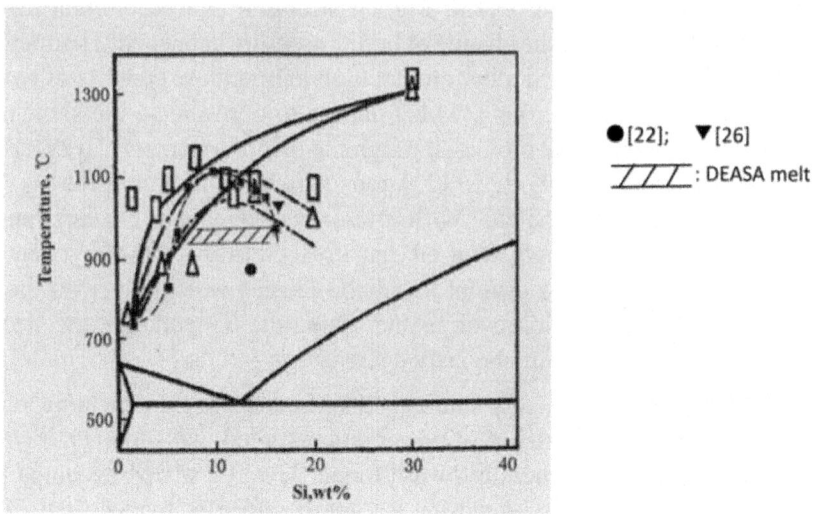

Figure 1. The dome of decay of metastable colloidal microheterogenity in Al-Si melts [21,23]

Superheating also prompts the morphological variety in iron-bearing compound in alloys [34]. As heating Al-7%Si-Mg alloy at temperature higher than 800°C, AlSiFe compound appeared in Chinese script form instead of coarse needle-like shape, increasing the impact strength of alloy. It is apparent that superheating is a powerful mean greatly affecting the feature of microstructure in Al-Si alloys.

It is worth noting that the overheating effects on the change in structure significantly depends upon the cooling rate in freezing in alloy [23, 35]. For hypereutectic Al-17%Si alloy even heated at temperature in the range of 1000-1050°C the primary silicon grows faceted in sand castings, where the freezing rate is less than 10°C/sec. By contrast the formation of more equiaxed, nearly globular silicon crystal can be observed if the melt is quenched with the cooling rate of higher than 100°C/sec.

The reason why superheating leads to a change in crystallographic characteristics of alloys is associated with the undercooling generated by a variety of structure in molten alloy, where the size and number of Si-Si clusters acting as a nuclei of eutectic silicon in solidification of alloy greatly affect the crystallization of alloy[36]. Higher superheating decreases Si-Si cluster in size and amount, depressing the liquid-to-solid transition temperature, as a result a deep undercooling occurs. A.Y.Gubinko[37] reported that superheating an Al-Si alloy melt to 100°C above its liquidus temperature offers an undercooling twice as great as for a melt superheated 35°C. The higher the superheating temperature, the greater the undercooling in freeze of alloy. Note that temperature in electrolysis cell is about 950°C lower than the critical temperature, above which structure of melt transits from microheterogeneous to homogeneous state (Fig.1) and DEASA is intrinsically homogeneous due to its reduction from oxides. Hence, it would be thought that the overheating in reduction cell does not affect the structure of melt, but holding DEASA melt in metal-mixer for long time causes the structural transition from homogeneous to heterogeneous state in some degree.

Role of Electric Field in the Crystallization of Al-Si Alloys

Over the past decades a lot of studies relating with the effect of electric field on the structure and properties of Al-alloys have been carried out [38-42]. Electric field either continuous current or pulse electric discharge deeply affects crystallographic character of alloy and its properties. In this chapter we will only focus our attention on the effect of direct current, which is related with electrolysis process.

By introducing the direct current into molten Al-10%Si alloy at 740°C for treating time of 50min, H.Li et al studied the effect of different current density on the structure and mechanical properties of alloy [43]. It was found that the electric field causes a morphological transition of eutectic silicon from flake to fibrous shape, accompanying with the reduction of second α-Al dendrite space. As increasing the electric current density to 30A/dm^2 silicon phase grows modified and finally primary Al-dendrite appears in near-nodular shape. As a result the elongation of alloy was raised by 100% and its tensile strength was improved by 15%. It is interesting that an increase in current density leads a rise in undercooling in freezing of alloy as shown in Fig.2. When increasing current density to 100A/dm^2 a deep undercooling of 15°C occurs, then undercooling grows slowly with current density. Undoubtedly, the deep undercooling is the reason of the change in morphology of silicon particles.

L.G.Huang et al introduced direct-current into melt poured into mould during solidification and investigated the effect of current density on the structural feature of Al-4%Si and Al-10%Si alloys, which were firstly heated at 700°C. It was found that silicon became finer with direct-current density and reached a limit as the density is increased to 283A/dm^2 and the size of α-Al dendrite arm space (DAS) also reduced with a minimum at density of 325A/dm^2. It is interesting that the effect of alternating current is the same to direct- current [44]. B.A. Timchenko et al [45, 46]studied the effect of high direct-current density (100-10 000A/dm^2) on the quality of casting made of eutectic Al-12%Si alloy. When a large current is passed through alloy during its solidification, the solubility of silicon in α-Al matrix is raised to 20%, and its distribution becomes more homogeneous with a reduced size of silicon particles. In addition, the mold filling ability (fluidity) of casting alloys is greatly improved accompanying with a less tendency to gas porosity and shrinkage. As a result the tensile strength and hardness are increased by 10%. Recently A. Prodhan[47] reported that molten eutectic Al-12.16%Si alloy, which firstly was superheated to 750°C, can be degassed by direct- current treatment during solidification (semisolid state). The initial hydrogen level in alloy made from the ingot is about 2.5ppm, and under current treatment within 10min the hydrogen content is reduced to near 1.7ppm, which is necessary for producing a casting without porosity [47]. However, a large current density will cause an increase of hydrogen concentration. It is obvious that electric field, which is introduced into melt at more or less higher temperature or during solidification, improves the casting properties with an increase in mechanical index. This is attributed to the structural rearrangement of alloy melt generated by electric field. However, we are unable to clarify how the electric field affects the properties of DEASA melt due to the absence of experimental results at high temperature of above 900°C.

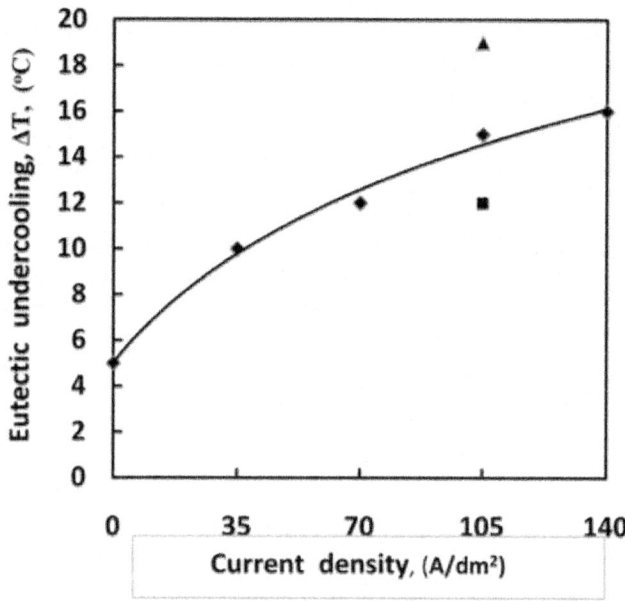

◆[22]; ■EZL101; ▲EZL109

Figure 2. Eutectic undercooling in freezing against current density in Al-Si melt.

The major effects induced by electric field on the behavior of alloy melt include Joule's effect and electron-transport. Electric current causes the input of heat due to Joule's heating effect, which leads to an increase in solidification time, resulting in the improvement of fluidity of alloy, and hence the reduce of shrinkage and porosity[47]. Obviously, Joule's heating effect doesn't affect the properties of DEASAs, which solidify without electric field in present study.

In electrolysis process Al-Si alloy melt is ionized to macroscopic homogeneous Al^{3+} and Si^{4+} ions, and conduct electrons, which are moving through the melt. Under electric potential, positive ions migrate to the cathode and the electrons move toward the anode. The so-called electron- transport, which depends on ionization potential of constituent elements and its mobility in the applied field, is most important factor that reduces the solute distribution co-efficient and influences the rearrangement of elements on the solid-liquid alloy boundary during solidification, therefore reducing the constitutional undercooling and changing the crystallographic behavior of alloys. Under the current potential, the conduction electrons surrounding the aggregates of Si^+-rich or Al^+-rich groups are readily to be transferred to unlike atoms, making the groups unstable[17,28]. When the electron-drag applied to the ions, the unstable groups either Si^+-rich or Al^+-rich are capable of splitting into smaller one. The smaller Si^+-rich aggregators, which act as a nuclear center of silicon phase in Al-Si alloys as reported in reference [36] will promote a larger undercooling in eutectic reaction, which strongly change the crystallographic characteristic of silicon phase. Our data showed that compared to commercial unmodified Al-Si alloys eutectic arrest temperature in DEASA ingot drops to about 15-18°C[15], which is sufficient to modify the microstructure of Al-Si alloys either eutectic[48] or hypereutectic.

Summarizing the experimental results in literatures mentioned above, it is apparent that the effect of overheating on the microstructure and properties of Al-Si alloys is more or less same as electric current that leads the same variety in arrangement of melt in some degree, resulting in a large undercooling in solidification of alloy. It is worth noting that the structure of liquid DEASA is homogeneous in electrolysis process, and therefore, the effect of both of superheating and current field is weakened compared to the existing alloys. It is thought that in the electrolysis process the combination of both factors (superheating and electric field) provides Al-Si alloys a circumstance, where the ability of melt to stabilize the homogeneous structure is enhanced, hence the morphological transition of constituents of DEASA easily undergoes either under lower cooling rate during solidification or upon remelting compared to the common alloy. DEASA is an excellent undercooled alloy, of which the crystallographic behavior is same to alloy treated with electric field at high temperature and rapid cooling rate during solidification. This inference has been evidenced in present and previous studies [14,15, 49].

CRYSTALLIZATION FEATURE OF DEASA

Morphology of Silicon Phase and its Inheritance upon Remelting

As well known, silicon is the major alloying element in Al-Si alloys and its morphology is primary important factor affecting the mechanical properties, castability, machinability and other physical properties. In 1950s it has been found that for Al-Si alloy the growth of silicon crystal is temperature-dependent and dictated by the undercooling in freezing [50]. Since then a number of investigations have been done to clarify the relationship between its morphology and undercooling in solidification [48,51-53]. In general, an eutectic temperature undercooling of 6-8°C is necessary for appropriate modification for hypoeutectic or eutectic Al-Si alloys. If the combination of undercooling induced by cooling rate during solidification and modifier is below the critical value, an unmodified structure is obtained.

The relationship between temperature / undercooling in freezing and morphological transition including eutectic, primary silicon and aluminum phase in Al-Si alloys containing different silicon content can be described in quasi-equilibrium Al-Si diagram (Fig.3.)[54].

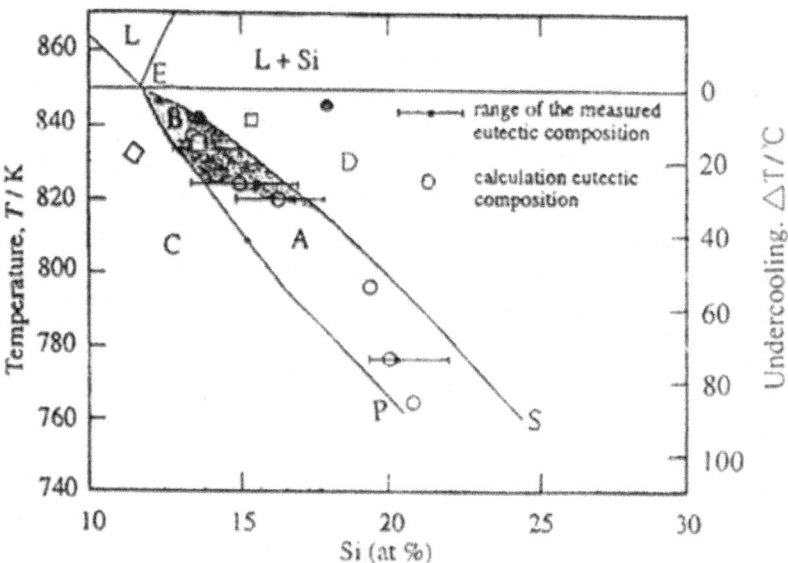

A,B: Quasi-eutectic   zone;   C:Al-dendrite+eutectic;   D:Primary silicon+eutectic;   ◇: Al-dendrtic + eutectic in present work; □: Coupled eutectic in present work; •: Primary Si + eutectic in present work.

Figure 3. Quasi-eutectic zone in the Al-Si system.[54].

Compared to equilibrium diagram, where eutectic reaction runs at a constant temperature and silicon content, the region of formation of quasieutectic structure exists, i.e.in a wide range of temperature/undercooling and silicon content the eutectic structure can be observed. For hypereutectic Al-Si alloys with an increase in silicon content the region shifts towards higher silicon concentration and depresses the eutectic temperature, implying that a higher undercooling is required to produce quasieutectic structure and, meanwhile, the silicon content in quasieutectic is much more than equilibrium. Whether hypereutectic alloy displays a quasieutectic structure or quasieutectic plus primary silicon grain depends upon undercooling. Obviously, the microstructure of eutectic alloy composes of eutectic plus primary α-Al dendrite in casting condition. On the other hand for hypoeutectic alloys due to eutectic shift toward higher silicon content the volume fraction of primary aluminum dendrite increases compared to the equilibrium Al-Si diagram with same silicon content, whereas with undercooling the volume fraction of Al-dendrite increases. In general, using the quasi-diagram the variety in crystallographic feature of Al-Si alloy with different silicon level and undercooling / temperature can be clearly explained.

In order to reveal this relationship between the crystallographic feature in DEASA and undercooling in freezing we observed the microstructure of DEASAs containing silicon content in the range 6- 18% and measured their cooling curves during solidification. Chemical analysis is listed in Table 1 and 2. The samples of eutectic (EZL102, EZL108 and EZL109) and hypoeutectic (EZL101 and ES9) alloys were cut from the center ingots. Hypereutectic alloys (EZL14, EZL16 and EZL18), of which the charge was composed of DEASA (EZL108)(Tab. 2.) ingot and Al-30%Si master alloy along with other master alloy additions., were melted in a 2 kg graphite crucible in an electric resistance furnace and heated to 850°C. After melting (Note: it is 1st remelting for EZL108) the molten alloy was held for 15 min to homogenizing the composition, then poured into a metallic mold, preheated to 250°C to form a casting 40x50x120mm^3 as shown in Fig.4. Pouring temperature is about 740°C for all alloys tested.

All tested alloys with different silicon content were repeatedly remelted to produce the unmodified structure with measured undercooling. This promotes to reveal the effect of undercooling on the structure in DEASA. Metallographic specimens were cut from the interiors of the casting near the site of a chromel-alumel thermocouple (Fig.4), by which the cooling curve was recorded. The cooling rate during solidification was about 1.0°C/sec.

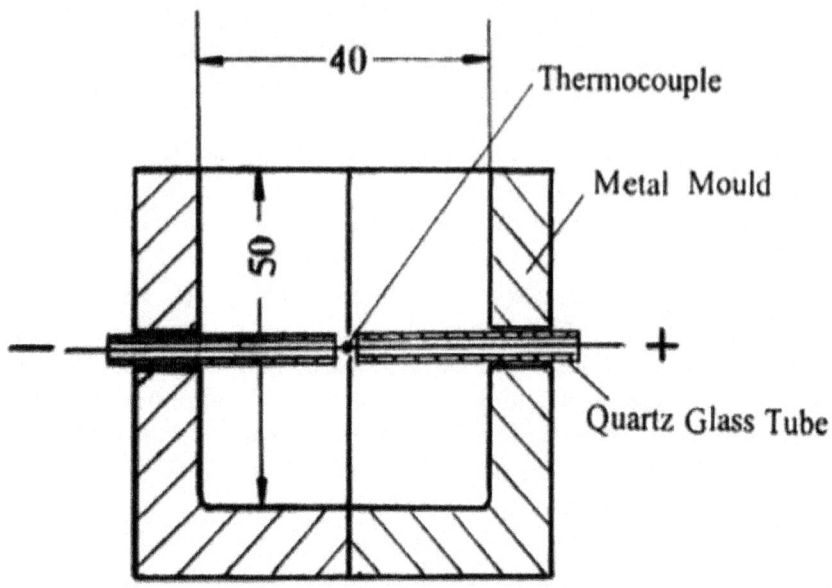

Figure 4. Mold and thermocouple.

Table 2. Chemical Analysis of DEASA tested

Alloy	Si	Mg	Cu	Mn	Ni	Fe	Ti	Sr	Ca	Zr	Remark
EZL108	11.60	0.65	1.00	0.60	0.25	0.25	0.10	<0.000	0.001	0.0070	DEASA
EZL14	13.70	0.55	0.80	0.31	<0.05	0.25	0.03	<0.0006	<0.001	0.0010	D*+AS30**
EZL16	15.70	0.59	1.00	0.34	<0.02	0.35	0.03	0.0006	0.001	0.0072	D+AS30
EZL18	17.60	0.39	0.75	0.25	<0.05	0.35	0.04	<0.0006	<0.001	0.0017	D+AS30

As well known, the alloying elements such as Mg, Cu, Mn, Ni, Fe and Zn lower the eutectic arrest temperature, T_E, in Al-Si alloy [55-57]. In general the following equation (1) is used to estimate the change of T_E in commercial alloys where the total of %Al +%Si is high, near 99% [57, 58].

$$T_E = 577 - (12.5 \ / \ \%Si) \left[\begin{array}{l} 4.43 \ (\%Mg) + 1.43 \ (\%Fe) + 1.93 \ (\%Cu) +1.7 \ (\%Zn) \\ + 3.0 \ (\%Mn) + 4.0 \ (\%Ni) \end{array} \right] \tag{1}$$

In present work the estimated eutectic arrest temperatures, $T_{E,}$ range from 569 °C to 573 °C depending upon the composition of alloys tested. Thus the undercooling, ΔT, will be

$$\Delta T = T_E - T_E{}' \tag{2}$$

where $T_E{}'$ is the measured eutectic temperature for given alloy.

Microstructure of eutectic DEASA (EZL102,EZL108 and EZL109) ingot is shown in Fig.5-8. A high volume proportion, 43-50%, of primary aluminum dendrite, which distributes evenly in modified eutectic matrix, can be found. and the eutectic undercooling is higher than 18 °C that is significant different from the commercial eutectic alloy and similar to the impurity-modified alloys although the silicon content is just near to the eutectic composition.

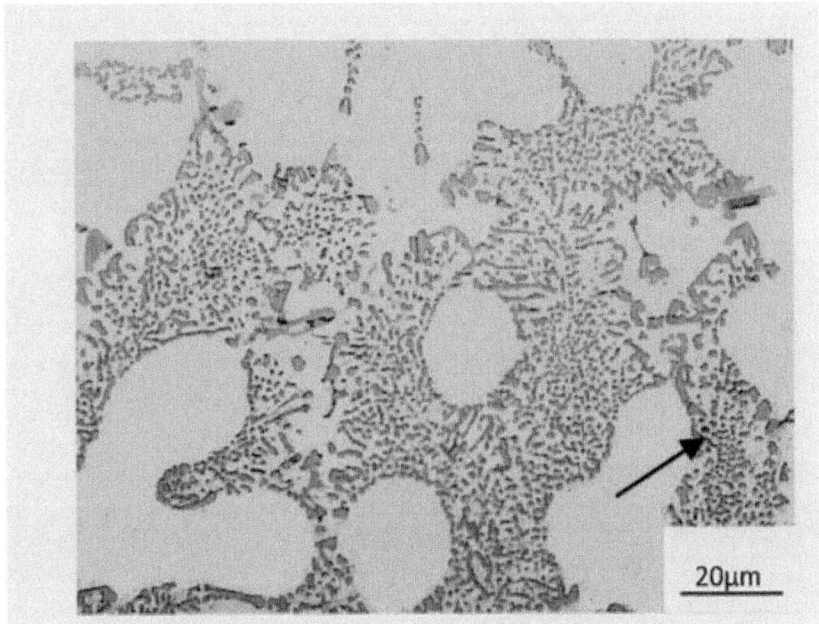

Figure 5. Optical micrograph of EZL102 ingot, showing as-cast self-modified structure. A few iron-rich crystal appears as a fine flake form as indicated by arrow.

The DEASA hypereutectic alloys, which contain 14% and 16%Si solidify with a completely modified eutectic microstructure (Fig.9 and 10) with undercooling of 12°C and 9°C, respectively. For the alloys with silicon content more than 17% (EZL18) the microstructure exhibits the coarse primary silicon crystals well distributed throughout the unmodified matrix as seen in Fig.11. In this case the eutectic temperature reached 568°C with undercooling of 5 °C. In the hypoeutectic electrolytic Al-7%Si ingot the volume proportion of α-Al dendrite reach 72% accompanying with modified eutectic silicon phase and undercooling of 12°C similar to Sr-or Na-modified Al-7%Si alloy (Fig.12). As increasing silicon level to 9% the fine silicon grows in modified mode with a high volume percentage of α-Al dendrite of 60% (Fig.13).

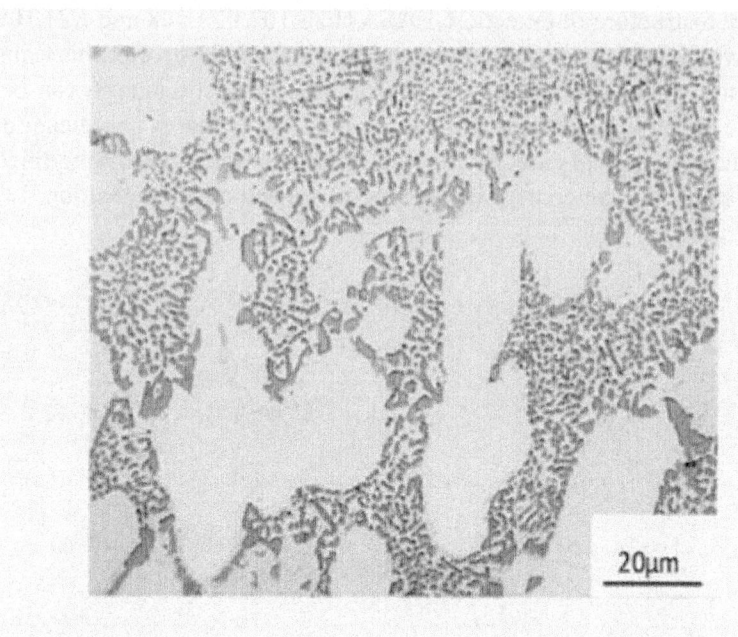

Figure 6. As-cast micrograph of EZL108 ingot, revealing self-modified structure. Optical.

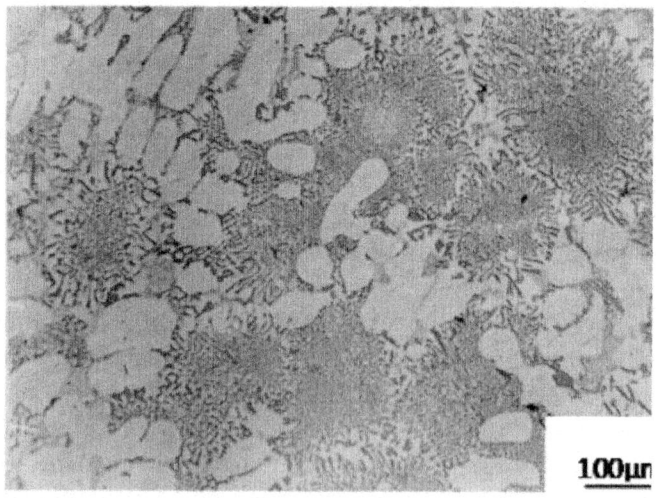

Figure 7. Microstructure in the top region of the electrolytic EZL109 ingot. The equiaxed coarse Al-Si eutectic cell appears, in which silicon grows in modification manner. Optical.

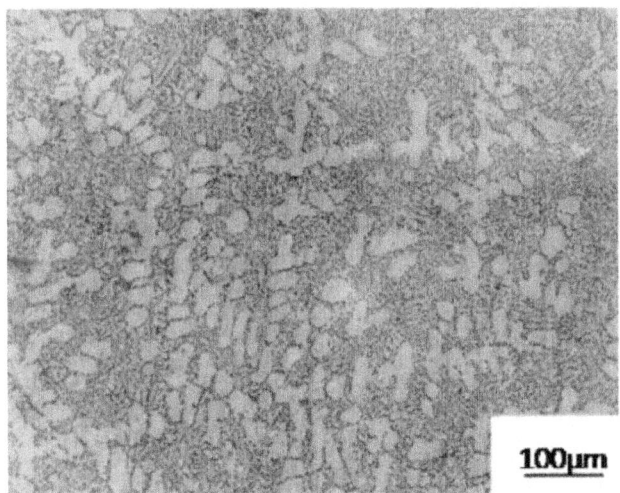

Figure 8. Optical micrograph in the bottom part of electrolytic EZL 109 ingot, indicating the fine self-modified structure.

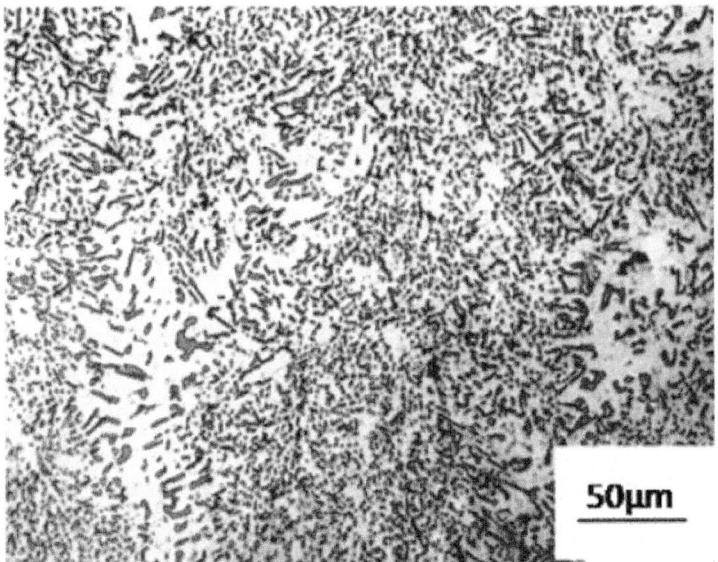

Figure 9. Self-modified microstructure of hypereutectic DEASA (EZL14), showing complete eutectic structure. On the boundary of eutectic cell some silicon flake can be observed. Optical.

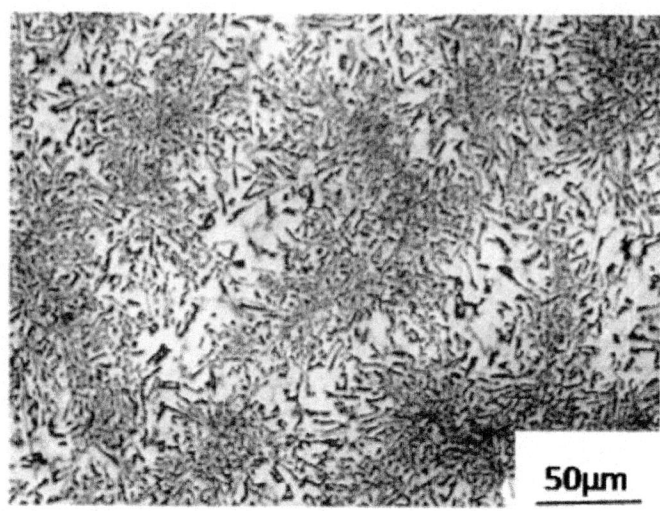

Figure 10. Optical self-modified structure of hypereutectic DEASA (EZL16). Complete eutectic structure appears accompanying some fine silicon flake on the boundary of eutectic cell.

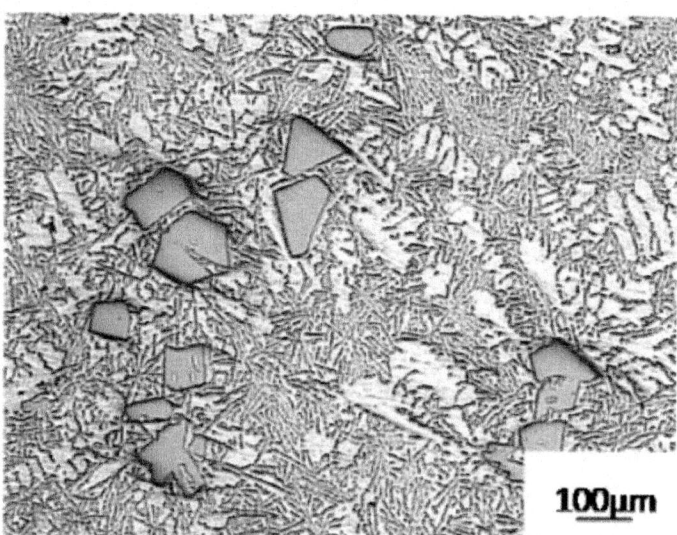

Figure 11. Microstructure of hypereutectic DEASA (EZL18). Coarse primary silicon distributes through the unmodified eutectic matrix. Some α-Al dendrites. occur. Optical.

By combining with the regularity of morphological transition of silicon phase in Al-Si alloys treated by electric field at high temperature in literatures and our experimental results it would be expected that the variety in crystallographic feature is attributed to the undercooling in freezing. This inference has been strongly supported by the experimental results in remelting DEASAs. The very fine self-modified structure in DEASA such as EZL101 and EZL109 is fully inherited upon first remelting with a deep undercooling of 9°C and 13°C, respectively, as shown in Fig.14. As undercooling is higher than critical value of 6-8°C, the alloys solidify in modified manner. In contrast, an unmodified structure in Al-17%Si alloy appears due to lower undercooling of 5°C

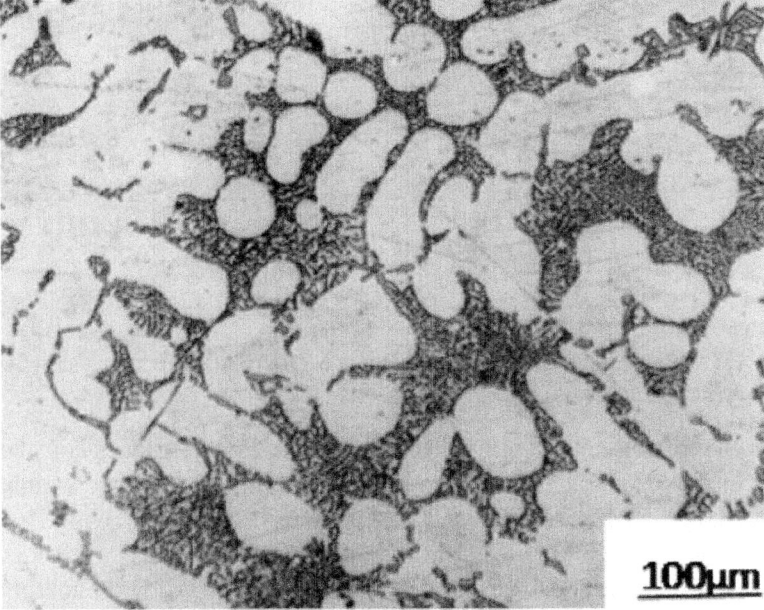

Figure 12. Microstructure of hypoeutectic DEASA(EZL101) ingot, demonstrating the self-modified structure with high volume fraction of α-Al dendrite. Optical.

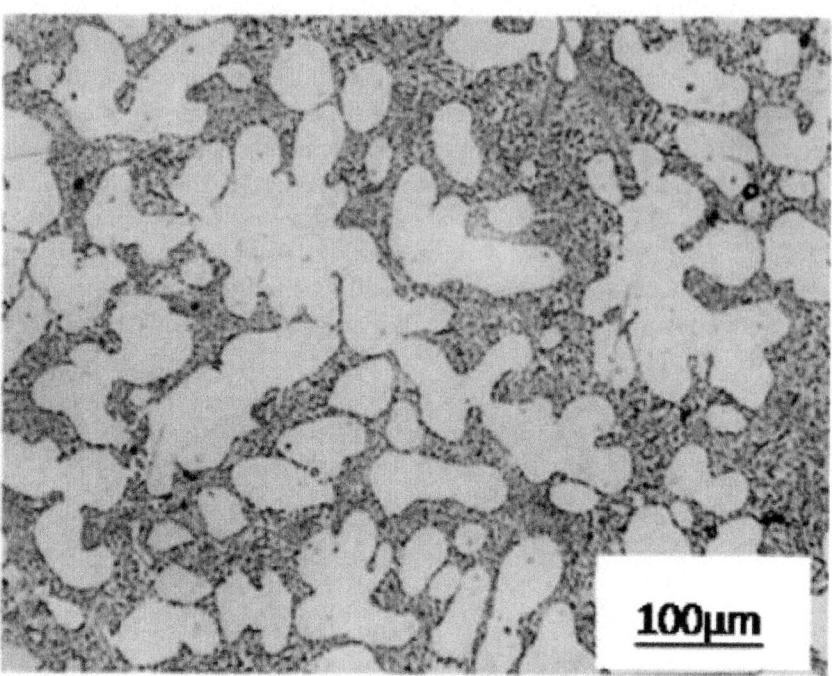

Figure 13. Optical micrograph of hypoeutectic DEASA(ES9) ingot, showing fine modified silicon phase with high volume percentage of α-Al dendrite.

It is interesting that the modified structure in EZL109 fades considerably slower and even upon 3-fold remelting the modified structure is inherited (Fig15) with an undercooling of 5°C. However, for EZL101with 7% Si or EZL102 having 12%Si after 3-fold remelting some silicon flake can be observed, displaying a decreased undercooling (Fig.16). It is thought that the alloying elements such as Cu, Mn, Ni and Mg prompt the occurrence of deep undercooling, strengthening the structural inheritance in EZL108,EZL109 and ES9 (Table I) In general, with 4-fold or more remelting the almost fully structural fading occurs and the undercooling disappears. In this case the microstructure in eutectic EZL108 and EZL10 is composed of eutectic with few, if any, α-Al dendrites. Note that upon first remelting the quasi-eutectic in DEASA hypereutectic alloys is subjected to fully fading, resulting in an appearance of coarse primary silicon grain distributed in unmodified eutectic matrix as shown inFig.17, while the undercooling cannot be found. Fig.18 shows the variety of undercooling with remelting for DEASA (EZL101, EZL109). When undercooling is lower than 5°C, the inheritance of self-modified structure of EZL101 is subjected to significantly fading. In contrast, as undercooling decreases to 5°C the self-modified microstructure of EZL109 remains unchanged. Thus, it is reasonable to consider that for hypoeutectic and eutectic DEASAs the critical undercooling is 5°C, which is lower than critical value of 6-8°Cfor commercial Al-Si alloys. This phenomenon is thought to be associated with the homogeneous characteristics of DEASA melt, which cause silicon to solidify in modification mode at lower undercooling and cooling rate [23].

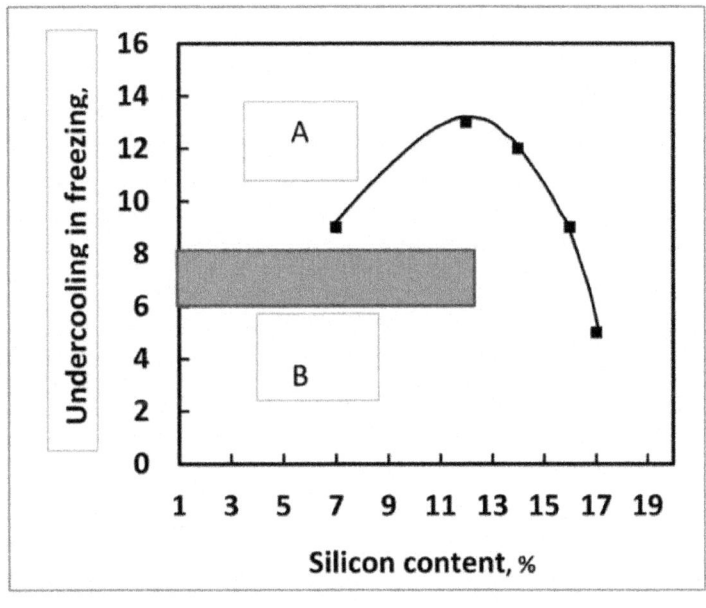

Critical region

A: modified silicon B: unmodified silicon

Figure 14. Relationship between eutectic undercooling and Si content in remelted DEASA.

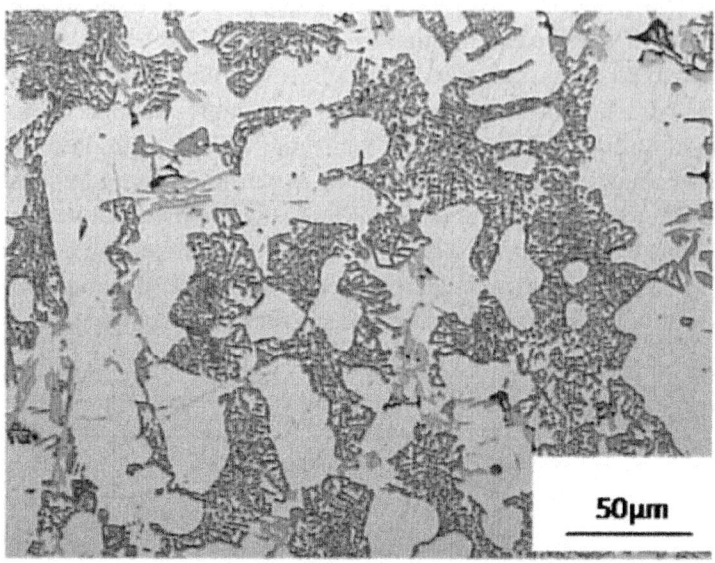

Figure 15. Optical structure of DEASA(EZL109) upon 3-fold remelting. Self-modified structure is fully inherited.

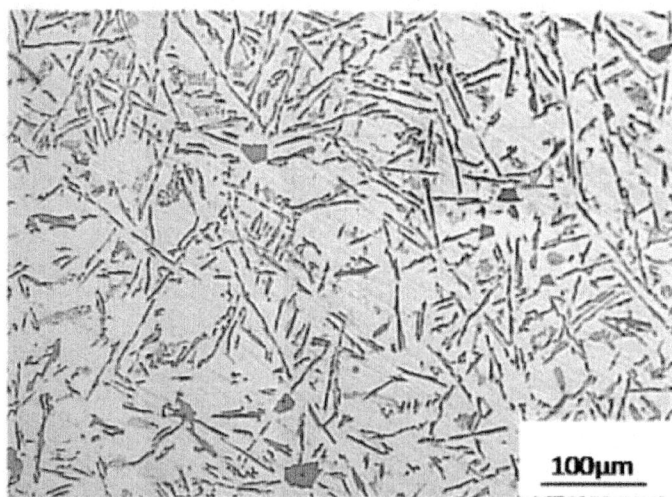

Figure 16. Microstructure of DEASA(EZL102) upon 3-fold remelting composed of unmodified structure. Some small faceted primary silicon appears. Optical.

It is worth noting that the self-modified structure is relatively insensitive to cooling rate as compared to commercial alloy. In general, the microstructure in top area of commercial ingot, where the cooling rate is very slow, displays coarse unmodified silicon flake but in the edge the fine silicon structure can be observed due to rapid cooling rate. Our observation reveals that there is no obvious difference in fineness of the eutectic between top and edge of eutectic DEASA ingot (EZL109) (Fig.7and 8). It is expected to be associated with the homogeneous melt, of which the stability is strengthened by the electric field in electrolytic process. That would be thought to be superiority over commercial alloy to produce complex castings.

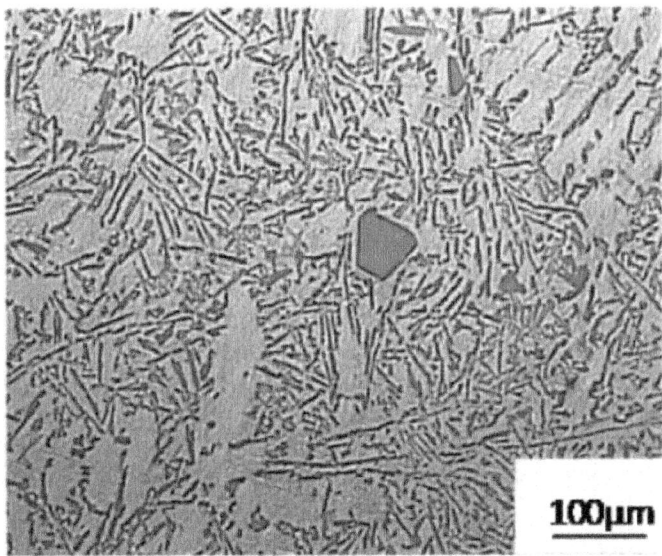

Figure 17. Optical micrograph of DEASA(EZL14) after first remelting. Self-modification is fully subjected to fading. Faceted primary silicon occurs.

The origin of the variety of undercooling of DEASA is associated with the homogeneous character of its melt. The original DEASA melt either hypoeutectic or eutectic is intrinsic homogeneous, causing silicon to solidify at a large undercooling due to the lack of large silicon-rich clusters acting as nuclei in freezing. The repeated re-melting of DEASA, in which the large undissolved silicon particles exist, causes a lower undercooling, accompanying with the fading of modified microstructure The homogeneous character in hypereutectic DEASA melting can be partially survives or fully lost depending upon the silicon composition, because with increasing silicon content the undissolved silicon particles dramatically increases, nucleating silicon in solidification with a lower undercooling.

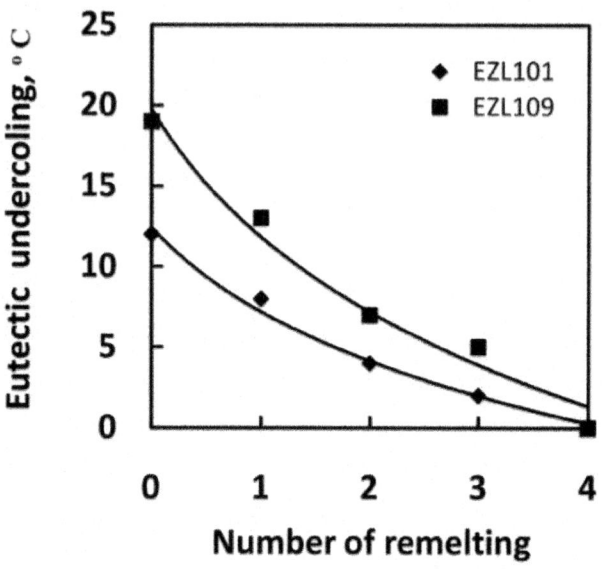

Figure 18. Relationship between undercooling and remelting under cooling and re-melting.

By combining the results in present and previous works we suggested the following growth mechanism of quasi-eutectic structure [49]. At initiation of the growth the silicon particle as nucleus would be assumed to be a nodule or irregular shape, with many different facets exposed in the melt [59-61]. Whether or not such a nucleus grows as polyhedron primary silicon crystal in freezing is determined by the degree of undercooling. As the nucleus grows, the boundary layer of eutectic composition starts to form around the growing nucleus and isolates it from the melting, thus preventing the further development of nucleus. With lower undercooling or higher silicon concentration the silicon atoms are capable of diffusing cross the layer to be trapped on the surface of the silicon nucleus, thus the silicon nucleus further grows, developing a primary silicon crystal and eutectic structure before the temperature of melt lowers down to the critical value shown by curve ES in Fig.3. Under high undercooling silicon atoms diffusion is limited, suppressing the primary silicon crystal to form. If the primary silicon cannot develop until the melt is cooled, reaching through the apparent eutectic temperature as curve ES shown in Fig.3, the quasi-eutectic structure occurs. In this case the silicon particles could act as the nucleus of eutectic, promoting the growth of eutectic structure.

As well known, whether the eutectic silicon grows in modified manner is attributed to the undercooling in solidification of alloy. This phenomenon is related with the entropy of melting and crystallographic structure as reported by A. Jackson in 1958 [62]. This relationship can be expressed as:

$$\alpha = \left(\Delta S / R \right) \left(N_S / N_V \right) \tag{3}$$

where α is Jackson criterion; ΔS is entropy of melting; R is gas constant; N_S and N_V are the number of an atom's nearest neighbors on the surface and within the body of a crystal. If α is less than 2 cal/°C, crystal grows isotropically with an atomically rough interface. By contrast, if α is greater than 2, crystal is faceted with an atomically smooth surface. It is very interesting that for silicon the Jackson criterion for principal crystallographic planes varies in range from 0.89 for (110) plane to 1.87 for (100) plane, to 2.67 for (111) plane. Thus silicon crystal is a borderline material, of which the growth mode can easily change from faceted to non-faceted when the undercooling increases [63]. The variety in undercooling, which is induced by cooling rate during freezing or impurity element or others, will significantly cause the change in morphology of silicon either eutectic or primary. Generally speaking, for hypoeutectic or near eutectic DEASAs undercooling of 5°C is considered as a critical value to change the growth mode of silicon(Fig.18). Recently H.S.Kang et al [64] reported that the critical undercooling is a linear function of silicon content. For the higher silicon content an increased undercooling is required to change the morphology of eutectic silicon phase. They revealed that for Al-13%Si alloy at undercooling of 14°C the eutectic silicon morphology changes from flake to fibrous shape. However, for hypereutectic Al-20%Si alloy an increase in undercooling to 73°C is required. The different critical undercooling reported in literatures is thought to be associated with the different structure in liquid state. In current study DEASA melt is homogeneous, but the melt heating treated at 720°C in study by Kang is microheterogeneous, for which a deep undercooling /high cooling rate is needed to achieve the modified eutectic structure as evidenced in study[23].

Other important variety in structural feature of DEASAs is that iron-rich phase appears in fine flaky form instead of needle-like shape in the center of ingot containing iron concentration of 0.25% as shown in Fig.19. It is interesting that as Fe-level is more than 0.5% in EZL102 ingot the morphology of Fe-bearing precipitate also remain unchanged as shown in Fig.5. That is also attributed to effect of superheating the melt in electrolysis pot on the crystallization of iron-rich composition as reported in reference [33,34].

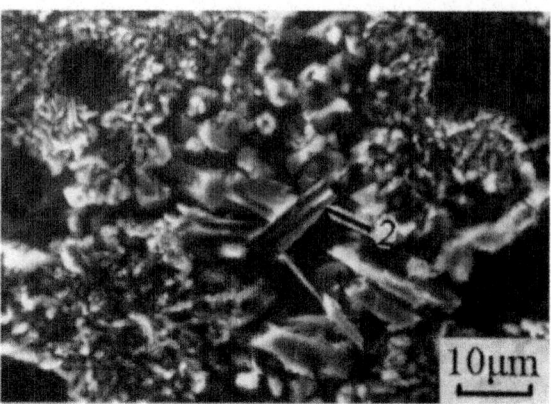

Figure 19. Fine flaky iron-rich phase indicated by arrow 2 appears in DEASA (EZL108) containing iron level of 0.25%.

Primary a-Al Phase

Primary α-Al phase is an important phase constituent, of which the volume fraction, grain size and morphology, dendrite arm space (DAS) and alloying element greatly affect the mechanical and foundry properties of hypoeutectic Al-Si alloys[65,66]. In DEASAs either eutectic such as EZL102, EZL108 and EZL109 or hypoeutectic such as EZL101, the volume proportion of primary α-Al dendrite is higher than unmodified Al-7%Si alloy (ZL101) (Fig.20), and with increasing silicon content the volume percentage of Al-phase decreases. Undoubtedly, the increase of primary α-Al dendrite greatly affects the properties of eutectic DEASA castings.

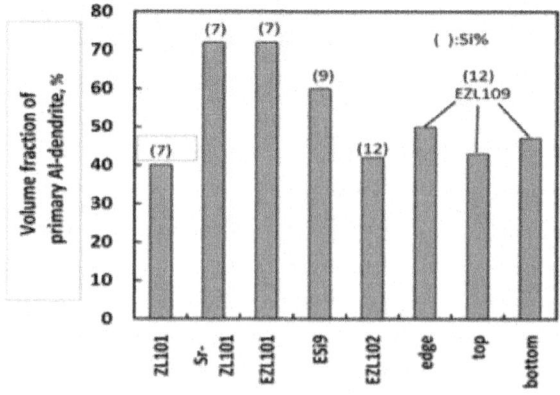

Figure 20. Volume fraction of α-Al dendrite in commercial and electrolytic Al-Si alloys.

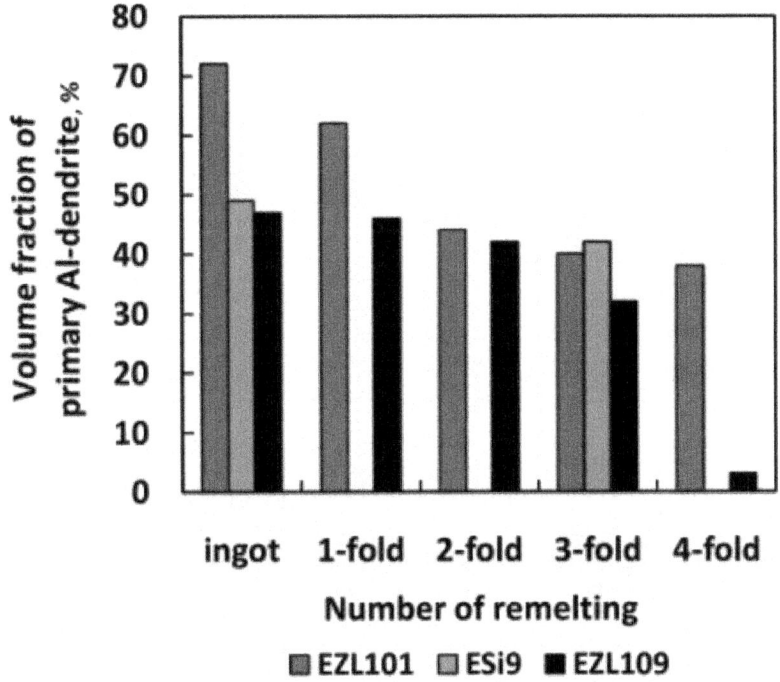

Figure 21. Volume percentage of α-Al dendrite in DEASA against the remelting.

The volume percentage of α-Al dendrite in the edge area of EZL109 ingot, where the cooling rate is much higher than bottom or top, is more or less larger than other areas (Fig.20). In addition, the volume proportion of aluminum dendrite decreases with remelting, which causes a decrease in undercooling in freezing (fig.18). After 3 or 4-fold remelting the volume percentage recovers to the value estimated from equilibrium Al-Si phase diagram (Fig.21) accompanying with an unmodified silicon structure. Apparently, the volume fraction of α-Al dendrite is a function of undercooling, which can be clarified using Al-Si quasi-diagram (Fig.3).

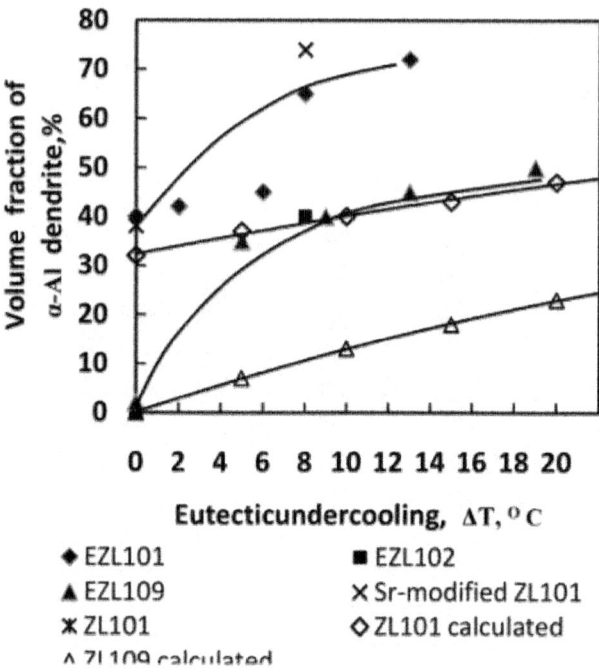

Figure 22. Volume percentage of α-Al dendrite in DEASA is a function of undercooling in freezing.

The fact that undercooling shifts the eutectic content toward to higher level during freezing and depresses the eutectic temperature, leading to an increase in temperature interval, in which the primary aluminum phase precipitates from melt, and, thus, the volume fraction of Al-phase is increased. Curve EP (Fig.3) represents the relation between undercooling and silicon content solved in Al-matrix. Therefore, we are able to estimate the volume fraction of α-Al dendrite in terms of undercooling in our tests (Fig.22). It is interesting that the volume fraction of α-Al dendrite measured in our study is much higher than the value calculated in terms of curve EP. By combining the successful achievement of quasi-eutectic structure in EZL16 with undercooling of 9°C, which is much smaller than the critical undercooling of 20°C shown on curve ES to obtain quasi-eutectic for commercial Al-16Si alloy (Fig.3), it is reasonably postulated that the region of formation of quasi-eutectic structure in DEASA moves toward higher silicon content and smaller undercooling due to the homogeneous DEASA melt.

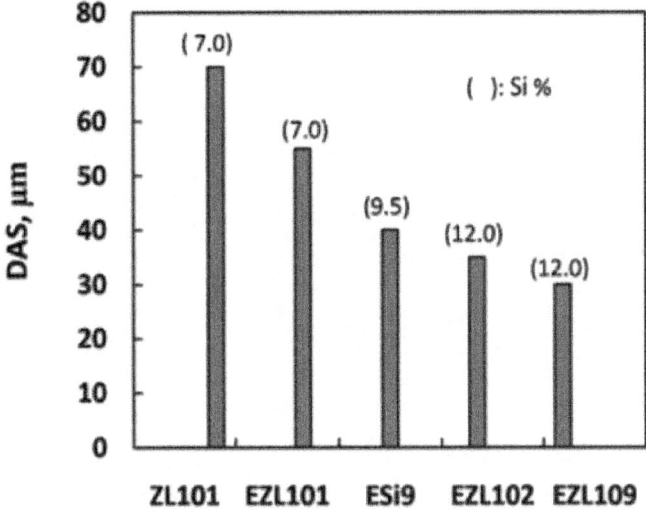

Figure 23. DAS of α-Al dendrite in DEASA and commercial ZL101(A356) ingot.

Dendrite arm space (DAS) is an important crystallographic feature in primary Al-phase, greatly affecting the mechanical properties of hypoeutetic Al-Si alloys. The DAS, which is not related in any way to the volume percentage of Al-dendrite, can be varied considerably by cooling rate. As far back as the 1960s it has been found that for commercial Al-Si alloy castings with cooling rate in range of 10^{-1}-10^2 °C/sec such as cast in sand and in a metal mould, and continuous castings the DAS value is a function of cooling rate as follows [67]:

$$d = A \cdot V^{-n} \qquad (4)$$

where d is DAS (μm), R is the cooling rate (°C/sec), A is related to the chemical composition and n=1/3-1/2.

In electrolytic EZL101ingot having 7% Si content the primary Al-dendrite displays the smaller DAS value than commercial Sr-modified ZL101 ingot as seen in Fig.23. Meanwhile the DAS decreases with silicon content. After 2 or 3-fold remelting DAS value doesn't change, if any. That is related with the same cooling rate in freezing of those samples [67].

Summarizing the structural characteristics of DEASAs we reach the conclusion that the electrolytic alloy castings either hypoeutectic or eutectic or hypereutectic exhibit very fine eutectic silicon grain with high volume fraction of α-Al dendrite, small DSA and small curved iron-bearing compound compared to commercial alloy. It is an advance superiority of DEASA over existing alloy for producing the high quality casting with excellent usage properties.

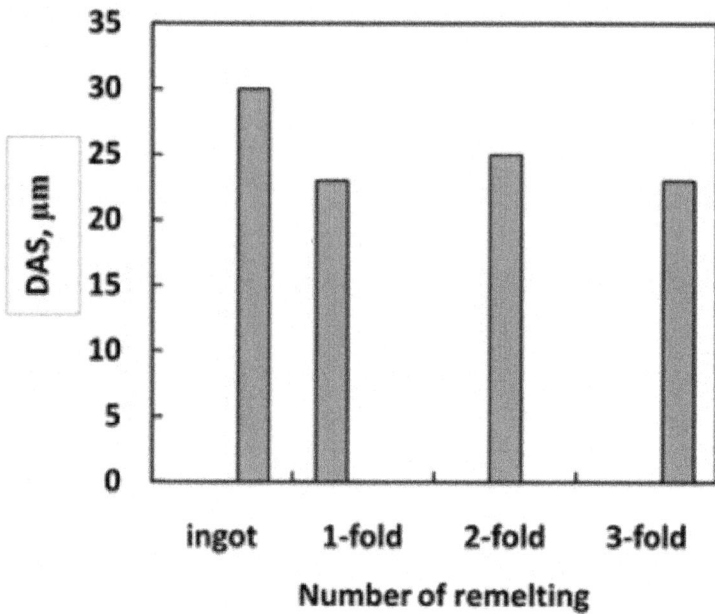

Figure 24. DAS in DEASA (EZL109) against remelting.

TECHNOLOGICAL PARAMETERS INFLUENCING SELF-MODIFICATION IN DEASA CASTINGS

In foundry several technological parameters including amount of DEASA ingot in metallic charge, furnace temperature and holding time, remelting and level of modifier added into melts, significantly affect the microstructure of castings and its mechanical properties. To optimize those parameters is an important event for producing high quality cast product with low cost.

In this chapter we have discussed the structural heredity of alloys upon remelting. It is concluded that remelting a fine metal easily produced a fine casting compared to a coarse metal at the same condition [68]. Experimental results demonstrated that at least 10% of a fine Al-Si ingot is required to achieve a casting with fine silicon grain [69]. Therefore, amount of DEASA ingot added in metal charge is an important factor for producing a casting with the fully modified microstructure.

In our test the metallic charge is composed of EZL101 ingot (Table 1.), pure aluminum ingot and Al-20%Si along with other master alloy additions. No modifier is added in melt. The percentage of DEASA ingot in metallic charge ranges from 10 to 50%. Table 3. lists the chemical analysis of alloys tested. Note that with either the holding time in furnace or remelting the strontium content is unchanged and much less than the critical level (0.004%) to create a modified eutectic in Al-Si alloys [70, 71]. Thus, the change in microstructure is associated with the amount of DEASA ingot used rather than strontium content in alloy.

Table 3. Technological parameters and chemical analysis of DEASA alloys wt%

Alloy	Amount of DEASA	Number of remelting	Holding time (min)	Chemical Analysis wt%								
	(%)			Si	Mg	Ti	Fe	Cu	Mn	Zn	Ni	Sr
EA50-0*	50	0	10	7.33	0.21	0.11	0.21	<0.01	0.01	0.003	0.016	0.0011
			10	7.08	0.41	0.19	0.27	<0.01	0.006	0.012	0.007	0.0012
EA30-0	30	0	120	6.90	0.34	0.17	0.30	<0.01	0.007	0.012	0.006	0.0013
			240	6.98	0.30	0.17	0.25	<0.01	0.007	0.012	0.006	0.0012
EA30-2		2	120	7.00	0.25	0.21	0.26	<0.01	0.006	0.012	0.009	0.0011
			10	6.71	0.39	0.053	0.25	<0.01	0.009	0.015	0.038	0.0016
EA10-0	10	0	120	6.84	0.28	0.050	0.25	<0.01	0.009	0.015	0.006	0.0016
			240	6.50	0.28	0.048	0.25	<0.01	0.009	0.015	0.006	0.0014
EA10-2		2	10	6.69	0.32	0.047	0.26	<0.01	0.006	0.013	-	0.0012

When 10% of metallic charge is DEASA silicon crystals grow in modified manner in as-cast microstructure of EA10-0 alloy, but a few silicon flakes can be found (Fig.25). However, increasing the amount of DEASA to 30% (EA30-0) or more (EA50-0) results in a fully modified microstructure as seen in Fig.26. It would be expected that the self-modified silicon crystal in DEASA acts as a modifier for the commercial Al-Si alloy. Like Na, Sr and other modifiers there is a critical amount of DEASA, below which the eutectic is not modifiable. For ZL101(A356) alloy 30% DEASA ingot in metallic charge is needed to obtain a full modified microstructure. Obviously, the higher the amount of DEASA used in charge, the stronger the trend to modification in alloy. Note that there is no overmodification with increasing the amount of DEASA ingot. It is a superior characteristic of DEASA castings to existing Al-Si alloy.

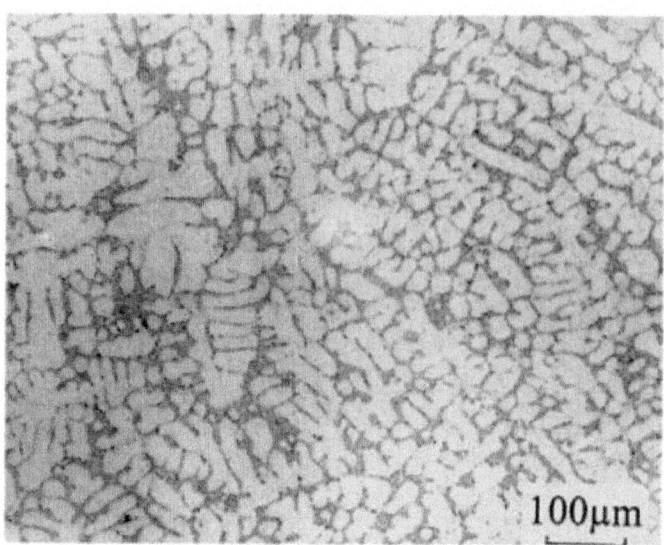

Figure 25. Modified microstructure of EA10-0 alloy with 10% of DEASA ingot in metallic charge and holding time of 10min. Some silicon flake can be found. Optical.

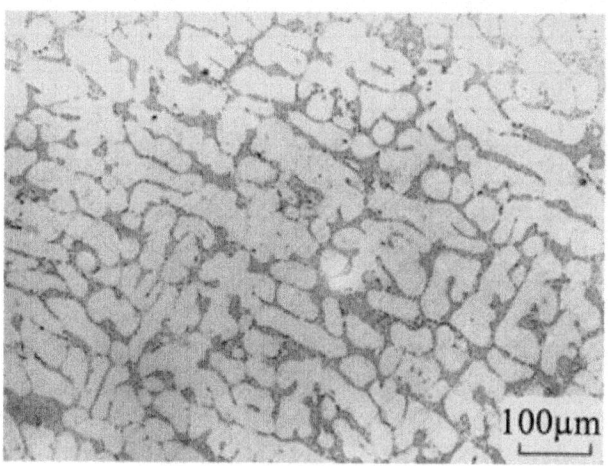

Figure 26. Optical microstructure of EA30-0 alloy with 30% of DEASA ingot in metallic charge and holding time of 10min, indicating a completely modified eutectic silicon.

Self-modified structure in EZL101(A356) alloys with either 10% or 30% of DEASA (EA10-0 and EA30-0) ingot strongly depends upon remelting and furnace holding time. When furnace holding time at 720°C is about 120min, both EA10-0 and EA30-0 alloys are subjected to partial structural fading as shown in Figs.27 and 28. Our test results reveal that for both alloys the inheritance of modified structure can be survived after first remelting. However, upon 2-fold remelting EA10-2 with 10% DEASA is subjected to partially fading in modified structure (Fig.29). As amount of DEASA increases to 30% (EA30-2), the modified structure can be maintained upon 2-fold remelting with holding time of 2 hrs, as shown in Fig.30, but when the holding time excesses 2 hours, some silicon flakes can be found.

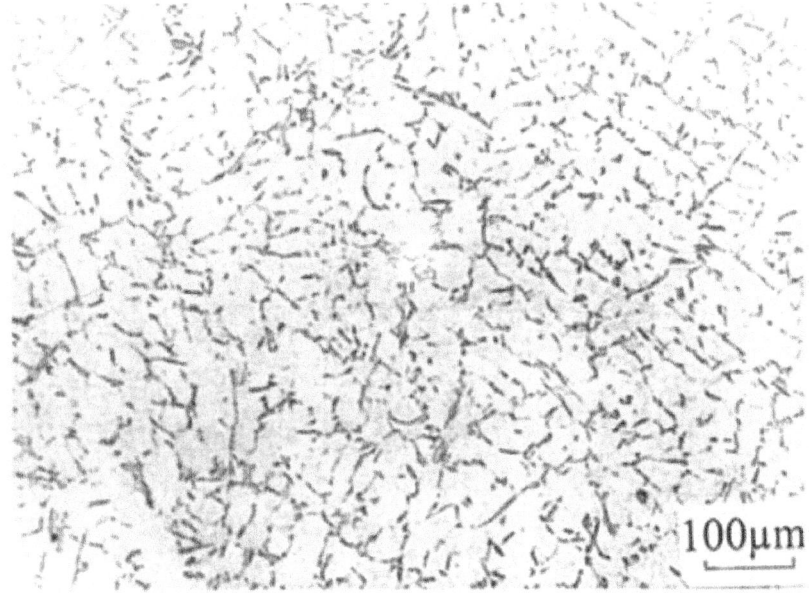

Figure 27. Optical micrograph of EA10-2 with 10% of DEASA ingot and holding time of 2hr at 720°C, indicating the unmodified structure.

In general, in foundries casting must be done within two hours after degassing aluminum melts in furnace. It is evident that strengthening the modification of DEASA alloys is needed for producing high quality castings. Below we will discuss the effect of small amount of strontium added into the molten alloy on the modified structure after different holding time. Table 4. lists the chemical analysis of Sr-modified DEASAs in tests.

Table 4. Chemical analysis of EZL101 alloys tested

Alloy	Amount of DEASA.%	Hold-ing time. min	Chemical Analysis wt%									
			Si	Mg	Cu	Mn	Fe	Ti	Cr	Ni	Zn	Sr
		10	7.08	0.43	0.03	0.006	0.29	0.064	<0.00	—	0.014	0.003
SEA10*	10%	120	6.84	0.28	0.04	0.009	0.35	0.088	0.004	0.006	0.015	0.002
		240	6.95	0.30	0.04	0.015	0.33	0.070	0.004	0.008	0.017	0.002
		10	6.81	0.39	0.032	0.007	0.36	0.16	0.002	0.004	0.012	0.003
SEA30	30%	120	6.66	0.34	0.032	0.007	0.35	0.17	0.002	0.006	0.012	0.002
		240	6.50	0.32	0.033	0.007	0.35	0.17	0.002	0.003	0.012	0.002

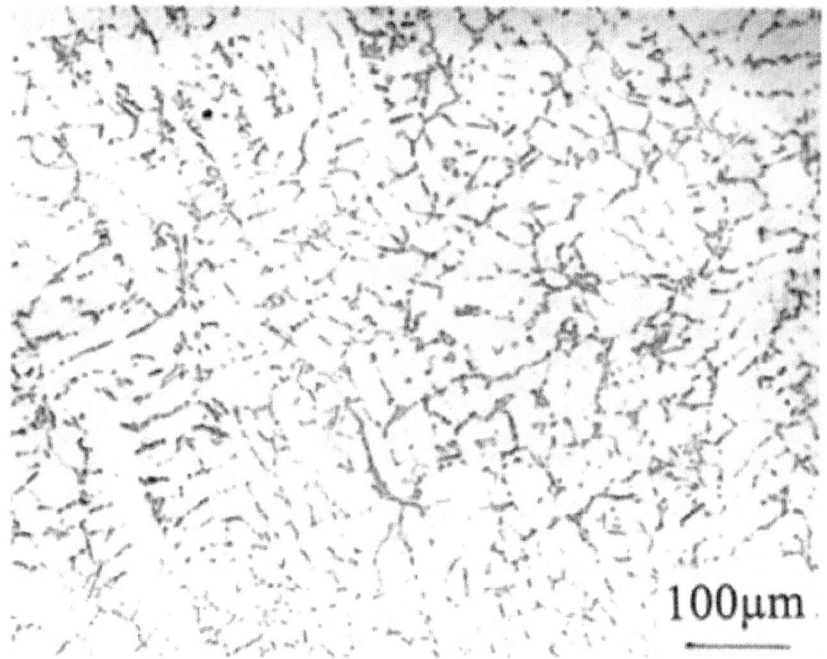

Figure 28. Micrograph of EA30-2 with 30% of DEASA ingot and holding time of 2hr at 720°C, indicating the unmodified structure. Optical.

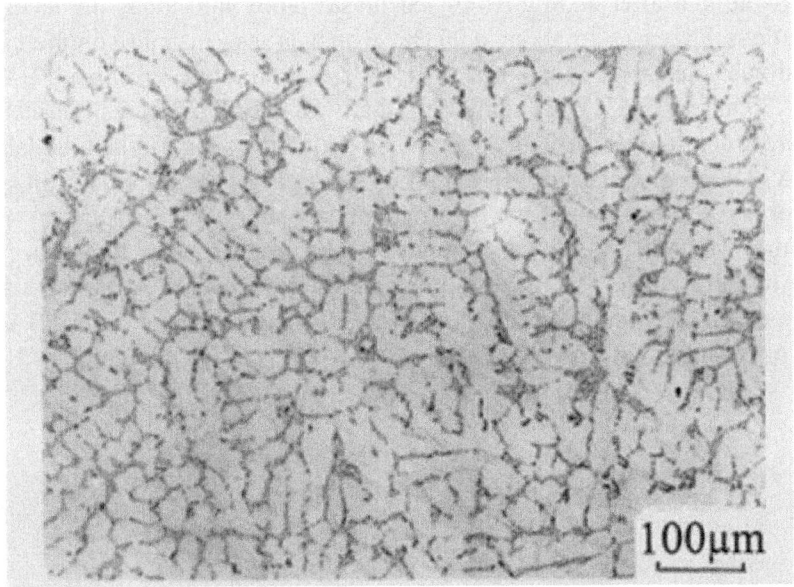

Figure 29. Optical micrograph of EA10-2 with 10% of DEASA ingot upon 2-fold remelting, indicating the partial fading of modified structure.

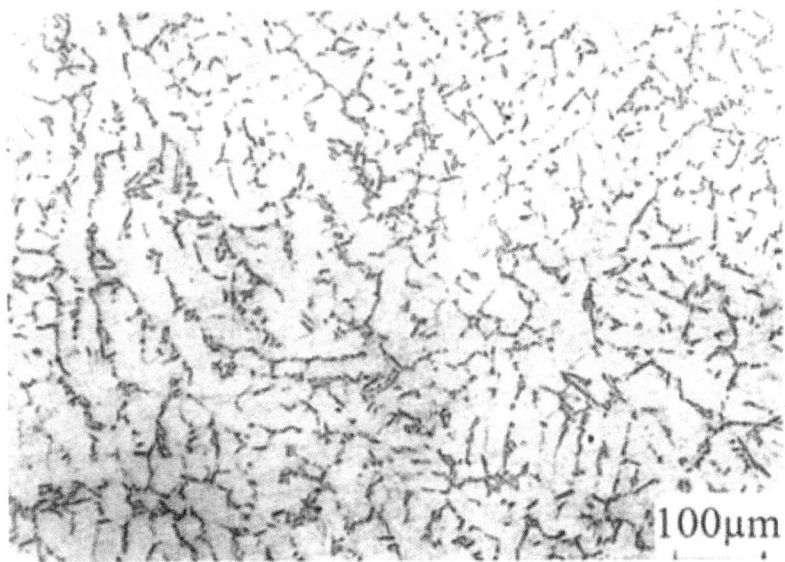

Figure 30. Micrograph of EA30-2 with 30% of DEASA ingot upon 2-fold remelting and holding time of 2 hrs, showing the modified structure. Optical.

Note that after adding Al-10%Sr master alloy into melts the level of Sr in alloys increases to about 0.003%, which is less than in commercial Sr-modified alloys, resulting in a fully modified structure in either SEA10 or SEA30 alloy (Fig.31). It is important that when holding time is four hours, in microstructure of SEA10 alloy some flaky silicon crystal can be found, but for SEA30 alloy the modification fading occurs in some degree as seen in Fig.32. Later is sufficient to meet the requirement in foundry. 30% of DEASA used in charge with 0.002-0.003% of Sr-level in alloy are necessary to produce high-quality Al-Si alloy. The process, in which the low level of strontium addition promotes a fully modified structure, is an advance advantage of DEASAs over Sr- or Na-modified Al-Si alloys.

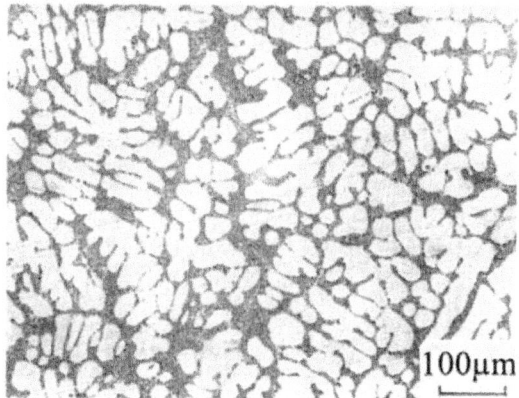

Figure 31. Micrograph of SEA30 alloy after furnace holding of 10min, showing the modified structure Optical.

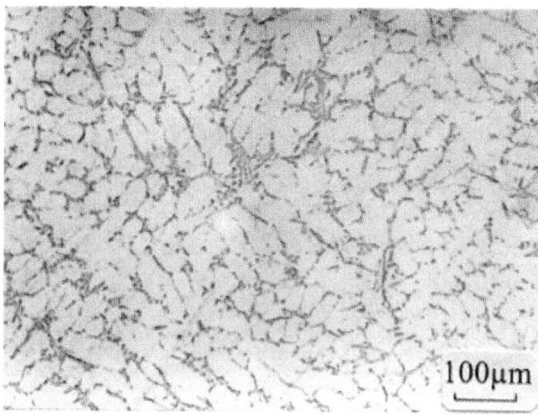

Figure 32. Partial modified microstructure of SEA30 alloy after furnace holding of 4hr at 720°C. Optical.

HIGH-QUALITY AUTOMOTIVE WHEELS MADE FROM DEASA

The wheel is an important part of a vehicle in terms of safety. The impact strength and fatigue life are on the top of the quality list of wheel characteristics. Al-7Si-0.3Mg alloy(A356), due to higher impact resistance and fatigue life, good castability and machinability, is a preferred choice to produce quality wheel. A lot of studies have demonstrated that the mechanical properties of A356 alloys are strongly affected by morphologies of eutectic silicon, iron content and porosity dispersed in castings, which is associated with the impurity modification and hydrogen level [71, 72]. For producing the quality wheel with higher impact resistance the maximum allowable iron level is limited to 0.20%. As mentioned above, DEASAs exhibit the excellent modified structure with a low level of modifier such as Sr, accompanying with the stringy Fe-rich precipitate of small size (Fig.19). Thus, the wheel made from DEASAs might be porous-free, resulting in higher impact resistance, ductility and tensile strength. At the end of last century we have examined the mechanical properties of wheel made from DEASA ingot (ES9, Tab.1) in the foundry [14].

A 600kg crucible was used to prepare the melts in an electric furnace. The metallic charge consisted of pure aluminum ingot, clean scrap of A356 alloy, other master alloy and DEASA (ES9) ingot pieces, of which the amount was a third of charge. Each melt was degassed with N_2 at temperature of 710°C. After degassing and holding for 15min, a small amount of Al-10%Sr master alloy was added into the melt to obtain Sr level in alloy below the critical value of 0.003%. Then the prepared melts were poured into a permanent mold to produce wheel casting and Y-shape plate castings with dimension of 22×150×220mm3. Finally, the castings were heat treated to a T_6 temper by solution at 535°C for 4 hrs, water quenching, and aging at 135°C for 6 hours. Table 5. lists the chemical analysis of DEASA (ES9) and EZL101 alloys, which have the different iron level, near or above the allowable value of 0.20% for wheel casting in order to clarify the effect of Fe-rich precipite on the mechanical properties of DEASA wheels.

Table 5. Chemical analysis of DEASAs for wheel tested wt%

Alloy	Si	Mg	Ti	Fe	Cu	Mn	Zn	Sr	RE
E S9	9.20-9.60	<0.01	0.40-0,60	0.44-0.65	<0.02	0.005	0.03	0.001-0.004	0.03
EZL101-17	6.63-6.80	0.28-0.30	0.12	0.16-0.17	0.03	0.1	0.01	0.0016-0.0024	—

EZL101-21	6.50-7.00	0.24-0.27	0.10	0.19-0.22	0.03	0.01	0.01	0.0022-0.0030	—
EZl01-27	6.90-7.30	0.28-0.30	0.10	0.26-0.27	0.03	0.008	0.01	0.0031-0.0034	—
ZL101 (A356)	6.90	0.30	0.12	0.10-0.13	0,02	0.01	0.01	0.0060-0.0080	0.0005

Table 6. lists the mechanical properties of conventional (ZL101A) and electrolytic Al-7Si-Mg alloy (EZL101). As iron content is less than or near the maximum allowable limit of 0.20%, the superiority of DEASAs over existing alloy is very evident. DEASA alloys offer the mechanical properties higher than existing alloy (ZL101A) with lower iron content of 0.12%. As increasing Fe-level from 0.21 to 0.27% there is a slight tendency to decrease the mechanical indexes. But the tensile strength remains to be higher than conventional alloy (ZL101A), the elongation and impact strength are lower than existing alloy.

Table 6. Mechanical properties of conventional and electrolytic Al-Si alloys with different Fe-level

Alloy	Fe content wt%	Tensile strength MPa	Elongation %	Impact strength J/cm2	Hardness HB
ZL101A	0.12	(213-238)/225	(7-16)/12	(15-52)/32	(76-80)/78
EZL101	0.16	(217-260)/239	(7-18)/13	(31-52)/37	(74-80)/76
	0.21	(211-241)/231	(11-20)/14	(28-52)/38	(75-85)/79
	0.27	(218-240)/232	(7-10)/8	(16-52)/29	(70-85)/79

Wheel impact strength test is carried out at wheel shock testing apparatus (Fig.33). In general, the critical impact strength for automobile wheel is 230mm (height) × 6000kN (weight) at 13degree of inclination. In production some 50% of the wheels made from ZL101 (A356) alloy exceed this minimum requirement by 10-20%. Addition of DEASA in charge exerts a significant improvement on the shock resistance. Testing results demonstrate that the wheels made of DEASA with different iron level offer the impact value exceeding 256mm×6000kN at 13degree of inclination, and most of them are higher than 276mm×6000kN, which exceeds the critical requirement by 20%. Moreover, in the extreme test at 30 degree of declination two third of DEASA wheels tested exceed the shock resistance of 230mm×1010kN. However, none of wheel made from ZL101(A356) could pass this limit.

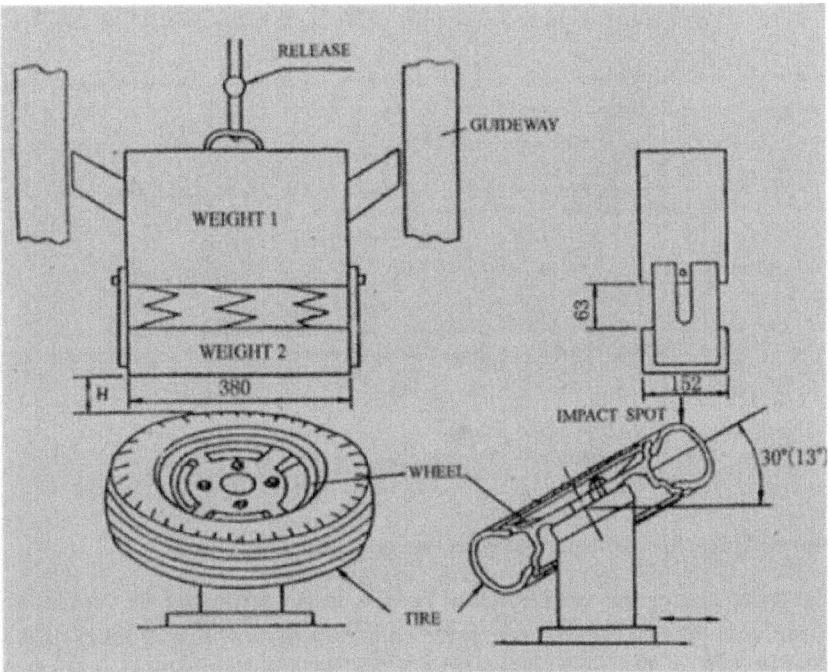

Figure 33. Schematic drawing of wheel shock testing apparatus.

Fig.34 shows the wheel fatigue test apparatus, which uses torque of 3000N-M for loading with rotating speed of 1500rpm of shift. For wheel of 14 or 15 inches diameter the design fatigue lifetime, which is expressed in terms of the number of cycles-to–failure, is 10^5 cycles. Usually the lifetime for wheel made of ZL101 (A356) ranges from 0.4 to 2.0×10^5 cycles. Some of them are not capable of exceeding the minimum lifetime. However, experimental data show that the dramatic improvement in impact resistance on DEASAs stated above is also evident in fatigue strength. DEASA wheels with iron content exceeding the allowable limit of 0.20% exhibit higher fatigue lifetime exceeding 2.0×105 cycles, except for E356-27 with higher iron content of 0.27% that has fatigue lifetime of 1.5×10^5 cycles.

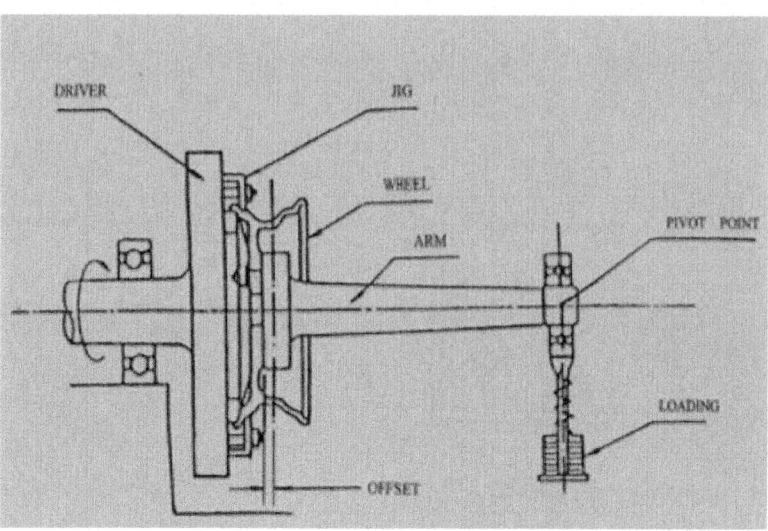

Figure 34. Schematic diagram of wheel fatigue testing apparatus.

Summarizing the experimental results it is reasonable to conclude that as iron content exceeds the maximum allowable limit of iron level of 0.20% in some degree, for example, reaching 0.27%, the mechanical properties of DEASAs, especially impact strength and fatigue resistance, significantly are improved. Therefore, it is expected that the allowable iron content would be limited to more than 0.20%, which would save the cost of wheel.

The reason why DEASA wheel containing different iron level exhibits an excellent impact strength and fatigue resistance compared to conventional alloy, is believed to be attributed to porosity, if the difference between morphologies of silicon crystals and Al-dendrites in DEASA and existing A356 is difficult to find in micrography of the wheel as demonstrated in Fig.35 and 36. Porosity is an undesirable feature of the cast structure because pores, either surface or internal, acting as stress raiser during loading, seriously degrade the mechanical properties [73-75]. This inference is strongly supported by leak test for wheel, revealing that all the DEASA wheels were leakproof, while 10% of ZL101 (A356) wheels were not. Moreover, visual inspection showed that no pinholes and microporosities could be found on the surface of DEASA wheels compared to common ZL101, implying that DEASA wheels exhibit much less porosity than existing alloy. Undoubtedly, the sound alloy made from DEASA has higher mechanical properties.

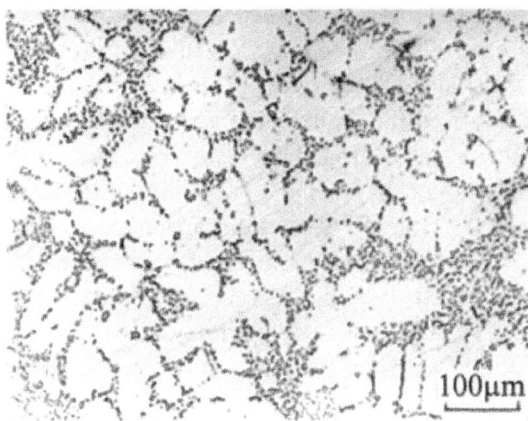

Figure 35. Opitical microstructure of hub in wheel made from DEASA EZL101-17(Table 6.) heat-treated by T_6

The origin of porosity is associated with two important, if not primary important, factors for Al-Si alloy in given casting condition, i.e. hydrogen dissolved in melt and amount of strontium or sodium added in molten alloy as modifier [71,76,77]. High hydrogen level causes an increase in porosity, resulting in decrease in mechanical properties [72-75]. Strontium or sodium increases the tendency to porosity of alloy [71]. In our study due to self-modification in DEASA ingot much less amount of modifier is required to be added into the DEASA molten alloy. Therefore, the tendency to porosity becomes weakened and sound castings are more easily obtained, resulting in higher impact resistance and fatigue strength in DEASA wheel.

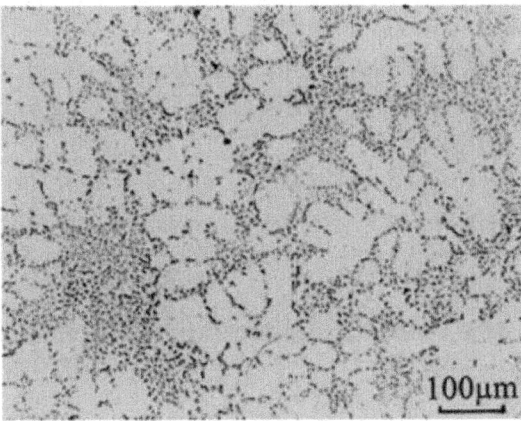

Figure 36. Microstructure of hub in wheel made of commercial A356 alloy heat treated by T_6 Optical.

Until now a few studies have been done on the behavior of hydrogen in aluminum in electrolytic process [78]. In electrolytic pot the surface of aluminum melt is usually crusted over by fused cryolite, acting as an insulator to isolate the liquid from the atmosphere and protecting melt from hydrogen pick-up. In this case the hydrogen level is very low. Prodhan [47] studied the behavior of hydrogen in Al-Si alloy during solidification under electric field, indicating that the hydrogen level can be decreased from 2.5ppm to 1.7ppm. This is within the acceptable limit of pore-free castings [47]. It would be expected that in electrolytic process the hydrogen in Al-Si alloy melt can be removed under electric field. Undoubtedly, the decreased hydrogen level in DEASAs strongly weakens the trend to porosity, enhancing the impact resistance and fatigue strength. It is evident that the electrolytic process would be a powerful mean to reduce the hydrogen level in alloy.

CONCLUSION

- DEASAs either hypoeutectic or eutectic display self-modified structure in ingot with excellent structural inheritance upon remelting. DEASA is self-undercooled alloy.

- Hypereutectic DEASAs with silicon in range from 13% to 17% exhibit completely self-modified eutectic structure, but are subjected to fully fading upon remelting due to disappearance of undercooling in freezing.

- DEASAs have high volume of α-Al dendrite that is associated with the high undercooling in freezing.

- DEASAs are insensitive to cooling rate in freezing.

- Iron-bearing precipitate in DEASAs appears in small curved shape as iron level increases to near 0.5%.

- 30% DEASA ingot in metallic charge with added Sr-level of 0.002-0.003% is necessary to produce high quality Al-Si casting with self-modified structure.

- Automobile wheel made of DEASA display high impact resistance and fracture strength that is associated to small amount of Sr-modifier added into melt, low hydrogen concentration and small curved shape of iron-rich compound.

- Electrolysis is a potential measure to produce high quality Al-Si casting.

REFERENCES

1.   Hayward C.R(1955An Outline of Metallurgical Practice.Third Edition. Toronto: D.Van Nostrand Company, Inc.285286p.

2.   J. Frilley, 1911Revue de Metallurgie. j. 87518523

3.   Bullough V.L1973US Patent 3Oct.16.

4.   Z. X. Qiu, Z. L. Zhang, K. Grjotheim, H. Kvande, 1987Aluminium.j. 631212471250

5.   Z. Orman, 1976Rudy Met. Niezelez. j. 215162164

6.   D. Weslan, 1984US Patent 4Jan.10, 1984.

7.   R. Keller, B. J. Weld, A. T. Tabereau, 1990In: Light metals 1990. TMS. 333340

8.   B. Moxnes, H. Gikling, H. Kvande, S. Rolseth, Straumsheim, K(2003) In: Crepeau P, editor. Light Metals 2003TMS. 329334

9.   Tabereaux A.T, McMinnC.J(1978In: Light Metals 1978.TMS.209222

10. C. J. Mc Minn, A. T. Tabereaux, 1976US Patent 3Sept.14, 1976.

11. Yang K.Q, Gu Q.S, Tian G.Y, LiQ.C(1994Chinese Patent. ZL94 11 6235.4, March 4th 1994. (In Chinese)

12. K. Q. Yang, S. Yang, 1997Foundry. j. No.14446In Chinese)

13. Yang D.X,.WangR.Y(1994Hot Working Technology. j. 221921In Chinese)

14. Wang R.Y, Lu W.H2001Light Metal Age. j. 59(5/6):.610

15. Wang R.Y,.Lu W.H,Hogan L.M2003Mat.Sci.Eng. j.A348289298

16. Bloomfield L.A, Freeman R.R, Brown W.L1954Phys. Rev. Letter. j. 5422462249

17. S. Mitsuo, (1997) Liquid Metals. London: Academic Press. 41 235-236.

18. T. Iida, R. I. L. Guthrie, 1988The Physical Properties of Liquid MetalsOxford: Clarendon Press. 1846

19. Gui M.C, Li Q.C, JunJ(1995Special Casting and Nonferrous Alloys.j.15158In Chinese)

20. Eskin D.G1996Z. Metallkunde.j. 894295301

21. Popel P.S, Tchikova O.A., Brodova J.G, Makeev V.V1992Nonferrous Metals. J. No.9: 53-56(In Russian).

22. Bian X.F, Wang W.M2000Materials letters.j. 4415458

23. Brodova J.G, Popel P.S,.Eskin G.I2002Liquid Metal Processing. Application to Aluminum Alloy Production. London: Taylor and Francis. 85145

24. Wang, L.D. Zhu D.Y, Wei Z.I, Chen Y.L, Huang L.G, Li Q.J, Wang Y.S2011Advance of Materials Research. J.146-147:79-89.

25. Nikitin V.I1991Foundry Production.j. 4In Russian)

26. Li P.J,. Xiong Y.H, Zhang Y.F, ZengD.B(2003Trans.Nonferrous Met. Soc.China.j. 13 (2):.329334

27. M. Singh, R. Kumar, 1973J. Mater. Sci. j. 8317323

28. Zhang L,.Bian X.F, MaJ.J(1995Foundry. j. 10In Chinese).

29. W. M. Wang, X. F. Bian, H. R. Wang, Z. Wang, Z. G. Lin, J. Liu, J.M(2001J. Materials Research. J. 16125923598

30. Xu C.L, JingQ.C(2006Mat.Sci.Eng.j. A4372451455

31. Gui M.C, Li Q.C, JiaJ(1995Special Casting and Nonferrous Alloys. j. 15 (1):.5-8 (In Chinese)

32. Zhang R,.Zhang L.M, Yang Z.H, LiuL(2011Materials Sci. Forum. J. 654-656:.14121415

33. Ri E.K, Ri K.S, Khimukhin S.N, Kalugin M.Y, Statsenko D.P, Kryuchkov I.V2011Foundry Production.j. 7In Russian).

34. Y. Awano, Y. Shimizu, 1987Imono.J.594233238In Japanese).

35. Li P.J,NikitinV.I, Kandalova E.G,   Nikitin   K.V(2002)   Mater.Sci. Eng.j.A332 : 371-372.

36. MüllerK(1998Metall. J.52 (1-2):29-35.

37. Gubinko A.Y1991Foundry Production. 4In Russian).

38. Z. L. Zhao, J. L. Wang, L. Lu, 2011Materials and Manufacturing Processing.j. 262249254

39. A. Prodhan, D. Sanyal, 1998Materials Science Letter.j. 16(11):.958961

40. A. Prodhan, C. S. Sivaramakrishnan, A. K. Chakrabariti, (200, 2001Met Mat. Trans B. j.32(2):.3723780

41. S. R. Yu, Y. H. Zheng, H. Feng, L. Cai, L.G(2010Hot working technology.j.39(9):47-50 (In Chinese).

42. H. Conral, 2000Mat. Sci. Eng. J.A2878205212

43. H. Li, X. F. Bian, X. F. Liu, J. J. , 1996Special Casting and Nonferrous Alloys. J.16(3): 8-10 (In Chinese).

44. HuangL.G,GaoZ.Y,ZhangZ.M2009FoundryTechnology.j.30(5):650-652 (In Chinese).

45. Temchenko S.L, Zadoroschnai N.A2005Foundry Production.j. 991213In Russian)

46. B.A.Rabkin,S.L.Temchenko(2003Foundry Production.j.10In Russian).

47. A. Prodhan, 2009AFS Trans. J.1176377

48. Loper C.R, Lu D.Y, Kang C.S1985AFS Trans. j. 93533543

49. Wang R.Y, Lu W.H2007AFS Trans. J.115241248

50. B. M. Thall, B. Chalmer, 1950J.Inst. Metals. j. 777997

51. C. B. Kim, R. W. Heine, 1963J. Inst. Met. j. 92367376

52. Jenkinson D.C, Hogan L.M1975J. Crystal Growth. j. 28171187

53. H. Fredriksson, 1991Scand. J. Metallurgy.J.204349

54. Fredriksson. H. Talaat-Benawy El, 2000Trans. J. of Japanese Institute of Light Metals. j. 41507515

55. G. Stuhldreier, K. W. Stoffregen, 1981Giesserei. j. 68404409

56. Stoffregen K.W1985Giesserei.j. 72545549

57. MondolfoL.F(1979Aluminum Alloys, Structures and PropertiesLondon: Butterworths.513515

58. D. K. Apelian, G. Sigworth, K. R. Whaler, 1984AFS Trans.j. 92297370

59. Wang R.Y, Lu W.H.Hogan L.M1995Mat. Sci.Tec. j. 11(5):.441449

60. Wang R.Y, Lu W.H, HoganL.M(1999J. Crystal Growth.j. 2074354

61. WangR.Y,LuW.H,HoganL.M(1997Trans.Metall.Mater.j.28A:12331243

62. JacksonK.A(1958Mechnism of growth, In: Liquid Metals and Solidification, ASM, Cleveland, OH.174186

63. Gilmer G.H,.Leaming H.J,.Jackson K.A1974Liquid Metals and Solidification. Proc.4th Int. Conf. on Crystal Growth, Amsterdam, North-holland. 495

64. Kang H.S.,.Yoon W.Y,.Kim K.H, Kim M.H. YoonE.2004Mater. Sci. Forum. j. 449-452: 169-172.

65. Oswalt K.J, MisraM.S(1980AFS Trans. J. 88845862

66. Mi J.W, Cheng J.N,.Yu Y.M1990In: C.Q.Chen, F.A.Starke.editors. Proceedings of the second international conference on aluminum alloys. Beijing: 566570

67. Dobatkin V.I, Eskin G.I.1990In: C.Q.Chen, F.A.Starke.editors. Proceedings of the second international conference on aluminum alloys. Beijing: 278282

68. Li S.S, Zhu Y.F, Zeng D.B1999Foundry. j. 8In Chinese).

69. Lui X.F, Bian X.F,.Ma J.J etal(1994Foundry.j. 10In Chinese)

70. W. H. Lu, R. Wang, R.Y(1995Special Casting and Nonferrous Alloys.j.15215In Chinese).

71. Gruzleski J.E., Closset B.M1990The Treatment of Liquid Aluminum-Silicon AlloysAFS Inc. Des Plaines, IL, US. 223chapter 4.

72. AnyalebechiP.N(2003In: P N. Crepeau. Editor. Light Metals 2003. TMS. 971981

73. Eady J.A,.Smith D.M1986Materials Forum.j. 9(4):.217223

74. Surappa M.K, Blank E.J, Jacquet.C1986Script Metallurgica.j. 20(9):.12811286

75. Caceeres C.H1995Script Metallurgica et Materialia.j. 321118511886

76. C. Mascre, M. Lefebvre, 1959Fonderie. j. 166166484497

77. ArbenzH(1962Giesserei.j. 49105110

78. Zhang M.J, Qiu Z.X, Di H.L1987Nonferrous Metals. 112731No.2:29-34 (In Chinese).

Chapter 6

ELECTROLYTIC EXTRACTION DRIVES VOLATILE FATTY ACID CHAIN ELONGATION THROUGH LACTIC ACID AND REPLACES CHEMICAL PH CONTROL IN THIN STILLAGE FERMENTATION

Stephen J. Andersen[1], Pieter Candry[1], Thais Basadre[1], Way Cern Khor[1], Hugo Roume[1], Emma HernandezSanabria[1], Marta Coma[1,2] and Korneel Rabaey[1]

[1]Laboratory of Microbial Ecology and Technology (LabMET), Ghent University, Coupure Links 653, Building A, Room A0.092, B9000 Ghent, Belgium

[2]Centre for Sustainable Chemical Technologies, University of Bath, Claverton Down, Bath BA2 7AY, UK.

ABSTRACT

Background

Volatile fatty acids (VFA) are building blocks for the chemical industry. Sustainable, biological production is constrained by production and recovery costs, including the need for intensive pH correction. Membrane electrolysis has been developed as an in situ extraction technology tailored to the direct recovery of VFA from fermentation while stabilizing acidogenesis without caustic addition. A current applied across an anion exchange membrane reduces the fermentation broth (catholyte, water reduction: $H_2O + e^- \rightarrow \frac{1}{2} H_2 + OH^-$) and drives carboxylate ions into a clean, concentrated VFA stream (anolyte, water oxidation: $H_2O \rightarrow 2e^- + 2 H^+ + O_2$).

Results

In this study, we fermented thin stillage to generate a mixed VFA extract without chemical pH control. Membrane electrolysis (0.1 A, 3.22 ± 0.60 V) extracted 28 ± 6 % of carboxylates generated per day (on a carbon basis) and completely replaced caustic control of pH, with no impact on the total carboxylate production amount or rate. Hydrogen generated from the applied current shifted the fermentation outcome from predominantly C2 and C3 VFA (64 ± 3 % of the total VFA present in the control) to majority of C4 to C6 (70 ± 12 % in the experiment), with identical proportions in the VFA acid extract. A strain related to *Megasphaera elsdenii* (maximum abundance of 57 %), a bacteria capable of producing mid-chain VFA at a high rate, was enriched by the applied current, alongside a stable community of *Lactobacillus* spp. (10 %), enabling chain elongation of VFA through lactic acid. A conversion of 30 ± 5 % VFA produced per sCOD fed (60 ± 10 % of the reactive fraction) was achieved, with a 50 ± 6 % reduction in suspended solids likely by electro-coagulation.

Conclusions

VFA can be extracted directly from a fermentation broth by membrane electrolysis. The electrolytic water reduction products are utilized in the fermentation: OH^- is used for pH control without added chemicals, and H_2 is metabolized by species such as *Megasphaera elsdenii* to produce greater value, more reduced VFA. Electro-fermentation displays promise for generating added value chemical co-products from biorefinery sidestreams and wastes.

BACKGROUND

The chemical industry requires a broad range of carbon-based building blocks and platform chemicals, many of which can be generated sustainably through microbial conversions from sugar, lignocellulosic biomass, and carbon dioxide [1–4]. Anaerobic microbial conversions increasingly contribute to the production of sustainable, non-fuel chemicals. In chemistry, greater value is associated with more reactive functional groups [4]. Non-fuel compounds, on average, are priced fifteen times higher per ton than fuels [3], though most anaerobic biotechnology success stories to date are those of bulk bio-fuels such as biogas and alcohols. These chemicals are generally recovered through gas separation or with petrochemical era separation technologies such as distillation and solvent extraction [5]. Though these technologies are mature and well developed for a broad range of chemicals, they are not broadly suited to biologically constrained titers in complex broths, particularly where extensive dewatering is required. The high energy and capital investment often fatally weaken the economics and sustainability of bulk biochemical production. This issue is compounded when production is constrained by a complex substrate, such as agro-industrial sidestreams and waste.

Production and recovery of compounds from a complex substrate of mixed organics require a strategy that includes both the conversion of non-ideal substrates and the targeted separation of products from fractions with overlapping physico-chemical characteristics (e.g. hydrophobicity, volatility) [6, 7]. Volatile fatty acids (VFA) are attractive yet challenging in both substrate conversion and separation, as VFA: (1) are ubiquitously formed as intermediates during the decomposition of organic matter, typically as a mixture of many VFA; (2) are hydrophilic at a more neutral pH and hydrophobic at low pH, allowing solvent extraction; and (3) have a charged carboxyl group under more neutral conditions, allowing electrical motility. Industrially, short-chain linear saturated VFA such as acetic (C2), propionic (C3) and butyric (C4) acid are mostly produced thermochemically. Targeting bio-production of VFA for use as bulk chemicals or chemical precursors (e.g. for conversion to solvents, fuels, polymers) is known as the Carboxylate Platform [8, 9], and is often associated with second generation biorefinery processes and sustainable substrates such as syngas, and agro-industrial residues and sidestreams. Separation and recovery technology is recognized as a major challenge within the Carboxylate Platform due to the aforementioned unit operation technological hurdles, but also due to product inhibition [10, 11]. Short to mid-chain VFA (C6–C8) acids are toxic at relatively low concentrations, but are commonly targeted for their high added value. Chain elongation to caproic acid (C6) has been demonstrated with mixed microbial communities on both synthetic feed and real streams, with *Clostridium kluyveri* and *Ruminococci* often identified as key players in the microbial community [12–17].*Megasphaera elsdenii* has been shown to generate caproic acid from sugars and lactate as a pure culture [18–21]. Active removal of caproic and heptanoic acid is critical in sustaining production [15, 17, 20].

Membrane electrolysis is an electrochemical extraction technique demonstrated for carboxylate recovery and concentration of short-chain VFA [22, 23] and phase separation of caproic acid [17]. In short, charged products flux across an ion exchange membrane, driven by the electrolysis of water in both the fermentation and the extraction compartment (Figure. 1). Hydrogen gas is produced at the cathode in membrane electrolysis, creating a surplus of biological reducing equivalents that may drive reverse β-oxidation and VFA chain elongation [24, 25]. Other than hydrogen (H_2), the electrolysis products of protons (H^+) and hydroxide ions (OH^-) can be utilized as acid and base without the addition of salts. Hydroxide can counter acidogenic fermentation, while protons generated at the anode acidify the extracted carboxylate, allowing acid accumulation [17, 22, 23]. Andersen et al. demonstrated acetate extraction in a synthetic broth with a high extraction efficiency for a concentration of around 10 g L^{-1}, with efficiency decreasing at lower concentrations [22]. Gildemyn et al. demonstrated combined acetate production (microbial electrosynthesis) and extraction with membrane electrolysis, using homoacetogens enriched for hydrogen metabolism and low applied current [23]. Xu et al. used membrane electrolysis to extract, acidify and phase separate caproic acid from a chain elongation reactor; however, the membrane electrolysis was separated from the broth by two units (solids separation, liquid–liquid membrane extraction) and did not interact directly with the fermentation [17].

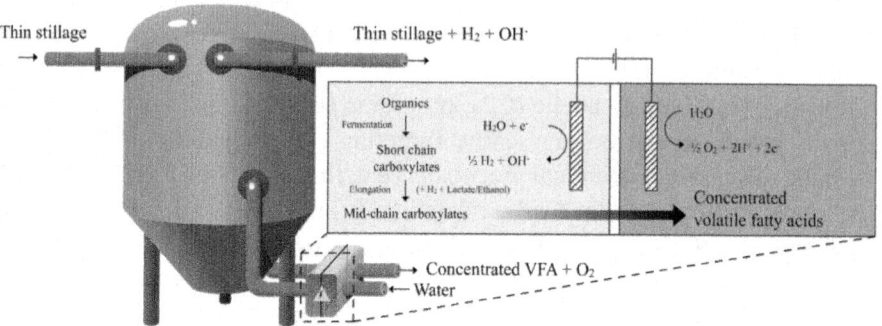

Figure. 1. Schematic of electro-fermentation and membrane electrolysis.

The unconverted organic fraction of bioethanol production from food crops is a rich, untapped source of complex organic compounds [26, 27], with as much as half or more of the carbon entering the system remaining unconverted [3]. This unconverted fraction is most commonly directed toward distillers grains (also known as 'dried distillers grains with solubles', DDGS), a low value agricultural feed product that is an integral co-product in modern bioethanol refineries [28,29]. In this study, we target thin stillage, the liquid fraction separated from the whole stillage (the distillation column bottoms in a bioethanol production). Though it varies depending on the size and operation of the plant, thin stillage is generated in the vicinity of 10^5 tons per year per plant and retains a high portion of solids, between 0.1 and 1 %, that are directed back to the production of DDGS after dewatering. Targeting the fermentation of thin stillage allows for a bio-production strategy on an organic rich, low impact stream, and avoids the embedded energy cost (e.g. distillation) in performing VFA chain elongation with recovered VFA and ethanol or lactic substrates.

Petrochemical era extractions typically sit downstream of the production process, while membrane electrolysis interacts directly with the fermentation for maximum utility of the input energy. It is therefore critical that the process implications of membrane electrolysis on the fermentation are understood in detail. Our study explored the impact of in situ membrane electrolysis on VFA fermentation of a real, untreated biorefinery stream. We characterized the impact of the electrolysis products (hydrogen, hydroxide) on thin stillage and the fermentative bacterial community native to this stream. We relate this to changes in the fermentation process, and discuss how this may impact sustainable VFA production.

RESULTS AND DISCUSSION

Stable Fermentation with Extraction and Electrochemical pH Control

Thin stillage was semi-continuously fermented under control and experimental (applied current) conditions, with membrane electrolysis extraction in the experimental case. No difference was observed between the total VFA produced at steady state under control conditions (Figure. 2a). The control is defined as the identical fermentation without an applied current, with sodium hydroxide supplied for pH control, while the experimental case had an applied current (100 mA, 3.22 ± 0.6 V, approximately 20–24 h) until the upper pH set point of pH 5.7 was reached, and then the current was lowered until the next feeding (20 mA, 2.36 ± 0.25 V, approximately 24–28 h) to prevent pH overshoot. Where the pH exceeded the upper set point, VFA rich acid from the anolyte was dosed back into the reactor, though this occurred on average at less than 1 mL day^{-1}. The average production rate of short-chain linear unsaturated C2–C7 VFA in the control and the experimental case were similar at 1.9 ± 0.8 g COD L^{-1} day^{-1} (2.5 ± 1.0 gC L^{-1} day^{-1}) and 1.9 ± 0.5 g COD L^{-1} day^{-1} (2.3 ± 0.6 gC L^{-1} day^{-1}), respectively, at the 6-day hydraulic retention time (HRT) condition. In both the control and experimental fermentation, a similar maximum amount of VFA was generated per total initial volume of thin stillage, with a total conversion of 31 ± 2 % sCOD for the control and 30 ± 5 % sCOD for the experiment (Figure. 3). If the sCOD that can be attributed to protein and oils is excluded, the conversion was 44 ± 2 % for the control and 43 ± 7 % for the applied current fermentation on a sCOD basis. The broth appears to have approached the maximum conversion to VFA under these reactor conditions. Membrane electrolysis extracted 28 ± 6 % of carboxylates generated (rate of extraction/rate of production). The extraction rate was not optimized in this study, which pertains more to electrochemical reactor design and operation (e.g. solids separation, membrane crossflow velocity, surface convection, optimized applied current scheme). The membrane flux rate is closely linked with the total molar concentration in the broth, as demonstrated in previous work involving extraction efficiency [22]. In this study, we focused on the effect of the electrolysis products on the fermentation rather than the effect of the extraction.

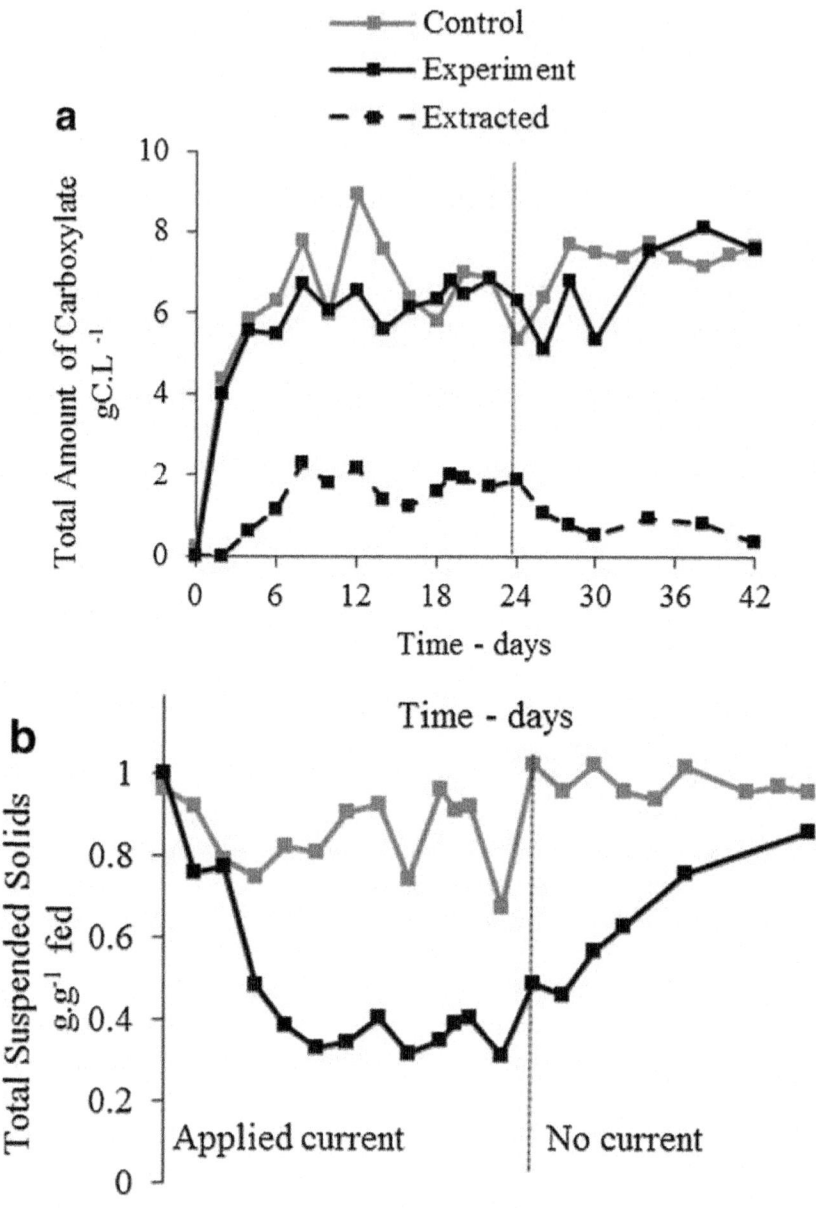

Figure. 2. Control and experimental fermentation over time; total amount of carboxylates and total suspended solids. In the experimental fermentation, current was applied prior to the vertical dotted line. **a** The total amount of measured carboxylates. Note the experimental case includes the amount extracted. **b** Total suspended solids measured represented by the proportion of total suspended solids in the fermenter relative to that of the feed

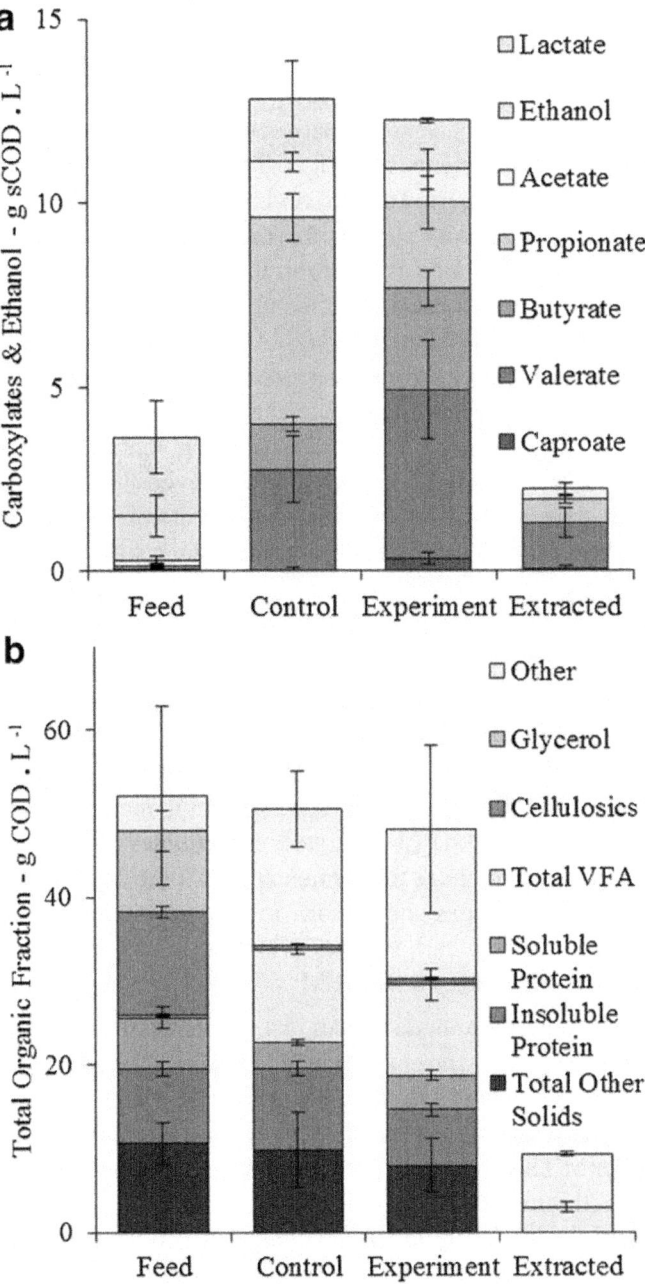

Figure. 3. Carbon oxygen demand (COD) balance in the feed, control fermentation, experiment fermentation (applied current) and in the extractant, measured during steady state (day 6–24). Note the extractant is also considered in the experiment col-

umn. **a** Total amount of measured carboxylates and ethanol. **b** Total amount of all identified components ($n = 9$ for carboxylates, $n = 4$ for other components, per stream).

The applied current for membrane electrolysis will cause a potential loss due to membrane fouling and solids between the electrode and the membrane. Current interrupt experiments investigated the change in resistance and the ohmic drop over a 5-day period due to solids buildup between the electrode and the membrane. The resistance attributed to the region between the electrode and the membrane was calculated at 3 Ω from the current interrupt experiments at the beginning of the experiment (new membrane), and increased to 10 Ω after 1 day. The resistance peaked at 19 Ω after 4 days. Cleaning the space between the membrane and the electrode decreased this resistance to 10 Ω. Solids buildup on the electrode and membrane can therefore account for almost 1 V of ohmic drop, representing a third of the applied potential under experimental conditions of 100 mA at 3.22 ± 0.6 V. An improved flow design or cleaning regime could translate into a power savings on membrane electrolysis, with increased cross flow velocity likely to improve anion flux.

Acidification of the fermentation broth was completely avoided by the cathodic generation of hydroxide for a zero chemical input pH control. Upper limit pH control was managed by dosing acid extractant back into the fermentation broth, at less than 1 mL day^{-1} on average, as the applied current scheme was designed to avoid excess electrolysis. In the control case, 24 ± 17 mL day^{-1} of 2 M NaOH was added to maintain the pH, equivalent to approximately 10 kg day^{-1} of caustic soda per cubic meter of thin stillage fed. For each kilogram of COD_{VFA} generated this equates to 0.83 kg of sodium hydroxide required to manage the fermentation. At the anode, the oxidation of water generates oxygen gas and protons in the anolyte (extractant), resulting in the protonation of the carboxylate and other anions (phosphate, chloride, sulfate, etc.) that have crossed the AEM.

The experimental reactor decreases in total suspended solids (TSS) relative to the control case after current is applied (Figure. 2b), stabilizing at 0.36 ± 0.04 g.g^{-1} TSS fed, relative to the control case of 0.86 ± 0.11 g.g^{-1} TSS fed in the control case during the steady state (solids content is reported here as grams per liter of $TSS_{Reactor}$/gram per liter of TSS_{Feed} fed due to inconsistency of solids in the fed thin stillage, i.e. minimum 14 g TSS L^{-1}, maximum 25 g TSS L^{-1}.) The total suspended solids in the control stabilized at a rate of 0.7 ± 1.4 g TSS L^{-1} day^{-1} compared to the applied current case which decreased at an average rate of 3.9 ± 1.9 g TSS L^{-1} day^{-1} in the steady state period from day 8 to day 24. This corresponds to a 4.5 times greater decrease in suspended solids, albeit with large variability. The current was stopped at day 24 to confirm the effect. The stoppage coincided with the suspended solids concentration returning to that of the feed. No equivalent decrease in the measured solids COD nor increase in measured soluble cellulosic fragments or VFA production was observed. This phenomenon is likely related to electro-coagulation, in which the applied current is neutralizing suspended particles, resulting in the formation of coagulated particles [30]. Some evidence to support this is revealed in nitrogen analysis, where 85 ± 6 % of total fed nitrogen remains in the broth in the control case, whereas 71 ± 5 % of total fed nitrogen is measured in the broth in the experimental case, with this difference accounted for mainly in the insoluble nitrogen compounds. Protein or lignin polymers can form colloids at high pH, which may adsorb at the high local pH on the cathode surface, or settle within the electrochemical cell. More research is required into the electro-coagulation phenomenon and the implication for the bacteria and their access to the substrate. If membrane electrolysis is applied in a similar broth in which conversion of solids are targeted, it is not clear if the apparent electro-coagulation will have an effect on the ability of the microbial community to metabolize these solids.

Fermentation and Chain Elongation of the Soluble Fraction to Carboxylates

The COD balance of the thin stillage fermentation revealed that membrane electrolysis resulted in a shift in the fermentation of VFA, from a majority C2 and C3 (64 ± 5 % C2 and C3, 36 ± 2 % C4–C6, as an average percentage of the total carboxylates on a COD basis during steady state, $n = 10$) to a majority of C4 to C6 (30 ± 5 % C2–C3, 70 ± 12 % C3–C6) without a change in the total amount of VFA on a COD basis (Figure. 3a). The proportion of acetate was lower in the applied current case than in the control, but only slightly and not significantly across all measured time points during the steady state (n = 10) (14 ± 2 % control, 8 ± 5 % with current). The proportion of propionate decreased from 51 ± 6 to 21 ± 7 % with current. The proportion of butyrate was greater under applied current (11 ± 2 % control, 25 ± 7 % with current), with a similar trend with valerate (25 ± 8 % control, 42 ± 11 % with current), and caproate (0 ± 1 % control, 3 ± 2 % with current). The extent of chain elongation can be compared as "chain elongation equivalents", the concentration of carbon (gC L^{-1}) at steady state that has been added through a chain elongation pathway on the theoretical assumption that all VFA start at either acetate (C2) or propionate (C3). The extent of VFA chain elongation at steady state in the applied current case was 2.6 ± 0.6 gC L^{-1} of chain elongation equivalents, significantly higher than the control case at 1.4 ± 0.4 gC L^{-1} (t test: $= 0.05, p = 1.9 \times 10^{-4}, n = 10$).

Only a fraction of the feed was converted to VFA. This fraction is referred to here as the "reactive fraction" (Figure. 3b). The reactive fraction consisted of soluble cellulosic fragments (or 'Sugars'), consisting of 4.4 ± 0.6 g L^{-1} glucose, 4.2 ± 0.5 g L^{-1}xylose and 2.9 ± 0.4 g L^{-1} arabinose (in total 12.3 ± 0.7 gCOD L^{-1}); and glycerol (9.8 ± 2.4 gCOD L^{-1}), lactate (2.1 ± 1.0 gCOD L^{-1}), and C1 to C8 carboxylates (0.3 ± 0.2 gCOD L^{-1}). The remainder of the thin stillage consisted of a solid fraction of proteins (8.9 ± 0.9 gCOD L^{-1}) and other lignocellulosic solids (10.6 ± 2.5 gCOD L^{-1}), a soluble COD fraction (sCOD) of protein (6.1 ± 1.3 gCOD L^{-1}), and an assumed balance fats, oils and other biomass (4.2 gCOD L^{-1}), in good agreement with a previous thin stillage characterization [27]. Membrane electrolysis did not alter any of the other main components characterized in this study (Figure. 3b). Zhou et al. studied glycerol fermentation with applied current, resulting in a mixed outcome with approximately 15–30 % of the carbon ending as VFA (mostly propionate), 20 % ending as biomass, 3–6 % as ethanol and 20–50 % as 1,3-propanediol [31]. Insignificant quantities of 1,3-propanediol were detected in this study, and propanol and butanol were detected at less than 1 gCOD L^{-1} each. Phenolic compounds were identified in the feed and fermentations at 1.2 g L^{-1}. Both the control and applied current case showed no net increase of ethanol in the broth from 1.2 ± 0.6 gCOD L^{-1} fed to 1.7 ± 1.0 gCOD L^{-1} and 1.3 ± 0.1 gCOD L^{-1}, respectively. Phenolic compounds were detected in the extractant in trace quantities.

Soluble cellulosic fragments, glycerol and lactate were consumed equally in both the control and experimental reactors (Figure. 3b). In the control, this resulted in predominantly acetate and propionate. Lactate is present in the fed thin stillage and can be used to elongate acetate to butyrate, as can ethanol [16]. Approximately 0.14 gC L^{-1} day^{-1} of lactate and 0.05 gC L^{-1}day^{-1} of ethanol enter the system by feeding. This fed lactate and ethanol can account for the chain elongation of C2 and C3 species to C4 and longer, assuming all carboxylates of C4 and longer are a result of chain elongation. The control case needs a total of 0.2 gC L^{-1} day^{-1} to elongate C2 and C3, compared with the applied current case which requires 0.7 gC L^{-1}day^{-1}.

Increased Organic Loading and Current Lead to Increased VFA Production

A brief, secondary experiment tested the system under doubled organic loading rate and a constant applied current of 100 mA, resulting in a 5.5 times increase in the VFA production rate to 10.4 ± 1.1 gCOD$_{VFA}$ L^{-1} day^{-1}. The disproportional increase in production can be partially attributed to removing the substrate limitation, in addition to the constant supply of hydrogen gas to the fermentation. The VFA production rate was 10.4 ± 1.9 g COD L^{-1} day^{-1}

$(3.3 \pm 0.6$ gC L^{-1} day$^{-1})$, at an extraction of 26 ± 5 % of produced VFA. The conversion rate of sCOD was also higher at 60 ± 11 % of the total sCOD fed, or 86 ± 16 % of the reactive fraction, with a higher proportion of C6 and now also C7 VFA. In the previous experiment, heptanoic acid (C7) was not detected (Additional file 1: Figure SI 1). The constant current appears to have increased the total conversion of the reactive fraction. The greater supply of hydrogen gas (per liter of reactor volume), 66 mmol L^{-1} day^{-1} compared with 25 mmol L^{-1} day^{-1} in the first experiment, resulted in a lower concentration of butyrate at 8 ± 0 % and valerate at 19 ± 1 % of the C2–C7 VFA. The proportion of caproate was 11 ± 6 %, compared to 3 ± 6 % in the previous experiment, and heptanoate at 1 ± 0 % (Additional file 1: Figure SI 1). The chain elongation equivalents increased relative to hydrogen generation, though not proportionally (Additional file 1: Figure SI 2). This suggests that the high organic loading rate and high current case (Additional file 1: Exp II, Figure SI 2) were either under excess hydrogen or the hydrogen was escaping the system before it could be utilized. Short-term batch tests at a range of applied current (Additional file 1: Figure SI 2B) showed a similar trend for total carbon chain elongation equivalents generated in the broth, where even though the applied current is doubled from 100 to 200 mA, the total chain elongation equivalents only increased incrementally.

In this high loading, constant current experiment, the volume of the extractant was halved and operated in batch to demonstrate acid accumulation and mimic a more realistic recovery strategy. A maximum concentration of 11.7 gC L^{-1} was reached in the anolyte, compared with the maximum of 2.3 gC L^{-1} in the previous experiment (Additional file 1: Figure SI 1). Phenolic acids were concentrated at up to 2.5 g L^{-1}. Phosphoric acid $(H_3PO_4$, pH <1) accumulated at up to a maximum of 4.1 g L^{-1} and hydrochloric acid to 1.6 g L^{-1}.

Membrane Electrolysis Favors Hydrogen Metabolizing Fermenters

The fermentation was initiated without an inoculum, and thus only organisms native to the thin stillage were cultivated. *Lactobacillus* spp. represented a relative abundance between 96 and 99 % of the bacteria at time zero and in the thin stillage fed throughout the experiment, with *Hallella* sp. and others making up the balance (Additional file 1: Figure SI 5, Table SI 1). *Lactobacillus* spp. abundance swiftly diminished in the control case to a maximum relative abundance of 1 %, and moved to a dominance of *Hallella* sp., with a relative abundance of 65 % after 2 days and a maximum relative abundance of 94 %, and an average of 75 ± 21 % across the whole experiment. The *Hallella* sp.

dominance was similar to the applied current case with an average abundance of 42 ± 26 % across the whole experiment. The next most abundant species in the control case were *Dialister* sp. and *Megasphaera* sp., both of the family Veillonellaceae. *Dialister* sp. had a relative abundance between 3 and 26 %, with an average across all measured time points of 13 ± 8 %. *Megasphaera* sp. had a relative abundance between 1 and 5 %, with an average of 3 ± 2 %. The applied current case contained *Dialister* sp. at a similar relative abundance to the control case at between 0 and 22 %, with an average of 11 ± 7 % across the whole experiment. The greatest difference between the control and experimental case arose from the abundance of *Lactobacillus* spp. and *Megasphaera* sp. The *Megasphaera* sp. was present with a relative abundance between 0 and 57 %, with an average of 15 ± 21 % (n = 9, at steady state). *Lactobacillus* spp. in the applied current case slowly decreased over the first 8 days and was then present between 6 and 17 % from day 8 to 24, in stark contrast to the control case of between 0 and 1 %. *Pectinatus sp.*, *Bifidobacterium* sp. and *Prevotella* sp. are also present in the applied current case at a relative abundance of up to at least 5 %, which is similar to the control case with the exception of *Pectinatus* sp. which never exceeded 0.7 %. The *Lactobacillus* spp. had a minimum abundance under applied current of 5.9 % and a maximum of 67.0 %, compared to the control case with a minimum of 0.0 % and a maximum of 1.3 % (ignoring $t = 0$, at which both had greater than 99 % abundance of *Lactobacillus* spp., identical to the feed). In the control case, the *Lactobacillus* spp. dropped from 99 % to a relative abundance of 1.3 % after 2 days. The time point of the greatest abundance of the *Megasphaera* sp. (57 %, Additional file 1: Figure SI 3 and SI 4) coincided with a slight increase in *Lactobacillus* spp. and an increase in chain elongation, albeit following a slump possibly related to competition with *Pectinatus* sp. (Additional file 1: Figure SI 3 and SI 4). Day 14 of the applied current fermentation coincided with a low point of chain elongation in the steady state fermentation, a high relative abundance of *Pectinatus* sp., and the *Hallella* sp. maximum. In a brief period where *Pectinatus* outcompeted *Megasphaera*, minimal C4–C6 carboxylates were produced and a peak of propionate was observed (Additional file 1: Figure SI 4).

Soluble cellulosic fragments, glycerol and lactate were consumed in both the control and experimental reactors (Figure. 3a). In the control, this resulted in predominantly acetate and propionate while with an applied current a greater concentration of mid-chain VFA was observed (Figures. 3a, 5). *Megasphaera* sp. can ferment glucose, lactate and short-chain VFA toward short to mid-chain VFA, alongside CO_2 and H_2 [18], while the majority of the other bacteria of high relative abundance produce short chain VFA and intermediates. *Hallella* sp, consistently the most abundant species, is known to produce acetate and succinate, that latter of which can be decarboxylated to propionate, or lactate [32]. *Dialister* is a non-fermenting bacillus [32] and its consistent abundance alongside *Hallella* suggests that it may have produced propionate from succinate. Neither *Hallella* nor *Dialister* have been associated with longer chain carboxylates [32, 33]. *Bifidobacterium* sp. produces lactate and acetate [34], *Prevotella* sp. mainly produces acetate and succinate, with a slight production of iso-butyrate and iso-valerate [35] while *Pectinatus* sp. is associated with the production of acetic and propionic acid [36, 37]. *Pectinatus* sp. has been observed to produce propionate from glycerol at a biological cathode [38].

Reverse β-oxidation VFA chain elongation with lactic acid has been described in *Megasphaera elsdenii* [16], and*Lactobacillus* spp., can produce lactic acid from a variety of substrates [39, 40]. The butyrate, valerate and caproate in both the control and applied current fermentation (Figure. 3a) were likely generated by *Megasphaera sp.* through the following pathway [18, 19]:

Lactate⁻+ Acetate⁻+H+→n-Butyrate⁻+ CO2+ H2O;ΔG0=−59.43 kJ mol−1at 37∘CLactate⁻+ Propionate⁻+ H+→Valerate− + CO2+ H2OLactate⁻+ Butyrate⁻+ H+→Caproate⁻+ CO2+ H2OLactate⁻+ Acetate⁻+H+→n-Butyrate⁻+ CO2+ H2O;ΔG0=−59.43 kJ mol−1at 37∘CLactate⁻+ Propionate⁻+ H+→Valerate− + CO2+ H2OLactate⁻+ Butyrate⁻+ H+→Caproate⁻+ CO2+ H2O

Megasphaera sp. stands out as one of few bacteria in these consortia with a high relative abundance that is known to generate mid-chain VFA, and there is evidence here to suggest that it was able to utilize hydrogen from membrane electrolysis to drive VFA chain elongation. *Megasphaera* sp. was the only species whose relative abundance correlated positively with the concentration of chain elongation equivalents (i.e. extent of VFA chain elongation) by the Pearson correlation test ($R = 0.63$, $p = 0.04$) in the applied current case, whereas no correlation can be seen in the control case ($R = 0.05$, $p = 0.89$). When compared in the RAST database, 100 % similarity was found with "Megasphaera NP3," a sequenced species closely related to *Megasphaera elsdenii* (RAST, http://rast.nmpdr.org/). Megasphaera NP3 contains genes for fatty acid production, glycolysis/gluconeogenesis and β-oxidation metabolism (for VFA chain elongation), along with genes for four NiFe hydrogenase mettalocenter assembly proteins, which makes it a good candidate for the ability to metabolize hydrogen [41]. Organisms that are both capable of reverse β-oxidation VFA chain elongation and hydrogen oxidation could gain energy by an increase of intracellular hydrogen which leads to an increase in NADH and NADPH, thus driving VFA reduction [2, 16, 24, 25, 41, 42, 43]. In this case, the majority of hydrogen was generated extracellularly in situ by membrane electrolysis and then transported into the cell to drive the so-called electro-fermentation.

A Redundancy Discriminant Analysis (RDA) was performed to examine the impact of applied current (experimental case) on the bacterial community, and their association with chain elongation over time, in comparison with a control (Figure. 4). We observed that the bacterial community in the applied current fermentation (Exp) was significantly different from that in the control case. The length of the chain elongation (CE) vector indicated a high relative association with the bacterial community in the Exp case ($r = 0.561$, RDA1 p value: 0.004). Within this community, Megasphaera sp. (Otu005) and Lactobacillus spp. (Otu006) are significantly correlated with chain elongation under the condition of the applied current (Exp), unlike in the control fermentation. In the control case, only time was significantly associated with the variations in the bacterial community ($r = 0.778$, RDA2 p value: 9.19×10^{-5}). The position of each arrow (time or CE) with respect to the plot axis represents its degree of correlation with particular OTUs. In this way, the differences in the relative abundance of *Lactobacillus* spp. may be associated with CE, while the variations in *Megasphaera* sp. may as well tend to be influenced by time. The species-environment correlation values confirmed these observations (RDA1 = 0.649, RDA2 = 0.778).

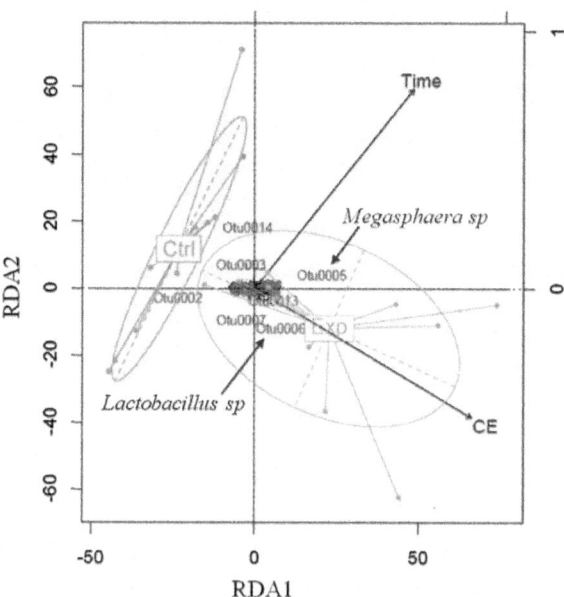

Figure. 4. Redundancy analysis highlighting the dissimilarities among the relative abundances of the bacterial communities in the control and experimental (applied cur-

rent) cases. "Ctrl" represents the community of the control reactor, which was described in RD2, while "Exp" indicates the community of the experimental (applied current) reactor, included in RD1. The *blue axis* represents time (days) and CE (extent of chain elongation).

The relationship between *Lactobacillus* spp's lactic acid production with the lactic, reverse β-oxidation and hydrogen utilizing capability of *Megasphaera* sp. enables the community to take advantage of the hydrogen generated in membrane electrolysis (Figure. 5). While *Lactobacillus* spp. are generally not recognized as hydrogen producers—or consumers—*Lactobacillus* spp. are often found in hydrogen producing consortia [44] and now here in a hydrogen consuming consortia. *Lactobacillus* spp. were always present in the applied current case during steady state at a relative abundance greater than 6 % at an average of 10.3 ± 3.4 % in the applied current case compared with an average relative abundance of 0.3 ± 0.4 % in the control (Additional file 1: Figure SI 3).

a Control

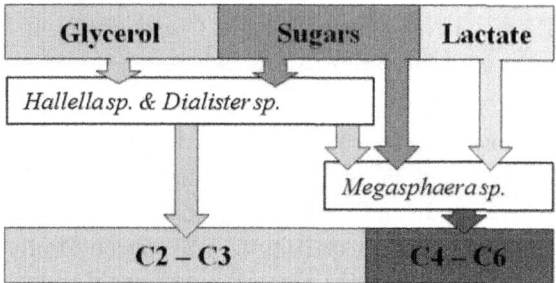

b Experiment (Applied Current)

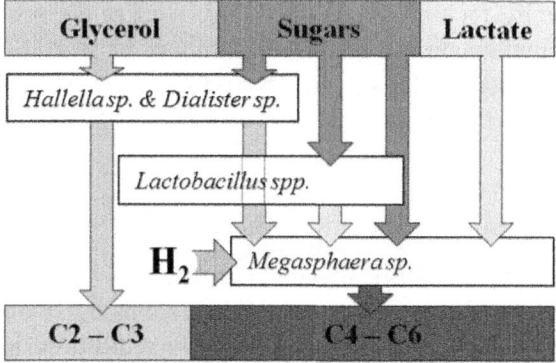

Figure. 5. Schematic of species with the top four greatest relative abundance and the proposed pathway of substrates, VFA intermediate and VFA product. **a** In the control case some acetate (C2) and propionate (C3) may be used as an intermediate

by *Megasphaera* sp. **b** In the experimental (applied current) case the *Megasphaera* sp. can metabolize electrolytically generated H_2 to gain additional energy and through generating more reduced, longer chain VFA by lactate elongation.

All species present in the fermentations were introduced from the feed (or whatever contamination followed). Phase contrast optical microscopy, flow cytometry and fluorescence in situ hybridization (FISH) confirmed that native species were present in the thin stillage despite the fact that thin stillage is retrieved between 75 and 80 °C, around the azeotrope of alcohol and water, downstream from the distillation column (Additional file 1: Figure SI 3). Enriching a native species from the target stream is attractive if the species fulfills the requirements of the process, though most chain elongation mixed culture studies use inocula with the intention of introducing *Clostridia*, more specifically *Clostridium kluyveri* [13, 16]. In our study *Clostridium* sp. are only observed in low abundance: less than 1 % in both the control and the experimental case. *Clostridium kluyveri* has been identified in other chain elongation studies, performing VFA chain elongation with an ethanol intermediate [12–17]. In one recent case, the *Clostridia* group IV was observed to generate caproic acid through a lactate substrate [45].

Understanding the principal actors in the community, be it in chain elongation or production of an intermediate, has implications in operational parameters such as temperature, residence time, pH and the substrate [46]. Selection of both substrate and strain are critical to ensure a targeted substrate conversion to the final product, be it in the conversion of oligosaccharides by *Lactobacillus spp.* [40], the production of caproic acid from sugars and lactic acid with *Megasphaera elsdenii* [18], or caproic production from acetic acid and ethanol with *Clostridium kluyveri* [12]. In the present state of technology, research generally focuses on product selectivity at the fermentation level, targeting optimum conditions toward the strain of interest [13, 15] with some examples of increased selectivity by developing extraction technology toward the product of choice, most notably with caproic acid due to its hydrophobicity [13, 17, 20].

Co-Products Support Biorefinery Economics

The VFA as presented in this study are intended as an added value co-product, similar to distiller grains in bioethanol production, though their production can incur considerable base and acid costs. Membrane electrolysis is an immature technology and a complete economic analysis is premature, but the hydroxide output can be compared against the potential cost associated with caustic soda dosing in biological VFA production, particularly from organic waste streams. The control fermentation in this experiment required 0.83 kg caustic per 1 kg COD_{VFA} produced. Consider caproic acid as an attractive target product at around 1000 USD t^{-1} unrefined, and up to 2000–3000 USD t^{-1} refined (price assumptions here are based on discussions with industry partners and may vary). Assuming a conservative price of 300 USD t^{-1} of caustic, if the 0.83 kg caustic per 1 kg COD_{VFA} ratio holds and all the COD_{VFA} generated in this study could be directed to caproic acid (2.21 tCOD per 1 t caproic acid), then 1.83 t of caustic soda (550 USD) would be required for 1 t caproic acid. For unrefined caproic, this is more than half the market price, and 18.3–27.5 % for the refined caproic acid price range. pH control may indeed be further optimized, but this control fermentation on real thin stillage demonstrates the clear and present issue of a high caustic dosing requirement. Moreover, the caustic dosing is required to maintain the fermentation in the biocompatible range for production, and to acquire an acid product, considerable acidification may be needed. The production of OH^- and H^+ by membrane electrolysis avoids caustic and acidic dosing, and also avoids salts entering the fermentation and extractant. The membrane electrolysis route has a cost associated with power input, which is approximately 2 kWh per kg COD_{VFA} based on these experiments. Assuming 0.05 USD kWh^{-1} electricity cost, this is around 100 USD t_{COD}^{-1}, though we stress this is specific to these experimental conditions. This includes the caustic correction and, unlike the control case, extraction and acidification. If the issue of solids buildup between the electrode and the membrane can be avoided and the theoretical power saving of 30 % holds, the power cost comes to 70 USD t_{COD}^{-1}. In principle this power input can come from renewables to improve the sustainability argument. Note that these calculations disregard downstream processing, which can be capital and energy intensive when separating compounds with similar properties, such as mixed VFA. For industrial, bulk chemicals, high product selectivity is critical to the economics of recovery and purification.

The reactive fraction of thin stillage is attractive as a target substrate as the cellulosic fragments and glycerol are readily convertible and have little value, though it is important not to detract from the contribution of these solids to existing production of DDGS. Avoiding degradation of solids is attractive in market in demand of livestock feed, as thin stillage is generally dewatered with the solids and syrup contributing to the production of DDGS, a low value but critical co-product whose value is linked to protein and fiber content. At this stage, it appears that membrane electrolysis allows most solids and proteins to pass without degradation. The apparent electro-coagulation effect may improve the prospects for solids recovery, though this is speculative, as the mechanism is scarcely documented.

CONCLUSIONS

In this work, we have demonstrated the impact of membrane electrolysis, a chemical-free extraction technology, on the fermentation of an under-utilized, real biorefinery stream to generate reduced VFA. This study has shown that membrane electrolysis can result in an increased abundance of *Megasphaera* sp. and *Lactobacillus* spp. in thin stillage, resulting in an increase in VFA chain elongation through lactic acid, driven by in situ hydrogen generation. Membrane electrolysis can provide a driving force to select for hydrogen metabolisers such as *Megasphaera elsdenii* and potentially *Clostridiumkluyveri* or others. In addition to hydrogen, membrane electrolysis generates hydroxide ions which can replace the extensive caustic dosing typical of VFA production.

A central tenet of this work is to move from chemical and heat intensive petrochemical processes, as chemicals and heat both imply embedded petrochemical energy. Membrane electrolysis is electricity driven, and can only be a sustainable technology if this electricity can be sourced sustainably. To move forward, an electro-fermentation should be performed at industrially relevant production and recovery rates. The overall sustainability with respect to the input power and the materials of the system must be critically assessed to confirm that membrane electrolysis can give an added value to the fermentation and chain elongation of sustainable VFA. Beyond sustainability, practicality and economics of scale are crucial to a new industrial biotechnology. Minimizing the electrode and membrane material relative to fermenter volume, and maximizing efficient use of power input will be crucial to transition to an industrial scale, with some lessons to be learned from full scale electrochemistry such as electrodialysis and the Chlor-Alkali process.

METHODS

Fermenter and Electrochemical Cell

In all experiments, semi-continuous fermentation was coupled with a continuous electrochemical membrane electrolysis. Multi-ported vessels of 1 L were used as fermenters (Glasgerätebau Ochs Laborfachhandel e.K., Germany) connected to electrochemical cells as described previously with a volume of 0.2 L per electrode chamber, internal dimensions 200 mm × 50 mm × 20 mm (depth) [23]. The total working liquid volume was 1.2 L, accounting for both the fermenter, the electrochemical cell and tubing. The anolyte was circulated from a 1 L schott bottle to the electrochemical cell, with an identical working volume of 1.2 L. The chambers were separated by an AEM (fumasep FAB, FumaTech GmbH, Germany). The cathode was an AISI Type 316L stainless steel wire mesh of 200 mm × 50 mm exposed working area with 564 μm mesh size, 140 μm wire thickness (Solana nv, Belgium), and the anode was an iridium mixed metal oxide coated titanium electrode (IrO_2/TaO_2: 0.65/0.35), 200 mm × 50 mm, with a centrally attached, perpendicular current collector (Magneto Special Anodes BV, The Netherlands). The catholyte and anolyte consisted in the fermentation broth and tap water (extractant), respectively. The electrolysis reactions consume water (catholyte, water reduction: H_2O + e^- → ½ H_2 + OH^-; anolyte, water oxidation: H_2O → $2e^-$ + 2 H^+ + O_2); however, assuming complete efficiency at the maximum current used here of 100 mA this would account for a 1.6 g of water per day from the catholyte and 0.8 g per day from the anolyte. Both were recirculated in their respective compartment at 6 L h^{-1} by peristaltic pump. In the control case, the pH was controlled between pH 5.4 (2 M NaOH dosing) and 5.7 (2 M H_2SO_4 dosing), whereas in the experimental case (with applied current), the pH was controlled between 5.4 and 5.7 by electrochemical water reduction and dosing with the acidified extractant in case of pH overshoot. The current was applied by a potentiostat (VSP, Biologic, France) in chronopotentiometry mode. The current was manually adjusted according to pH of the fermentation broth. Following feeding, an applied current of 100 mA (i.e. 10 A m^{-2}) was applied for the first 20–24 h, or until pH 5.7 was reached. The pH controller would automatically dose acidic extractant from the previous period to return the pH below the set point. A maintenance period followed at which 20 mA was applied until the next feeding event as to minimize dosing of acidified extractant into the broth. All experiments were performed in a 35 °C temperature controlled room.

In the first experiment, 400 mL of the reactor volume was replaced with thin stillage at 2-day intervals for an equal HRT and solids retention time (SRT) of 6 days. The 6-day HRT/SRT was chosen to allow good conversion of substrate and sufficient time for adaption of the community to the conditions. The fermentation broth and the extractant were replenished at the same rate and the reactors were sampled every 2 days before feeding. After a steady state was reached the experimental cell was operated for four HRTs. The applied current was removed after 24 days to confirm the negation of the effect.

In the second experiment, the current was applied and the residence time of the fermentation broth was decreased to 3 days by replacing 400 mL of reactor volume daily, and sampling daily. The extractant volume was decreased to 600 mL from 1200 mL and operated in batch mode during this experiment, to mimic more realistic operation of extractant concentration and accumulation.

Supplementary batch tests explored a range of applied currents in which the reactor and extraction cell were filled with thin stillage and run for 6 days for each run. The reactors were emptied and refilled between each test. The applied current was set at 0.1 mA for the no current case, 50 mA (5 A m⁻¹), 100 mA (10 A m⁻¹) and 200 mA (200 A m⁻¹) and the pH was controlled between pH 5.4 and 5.7 with 2 M NaOH and 2 M H_2SO_4. The AEM was replaced between each test and reactor components were scrubbed with a brush and tap water only to limit cross-contamination.

Microbial Characterization and Community Analysis

DNA extraction was performed using a PowerSoil DNA extraction kit (MO BIO Laboratories, USA) according to Roume et al. [47]. Biomass has first been concentrated by centrifugation in sterile 2 mL bead beating Micrewtubes (Simport, Canada) for 1 min at 20,238g. The protocol consists of cell lysis by bead beating in 2 mL Microtubes bead beating tubes and lysis buffer, in a Fast Prep-96 instrument (2 times 40 s at 1600 rpm) followed by removal by precipitation of diverse polymerase chain reaction (PCR) inhibitors according to the manufacturer's instructions. Total DNA was captured on a silica membrane incorporated into a chromatographic spin column, washed and then eluted in the dedicated buffer. Concentration of double-stranded DNA was quantified using the QuantiFluor dsDNA system and measured with a GloMax 96 Microplate Luminometer (Promega GmbH, Germany). For quality control, the isolated DNA was amplified by PCR using Illumina sequencing primers and separated by electrophoresis.

The V3–V4 region of the bacterial 16S rRNA gene was sequenced with Illumina sequencing Miseq v3 Reagent kit (http://www.illumina.com/products/miseq-reagent-kit-v3.ilmn, by LGC Genomics GmbH, Berlin, Germany) using 2 × 300 bp paired-end reads and primers 341F-785R described by Stewardson et al. [48]. The PCRs included about 5 ng of DNA extract, 15 pmol of each forward primer 341F 5′-NNNNNNNNTCCTACGGGNGGCWGCAG and reverse primer 785R 5′-NNNNNNNNTGACTACHVGGGTATCTAAKCC in 20 µL volume of MyTaq buffer containing 1.5 units MyTaq DNA polymerase (Bioline) and 2 µL of BioStabII PCR Enhancer (Sigma). For each sample, the forward and reverse primers had the same 8-nt barcode sequence. PCRs were carried out for 30 cycles using the following parameters: 2 min 96 °C predenaturation; 96 °C for 15 s, 50 °C for 30 s, 72 °C for 60 s. DNA concentration of amplicons of interest was determined by gel electrophoresis. About 20 ng amplicon DNA of each sample was pooled for up to 48 samples carrying different barcodes. PCRs showing low yields were further amplified for 5 cycles. The amplicon pools were purified with one volume AMPure XP beads (Agencourt) to remove primer dimer and other small mispriming products, followed by an additional purification on MinElute columns (Qiagen). About 100 ng of each purified amplicon pool DNA was used to construct Illumina libraries using the

Ovation Rapid DR Multiplex System 1-96 (NuGEN). Illumina libraries were pooled and size selected by preparative gel electrophoresis. Sequencing was done on an Illumina MiSeq using v3 Chemistry (Illumina).

Bioinformatics was executed with the mothur community analysis pipeline [49]. The analysis started from the primers clipped 16Sr RNA sequences; containing sequences where primers were detected and removed (allowing 2 mismatches) and turn into forward and reverse primer orientation which was later combined using the make.contigs command. The use of this open-source software package involved three sequential steps. The first step consists of the preparation and denoising of sequences, and extraction of the V3–V4 region. Low-quality sequences were removed and the frequency of sequencing and PCR errors reduced. The sequences were first trimmed using screen.seqs command (allowing no base name ambiguity and a maximum length of 427 bases). Sequences showing a weak alignment (allowing a maximum of four bases in homopolymer) with a V3–V4 customized SILVA database (v119) were removed as well as overhangs at both end of each sequence. The sequences were pre-clustered (allowing a maximum of 4 mismatches per sequence) and chimeric sequences were removed using UCHIME software [50] and sequences have been classifying using RDP v10 database (allowing at least a bootstrap value of 65 %) and sequences not identified as bacteria have been removed. The second step of the pipeline consisted of clustering sequences into operational taxonomic units (OTUs). OTU binning was completed using a hierarchical clustering algorithm implemented within mothur and considering a cutoff of 0.03. The third step involved assignment of the taxonomic information to sequences and OTUs. The analysis was carried out on randomly subsampled OTUs such that each file contained the same number of sequences (5401). The ade4, stat and psych packages of the R software (R version 2.13.2, http://www.r-project.org/) were used, respectively, for a principal component analysis (PCA; dudi.pca function), representation of the heatmap (heatmap function) and Pearson coefficient calculation (corr.test function).

For microbial activity analysis from the thin stillage, raw samples were evaluated under an optical microscope (Axioskop, Zeiss) by phase contrast and fluorescence. Fluorescence in situ hybridization (FISH) was performed as described by Anton et al. [52]. Samples were fixed with 4 % paraformaldehyde and stained with probes for general bacteria domain EUB 338 I, II and III with FLUO fluorophores.

Stream Characterization and Analysis

In this study we refer to the linear, saturated monocarboxylate C2–C8 fatty acid conjugates collectively as VFA or individually by the common name of the dissociated anion, e.g. acetate, caproate, when discussing the compound in the more neutral fermentation broth. All carboxylate fermentation is from thin stillage of Alco Bio Fuel NV (Ghent, Belgium) (stored at 4 °C). No inoculum was added, the fermentation proceeded according to the bacterial community already present in the broth. New batches of stillage were periodically retrieved, with one batch requiring dilution to the appropriate COD range to maintain consistent organic loading. Stream characterization confirmed that after dilution the stream remained sufficiently consistent for these experiments, with some variation in solids content. Reported feed concentrations are averages of all feed streams including the diluted stream.

All samples were tested for TSS and VSS according to Standard Methods 2540D and E [51]. The C2–C8 fatty acids (including isoforms C4–C6) were measured according to Andersen et al. [22]. Management of pH was tracked by mass of acid or base dosed, and gas production was quantified with an external gas trap and assessed with a Compact Gas Chromatograph (Global Analyser Solutions, Breda, The Netherlands), equipped with a Molsieve 5A pre-column and Porabond column (CH_4, O_2, H_2 and N_2) and a Rt-Q-bond pre-column and column (CO_2, N_2O and H_2S). Concentrations of gases were determined by means of a thermal conductivity detector. Eight samples were selected at random from the thin stillage feed samples for further characterization: four samples from the control effluent, four from the experimental effluent and four from the extractant during the steady state period with the average reported of the following analyses: total and soluble COD by Nanocolor® kits (Macherey–Nagel GmbH, Germany); lactate, glycerol, 1,3-propanediol, ethanol, propanol and butanol by ion chromatography (Dionex DX 500); hemicellulosic and cellulosic fragments in the soluble phase (reported in this text as "soluble cellulosic fragments", as to differentiate from (hemi)cellulosic material in the solids) by the NREL procedure according to Sluiter et al. [53], measured with high-performance liquid chromatography [Agilent Varian ProStar 220 SDM, USA; 5 mM H_2SO_4 mobile phase, 0.6 mL min^{-1} and 60 °C column temperature with a refractive index detector and Rezex H + column (Aminex)]; soluble and insoluble proteins by Kjeldahl nitrogen measurements according to Standard methods (4500-Norg B; APHA, 1992) [51], and calculated to protein COD content based on an assumed carbon to nitrogen ratio of 5:1. The protein measurement is intended only as an indication of changing protein concentrations and not intended as a strict protein quantification. The "Total Other Solids" is based on the difference between the measured total COD and the soluble COD, and its difference from the calculated insoluble protein COD. Non-organic anions chloride, nitrite, nitrate, sulfate and phosphate were determined on a 761 Compact Ion Chromatograph (Metrohm, Switzerland) equipped with a conductivity detector. The total phenolic content was assessed with the Folin Ciocalteu Assay method [54].

Electrochemical Analysis

The resistance of the whole cell was assessed by Current Interrupt [55]. 10 mA of current was applied at a period of 100 ms over 10 cycles successively, and the resulting voltage recorded at 0.2 ms intervals. The cell voltage change during the first interval of 0.2 ms is the ohmic drop of Current × Resistance, assuming that faradaic and diffusional processes present much slower relaxation times and therefore do not impact the voltage amplitude.

ABBREVIATIONS

VFA: volatile fatty acid

COD: chemical oxygen demand

sCOD: soluble chemical oxygen demand

DDGS: dried distillers grains with solubles

gC: grams of carbon, i.e. 1 mol of Caproic (C6) Acid = 116.16 g = 6 × 12.01 gC = 72.06 gC

HRT: hydraulic retention time

TSS: total suspended solids

PCR: polymerase chain reaction

AUTHORS' CONTRIBUTIONS

SA designed and executed experiments, interpreted data and drafted the manuscript. PC and TB executed experiments, performed analysis and contributed to data interpretation. WCK was involved in stream characterization and contributed to data interpretation. HR and EHS designed, executed and interpreted the microbial community analysis and performed statistical analysis. MC contributed to experimental design and interpretation of data. KR conceived of the study, participated in its design and coordination and contributed to the drafting of the manuscript. All authors read and approved the final manuscript.

ACKNOWLEDGEMENTS

SJA, MC, and KR are supported by Ghent University Multidisciplinary Research Partnership (MRP)—Biotechnology for a sustainable economy (01 MRA 510 W). WCK is supported by Het Bijzonder Onderzoeksfonds (BOF,

DEF13/AOF/010). EHS is a postdoctoral fellow supported by the Research Foundation of Flanders (Fonds Wetenschappelijk Onderzoek-Vlaanderen, FWO). KR and HR are supported by European Research Council Starter Grant ELECTROTALK. Prof Lars Angenent is heartily thanked for his insights and stimulating discussion. Tim Lacoere is warmly thanked for his graphical contribution. Antonin Prévoteau is thanked for the electrochemical analysis and insight, and a critical read of the manuscript, with thanks for the latter also extended to Alberto Scoma and Ruben Props. We thank Alco Bio Fuel nv for generously providing thin stillage.

ADDITIONAL FILE

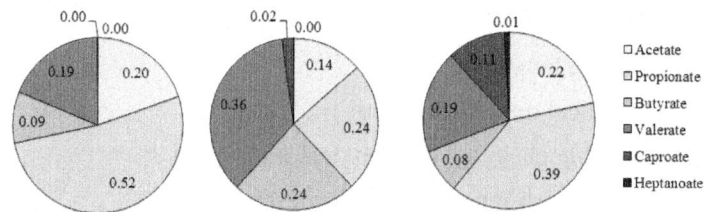

	Control	Experiment	Experiment II: High Loading Rate, High Current
Production Rate	1.9 ± 0.8 gCOD$_{VFA}$ L^{-1} d^{-1}	1.9 ± 0.5 gCOD$_{VFA}$ L^{-1} d^{-1}	10.4 ± 1.1 gCOD$_{VFA}$ L^{-1} d^{-1}
Maximum Anolyte Concentration		4.0 gCOD$_{VFA}$ L^{-1} 2.3 gC$_{VFA}$ L^{-1}	18.9 gCOD$_{VFA}$ L^{-1} 11.7 gC$_{VFA}$ L^{-1}

Figure SI 1. Comparison of the average outcome of VFA and production for the control case, experimental case (applied current) and experimental case with increased loading rate and current.

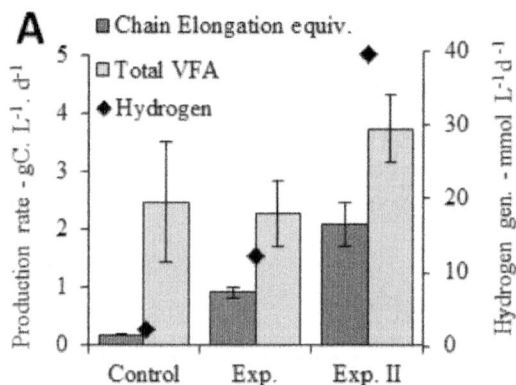

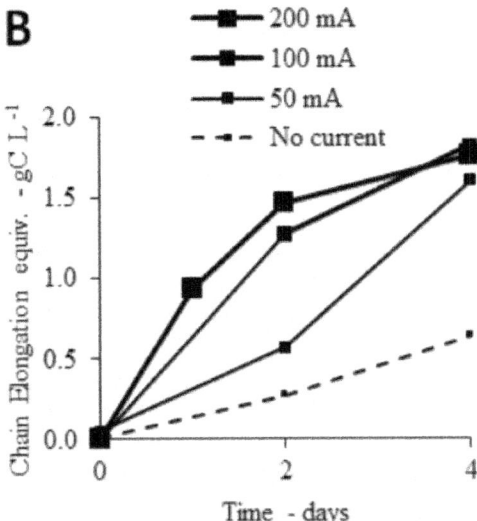

Figure SI 2. Chain elongation equivalents vs. electrolytic hydrogen. **A.** Production rate of total VFA compared against production of ethanol equivalents required for chain elongation to C4 to C7 carboxylates. Hydrogen gas reported is the average of that measured from the headspace and gas column. **B.** Chain elongation equivalents as measured in batch tests over time, for 0, 50, 100 and 200 mA.

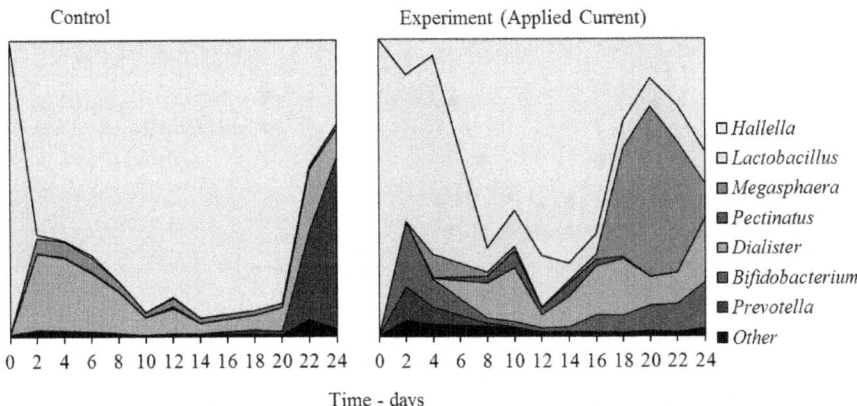

Figure SI 3. Cumulative line plot of species abundance over time. Data points at day 16 for the control and day 6 for the experiment are assumed values due to data collection error. Feed abundance is consistently in excess of 99% *Lactobacillus*.

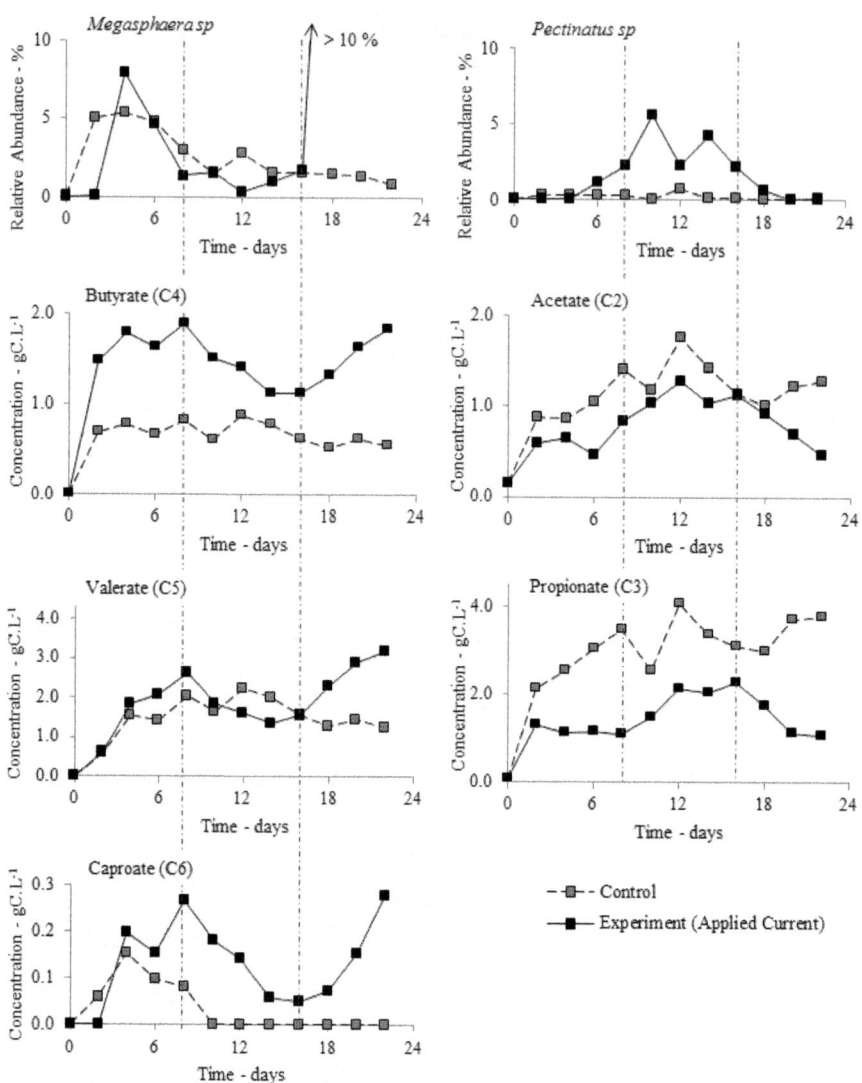

Figure SI 4. Higher resolution perspective of *Megasphaera sp* and *Pectinatus sp* relative abundance and the short chain carboxylates concentration in the fermentations.

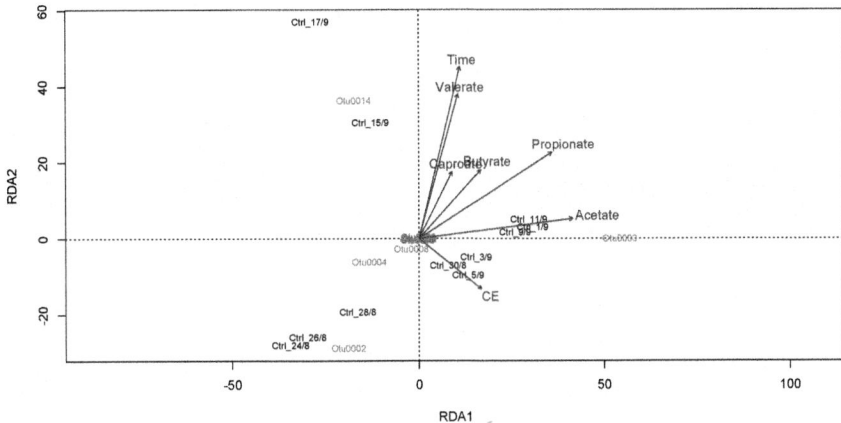

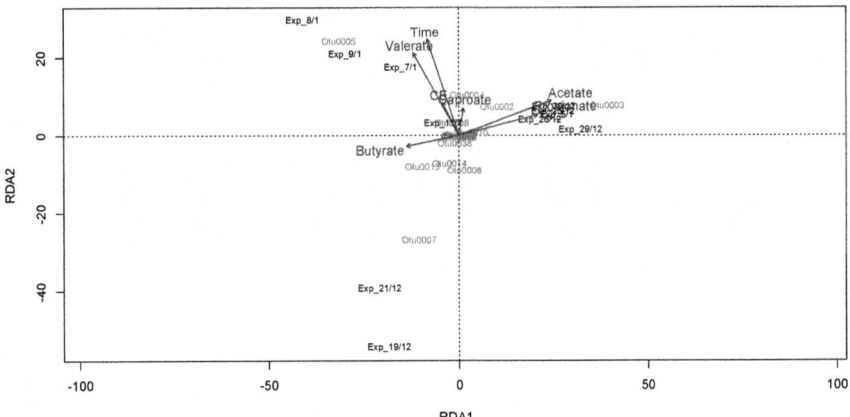

Figure SI 5. Redundancy discriminant analysis of the control and experiment, C1 to C6 VFA.

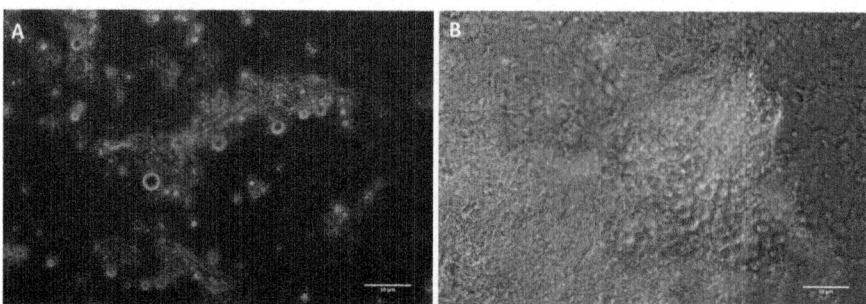

Figure SI 6. Micrographs from raw thin stillage. **A.** Contrast light image at 1000x. **B.** FISH image overlaid with transmitted light image. Green color stands for hybridization with EUB 338 I-II-II RNA probe. The scale bar signifies 10μm in both cases.

Table SI 1. Measured and calculated chemical oxygen demand. All values are gCOD L^{-1}

	Feed		Control		Experiment		Extract-ed	
	Average	StDev	Average	StDev	Average	StDev	Average	StDev
COD	**55.50**	**10.60**	**52.30**	**4.50**	**43.30**	**10.10**	**9.32**	**0.30**
sCOD	36.00	10.10	32.80	4.50	22.33	3.10	9.32	0.30
Total VFA	**0.29**	**0.17**	**11.11**	**0.57**	**7.94**	**1.87**	**2.96**	**0.55**
Acetate	0.16	0.14	1.50	0.27	0.64	0.54	0.27	0.15
Butyrate	0.01	0.03	1.22	0.20	2.00	0.50	0.75	0.21
Caproate	0.00	0.00	0.02	0.07	0.27	0.17	0.06	0.07
Propionate	0.08	0.26	5.63	0.64	1.67	0.72	0.65	0.10
Valerate	0.04	0.16	2.74	0.90	3.36	1.35	1.23	0.40
Reactive Fraction	**25.38**		**2.19**		**1.94**		**0.00**	
Ethanol	1.22	0.56	1.72	1.04	1.33	0.06	0.00	0.00
Sol. Cellu. Fragm.	12.30	0.70	0.47	0.03	0.61	0.14	0.00	0.00
Glycerol	9.74	2.43	0.00	0.00	0.00	0.00	0.00	0.00
Lactate	2.12	0.97	0.00	0.00	0.00	0.00	0.00	0.00
Solids	**19.50**	**2.53**	**19.50**	**4.48**	**20.97**	**3.14**	**0.00**	**0.00**
Insoluble Protein	8.87	0.90	9.64	0.80	6.62	0.80		

Total Other Solids	10.63		9.86		14.35			
Other Solubles	**10.62**		**19.51**		**13.45**		**6.36**	
Soluble Protein	6.13	1.30	3.18	0.38	4.09	0.61		
Other Solubles	4.19		16.32		9.36			

Table SI 2. Other physico-chemical data. (TSS, Kjeldahl Nitrogen, pH, Temperature, Conductivity)

	Feed		Control		Experiment	
	Average	StDev	Average	StDev	Average	StDev
Total Suspended Solids, g L^{-1}	20.33	2.84	18.62	1.86	11.17	1.70
Total Kjeldahl Nitrogen, gN L^{-1}	1.50	0.28	1.28	0.09	1.07	0.09
Soluble Kjeldahl Nitrogen, gN L^{-1}	0.64	0.04	0.32	0.04	0.41	0.07
pH	4.63	0.3	5.5		5.5	
Temperature, °C	4		35		35	
Conductivity, mS/cm	6.10	0.46				

Table S3. The top 11 most abundant bacterial species

Bacterial species (*sp.*, %)	Start of experiment			
	Control Feed	Control Eff	Exp Feed	Exp Eff
Lactobacillus	99.09	0.17	99.65	8.29
Hallella	0.07	93.70	0.00	70.25
Dialister	0.00	3.50	0.00	11.81
Megasphaera	0.07	1.59	0.00	1.35
Pectinatus	0.02	0.09	0.00	2.22
Clostridium_sensu_stricto	0.06	0.17	0.00	0.17
Bifidobacterium	0.04	0.00	0.00	1.74
Prevotella	0.04	0.15	0.00	1.11
Propionibacterium	0.00	0.07	0.00	0.00
Desulfovibrio	0.00	0.00	0.00	2.28
Oscillibacter	0.00	0.00	0.00	0.07
Other	0.61	0.56	0.35	0.70

Bacterial species (*sp.*, %)	End of experiment			
	Control Feed	Control Eff	Exp Feed	Exp Eff
Lactobacillus	99.74	0.07	97.32	10.72
Hallella	0.06	28.51	2.15	38.10
Dialister	0.04	9.72	0.20	21.72
Megasphaera	0.00	1.43	0.02	11.33
Pectinatus	0.00	0.07	0.00	0.06
Clostridium_sensu_ stricto	0.00	0.07	0.00	0.37
Bifidobacterium	0.00	0.00	0.00	15.44
Prevotella	0.00	57.79	0.00	0.13
Propionibacterium	0.00	0.02	0.09	0.00
Desulfovibrio	0.00	0.00	0.00	0.52
Oscillibacter	0.00	0.00	0.00	0.22
Other	0.17	2.31	0.22	1.39

REFERENCES

1. Gavrilescu M, Chisti Y. Biotechnology—a sustainable alternative for chemical industry. Biotech Adv. 2005;23:471–99. doi:10.1016/j.biotechadv.2005.03.004.

2. Rabaey K, Ronzendal RA. Microbial electrosynthesis—revisiting the electrical route for microbial production. Nat Rev Microbiol. 2010;8:706–16. doi:10.1038/nrmicro2422.

3. Deloitte. Opportunities for the fermentation-based chemical industry: an analysis of the market potential and competitiveness of North-West Europe. September 2014; http://www2.deloitte.com/global/en/pages/manufacturing/articles/opportunities-for-fermentation-based-chemical-industry.html. Accessed 4th Aug 2015.

4. Wagemann K. Herstellung von Grundchemikalien auf Basis nachwachsender Rohstoffe als Alternative zur Petrochemie?/Production of basic chemicals on the basis of renewable resources as alternative to petrochemistry? Chelie Ingenieur Technik. 2014;86(12):2115–34. doi:10.1002/cite.201400108.

5. Ramaswamy S, Huang HJ, Ramarao BV. Separation and purification technologies in biorefineries. Hoboken: Wiley. 2013. doi:10.1002/9781118493441.

6. Percival Zhang YH. What is vital (and not vital) to advance economically-competitive biofuels production. Process Biochem. 2011;46(11):2091–110. doi:10.1016/j.procbio.2011.08.005.

7. Fava F, Totaro G, Diels L, Reis M, Duarte J, Carioca OB, Poggio-Varaldo HM, Ferreira BS. Biowaste biorefinery in Europe: opportunities and research and development needs. N Biotechnol. 2015;32(1):100–8. doi:10.1016/j.nbt.2013.11.003.

8. Agler MT, Wrenn BA, Zinder SH, Angenent LT. Waste to bioproduct conversion with undefined mixed cultures: the carboxylate platform. Trends Biotechnol. 2011;29(2):70–8. doi:10.1016/j.tibtech.2010.11.006.

9. Holtzapple MT, Granda CB. Carboxylate platform: the MixAlco process Part 1: comparison of three biomass conversion platforms. Appl Biochem Biotechnol. 2009;156:525–36. doi:10.1007/s12010-008-8466-y.

10. Yang ST. Bioprocessing for value-added products from renewable resources: new technologies and applications. USA: Elsevier BV; 2011.

11. López-Garzón CS, Straathof AJJ. Recovery of carboxylic acids produced by fermentation. Biotech Adv. 2014;32(5):873–904. doi:10.1016/j.biotechadv.2014.04.002.

12. Barker HA, Taha SM. Clostridium kluyverii, an Organism concerned in the formation of caproic acid from ethyl alcohol. J Bacteriol. 1942;43(3):347–63.

13. Agler MT, Spirito CM, Usack JG, Werner JJ, Angenent LT. Chain elongation with reactor microbiomes: upgrading dilute ethanol to medium-chain carboxylates. Energ Environ Sci. 2012;5(8):8189–92. doi:10.1039/C2EE22101B.

14. Steinbusch KJJ, Hamelers HVM, Plugge CM, Buisman CJN. Biological formation of caproate and caprylate from acetate: fuel and chemical production from low grade biomass. Energ Environ Sci. 2011;4(1):216–24. doi:10.1039/C0EE00282H.

15. Grootscholten TIM, Steinbusch KJJ, Hamelers HVM, Buisman CJN. High rate heptanoate production from propionate and ethanol using chain elongation. Bioresour Technol. 2013;136:715–8. doi:10.1016/j.biortech.2013.02.085.

16. Spirito CM, Richter H, Rabaey K, Stams AJM, Angenent LT. Chain elongation in anaerobic reactor microbiomes to recover resources

from waste. Curr Opin Biotechnol. 2014;27:115–22. doi:10.1016/j. copbio.2014.01.003.

17. Xu J, Guzman JJ, Andersen SJ, Rabaey K, Angenent LT. In-line and selective phase separation of medium chain carboxylic acids using membrane electrolysis. Chem Commun. 2015;51:6847–50. doi:10.1039/ C5CC01897H.

18. Rogosa M, Transfer of Peptostreptococcus elsdenii Gutierrez, et al. to a new genus, Megasphaera [M. elsdenii (Gutierrez et al.) comb. nov.]. Int J Syst Bacteriol. 1971;21:187–9.

19. Hino T, Shimada K, Maruyama T. Substrate preference in a strain of Megasphaera elsdenii, a ruminal bacterium, and its implications in propionate production and growth competition. Appl Environ Microbiol. 1994;60(6):1827–31.

20. Choi K, Jeon BS, Oh MK, Sang BI. In situ biphasic extractive fermentation for hexanoic acid production from sucrose by Megasphaera elsdenii NCIMB 702410. Appl Biochem Biotechnol. 2013;171(5):1094–107. doi:10.1007/s12010-013-0310-3.

21. Weimer PJ, Moen GN. Quantitative analysis of growth and volatile fatty acid production by the anaerobic ruminal bacterium Megasphaera elsdenii T81. Appl Microbiol Biotechnol. 2013;97(9):4075–81. doi:10.1007/ s00253-012-4645-4.

22. Andersen SJ, Hennebel T, Gildemyn S, Coma M, Desloover J, Berton J, Tsukamoto J, Stevens C, Rabaey K. Electrolytic membrane extraction enables production of fine chemicals from biorefinery sidestreams. Environ Sci Technol. 2014;48(12):7135–42. doi:10.1021/es500483w.

23. Gildemyn S, Verbeeck K, Slabbinck R, Andersen SJ, Prevotéau A, Rabaey K. Integrated production, extraction and concentration of acetic acid from CO2 through microbial electrosynthesis. Environ. Sci. Technol. Lett. 2015;2(11):325–8. doi:10.1021/acs.estlett.5b00212.

24. Steinbusch KJJ, Hamelers HVM, Buisman CJN. Alcohol production through volatile fatty acids reduction with hydrogen as electron donor by mixed cultures. Water Res. 2008;42(15):4059–66. doi:10.1016/j. watres.2008.05.032.

25. Angenent LT, Rosenbaum MA. Microbial electrocatalysis to guide biofuel and biochemical bioprocessing. Biofuels. 2013;4(2):131–4. doi:10.4155/bfs.12.93.

26. Kim S, Dale BE. Global potential bioethanol production from wasted crops and crop residues. Biom Bioenerg. 2004;26(4):361–75. doi:10.1016/j. biombioe.2003.08.002.

27. Kim Y, Mosier NS, Hendrickson R, Ezeji T, Blaschek H, Dien B, Cotta M, Dale B, Ladisch MR. Composition of corn dry-grind ethanol by-products: DDGS, wet cake and thin stillage. Biores Technol. 2008;99:5165–76. doi:10.1016/j.biortech.2007.09.028.

28. Liu K, Rosentrater KA. Distillers grains: production, properties and utilization. Boca Raton: CRC Press; 2012.

29. Lupitskyy R, Staff C, Satyavolu J. Towards integrated biorefinery from dried distillers grains: Evaluation of feed application for Co-products. Biom Bioenerg. 2015;72:251–5. doi:10.1016/j.biombioe.2014.10.029.

30. Mollah MYA, Morkovsky P, Gomes JAG, Kezmez M, Parga J, Cocke DL. Fundamentals, present and future perspectives of electrocoagulation. J Hazard Mater B. 2004;114:199–210. doi:10.1016/j.jhazmat.2004.08.009.

31. Zhou M, Chen J, Freguia S, Rabaey K, Keller J. Carbon and electron fluxes during the electricity driven 1,3-propanediol biosynthesis from glycerol. Environ Sci Technol. 2013;47:11199–205. doi:10.1021/es402132r.

32. Moore LVH. Moore WEC. Oribaculum catoniae gen. nov., sp. nov.; Catonella morbi gen. nov., sp. nov.; Hallella seregens gen. nov., sp. nov.; Johnsonella ignava gen. nov., sp. nov.; and Dialister pneumosintes gen. nov., comb. nov., nom. rev., Anaerobic Gram-Negative Bacilli from the Human Gingival Crevice. Int J Syst Evol Microbiol. 1994;44:187–92. doi:10.1099/00207713-44-2-187.

33. Jumas-Bilak E, Hélene JP, Carlier JP, Teyssier C, Bernard K, Gay B, Calpos J, Morio F, Marchandin H. Dialister micraerophilus sp. nov. and Dialister propionicifaciens sp. nov., isolated from human clinical samples. Int J Syst Evol Microbiol. 2005;55:2471–8. doi:10.1099/ijs.0.63715-0.

34. Sgorbati B, Biavati B, Palenzona D. The genus Bifidobacterium. The genera of lactic acid bacteria; the lactic acid bacteria, vol 2. USA: Springer. 1995. pp. 279–306. doi:10.1007/978-1-4615-5817-0_8.

35. Shah HN, Collins DM. Prevotella, a new genus to include Bacteroides melaninogenicus and related species formerly classified in the genus bacteroides. Int J Syst Bacteriol. 1990;40(2):205–8.

36. Lee SY, Mabee MS, Jangaard NO. Pectinatus, a new genus of the family Bacteroidaceae. Int J Syst Bacteriol. 1978;28:582–94.

37. Caldwell JM, Juvonen R, Brown J, Breidt F. Pectinatus sottacetonis sp. nov., isolated from a commercial pickle spoilage tank. Int J Syst Evol Microbiol. 2013;63:3609–16. doi:10.1099/ijs.0.047886-0.

38. Denis PG, Harnisch F, Yeoh YK, Tyson GW, Rabaey K. Dynamics of cathode-associated microbial communities and metabolite profiles in

a glycerol-fed bioelectrochemical system. Appl Environ Microbiol. 2013;79(13):4008–14. doi:10.1128/AEM.00569-13.

39. Beijerink MW. Sur les ferments lactiques de l'industrie. Arch Neerl Sci. 1901;II(7):212–43.

40. Gänle MG, Follador R. Metabolism of Oligosaccharides and starch in Lactobacilli: a review. Front Microbiol. 2012;3:340. doi:10.3389/fmicb.2012.00340.

41. Shafaat HS, Rüdiger O, Ogata H, Lubitz W. [NiFe] hydrogenases: a common active site for hydrogen metabolism under diverse conditions. BBA-Bioenergetics. 2013;1827(8–9):986–1002. doi:10.1016/j.bbabio.2013.01.015.

42. Velt A, Akhtar MK, Mizutani T, Jones PR. Constructing and testing the thermodynamic limits of synthetic NAD(P)H:H2 pathways. Microb Biotechnol. 2008;1(5):382–94. doi:10.1111/j.1751-7915.2008.00033.

43. Kracke F, Krömer JO. Identifying target processes for microbial electrosynthesis by elementary mode analysis. BMC Bioinform. 2014;15(1):410. doi:10.1186/s12859-014-0410-2.

44. Sikora A, Zielenkiewicz U, Blaszczyk M, Jurkowki M. Lactic acid bacteria in hydrogen-producing consortia: on purpose or by coincidence?, Lactic Acid Bacteria—R & D for Food, Health and Livestock Purposes, Dr. J. Marcelino Kongo (Ed.), InTech, 2013. doi:10.5772/50364.

45. Zhu X, Tao Y, Liang C, Li X, Zhang W, Zhou Y, Tank Y, Bo T. The synthesis of n-caproate from lactate: a new efficient process for medium-chain carboxylates production. Sci Rep. 2015;5:14360. doi:10.1038/srep14360.

46. Vanwonterghem I, Jensen PD, Rabaey K, Tyson G. Temperature and solids retention time control microbial population dynamics and volatile fatty acid production in replicated anaerobic digesters. Sci Rep. 2015;5:8496. doi:10.1038/srep08496.

47. Roume H, Muller EEL, Cordes T, Renaut J, Hiller K, Wilmes P. A biomolecular isolation framework for eco-systems biology. ISME J. 2013;7:110–21.

48. Stewardson AJ, Gaïa N, François P, Malhotra-Kumar S, Delémont C, Martinez de Tejada B, Schrenzel J, Harbarth S, Lazarevic V, SATURN WP1 and WP3 Study Groups. Collateral damage from oral ciprofloxacin versus nitrofurantoin in outpatients with urinary tract infections: a culture-free analysis of gut microbiota. Clin Microbiol Infect. 2015;21(4):344. e1–11. doi:10.1016/j.cmi.2014.11.016.

49. Schloss PD, Westcott SL, Ryabin T, Hall JR, Hartmann M, Hollister EB, Lesniewski RA, Oakley BB, Parks DH, Robinson CJ, Sahl JW, Stres B, Thallinger GG, Van Horn DJ, Weber CF. Introducing mothur: open-source, platform-independent, community-supported software for describing and comparing microbial communities. Appl Environ Microbiol. 2009;75(23):7537–41. doi:10.1128/AEM.01541-09.

50. Edgar RC, Haas BJ, Clemente JC, Quince C, Knight R. UCHIME improves sensitivity and speed of chimera detection. Bioinformatics. 2011;27(16):2194–200. doi:10.1093/bioinformatics/btr381.

51. APHA. Standard methods for the examination of water and wastewater. Washington DC: American Public Health Association; 2005.

52. Anton JE, Llobet-Brossa E, Rodriguez-Valera F, Amann R. Fluorescence in situ hybridization analysis of the prokaryotic community inhabiting crystallizer ponds. Environ Microbiol. 1999;1:517–23.

53. Sluiter A, Hames B, Ruiz R, Scarlata C, Sluiter J, Templeton D. Determination of sugars, byproducts, and degradation products in liquid fraction process samples: laboratory analytical procedure 2006. National Renewable Energy Laboratory; U.S. Department of Energy; NREL/TP-510-42623.

54. Swain, T., Goldstein, J.L. Chapter 11—the quantitative analysis of phenolic compounds. In: Methods in polyphenol chemistry, Proceedings of the Plant Phenolics Group Symposium, Oxford, 1964.

55. Bard AJ, Faulkner LR. Electrochemical methods: fundamentals and applications. 2nd ed. New York: Wiley; 2001.

Chapter 7

GENERATION OF HYDROGEN, LIGNIN AND SODIUM HYDROXIDE FROM PULPING BLACK LIQUOR BY ELECTROLYSIS

Guangzai Nong, Zongwen Zhou and Shuangfei Wang

Institute of Light Industry and Food Engineering, Guangxi University, Nanning 530004, China

ABSTRACT

Black liquor is generated in Kraft pulping of wood or non-wood raw material in pulp mills, and regarded as a renewable resource. The objective of this paper was to develop an effective means to remove the water pollutants by recovery of both lignin and sodium hydroxide from black liquor, based on electrolysis. The treatment of a 1000 mL of black liquor (122 g/L solid contents) consumed 345.6 kJ of electric energy, and led to the generation of 30.7 g of sodium hydroxide, 0.82 g of hydrogen gas and 52.1 g of biomass solids. Therefore, the recovery ratios of elemental sodium and biomass solids are 80.4% and 76%, respectively. Treating black liquor by electrolysis is an environmentally friendly technology that can, in particular, be an alternative process in addressing the environmental issues of pulping waste liquor to the small-scale mills without black liquor recovery.

INTRODUCTION

Black liquor is a wastewater that is generated in pulp mills during Kraft pulping of wood or non-wood raw materials [1]. Black liquor contains approximately 25%–41% lignin [2,3] and 18%–23% sodium [4] and other dissolved organics (hemicelluloses, cellulose, extractives, *etc.*), and inorganics. Due to the presence of large amounts of organics, black liquor can be considered as a renewable resource [5]. In a modern pulp mill, black liquor is processed in the so-called black liquor recovery process [6], where the original thin black liquor is concentrated to 65%–75% (wt %) and is then combusted in a recovery boiler so that the organics are recovered in the form of steam and electricity while the sodium salts are converted to sodium carbonate and sodium sulfide. For small scale operations, for example, those typically found in Asia, the conventional black liquor recovery process is not economic; instead black liquor is treated following standard industrial waste water treatment technologies, such as aerobic and/or anaerobic systems.

Another process to treat black liquor is the gasification technology so that syngas, and other gases are produced. Such a technology has been under development for the past three decades [7]. The process of black liquor gasification feeds black liquor with a 65%–75% solid content and limited amount of oxygen/air into the gasifier, so that a series of reactions lead to the conversion of the black liquor into CO, H_2, CO_2 and sodium carbonate at high temperatures [8]. Biomass syngas containing CO and H_2 can then be used as the base material to synthesize methanol and dimethyl ether (DME) [9,10] or as a fuel for combustion in a gas turbine to generate power [11]. The generated sodium carbonate is converted into sodium hydroxide via alkalization with slaked lime.

Lignin can be used to generate liquid fuels for internal combustion engines [12,13] and to generate other useful materials [14,15]. One way to utilize lignin for the above is to acidify the black liquor to pH = 2–4 to precipitate the lignin; subsequently, the wet lignin is obtained by filtering the mixture and then further processed [16].

Treating black liquor by combustion and gasification treatments, sodium hydroxide and heat can be recovered; however, lignin is converted into energy and gas pollutants are discharged using these treatments. Treating black liquor by the acidification treatment, recovering the lignin consumes sulfuric acid and loses sodium hydroxide. To date, no method has been found to fully recover both lignin and sodium hydroxide from black liquor. Therefore, we reported herein our results on the generation of hydrogen, lignin and sodium hydroxide from black liquor using the electrolysis concept.

METHODS AND MATERIALS

Black Liquor

Black liquor with a concentration of 122 g/L of solids and a chemical oxygen demand (COD_{Cr}) of 119,198.4 mg/L was obtained from the Guangxi Huajing Co., Ltd. (Nanning, China). The black liquor was generated from gumwood/

bamboo by sulfite process containing 18% (wt %) sodium in the dry black liquor solids. By pyrolysis at 700 °C for 5 h, the organic compositions were converted completely, remaining mass of residue accounting for 43.8% of the dry solids. Thereby, the organic components in the dry black liquor are estimated to be 56.2%.

Electrolysis of Black Liquor

The set-up primarily consisted of a cationic electrolytic reactor and a membrane filter, as shown in Figure 1. The cationic electrolytic reactor is the key device in the system [17]; the reactor consisted of a large anode chamber, which is equipped with a cathode plate. A 20 cm^2 cationic exchange membrane (CEM) was used as the separation membrane. The working voltage was 4 V, which produced an average 100 mA of current.

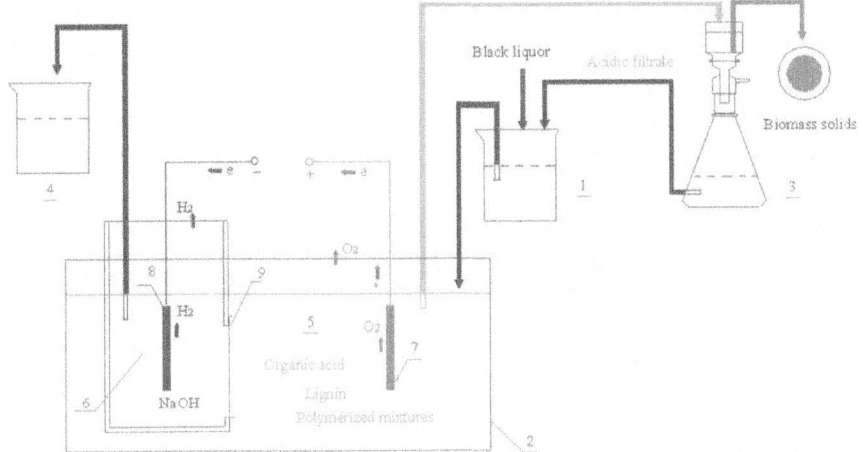

1. Beaker; 2. Electrolytic reactor; 3. filter; 4. NaOH stored; 5. Anode chamber; 6. Cathode chamber; 7. Anode plate; 8. Cathode plate; 9. Cationic membrane.

Figure 1. Experimental set-up of the novel black liquor processing system by electrolysis.

The electrolysis reactor is utilized in the chlor-alkali and other industries. In the process of water electrolysis, water (with some added electrolyte) is fed into the reactor, producing hydrogen gas and oxygen gas in the cathode and anode chambers, respectively [18]. In the sodium chloride electrolysis process, saturated salt water is fed into the reactor, generating hydrogen gas and sodium hydroxide in the cathode chamber and chlorine gas in the anode chamber. Here, the electrolysis reactor was utilized to treat black liquor, hydrogen gas and sodium hydroxide are generated in the cathode chamber; at the same time, oxygen gas, lignin and other organic precipitates were generated in the anode chamber. A filter was used to remove them from the reactor to maintain the reactor working continually. In addition, the conductivity of diluted black liquor is rather lower; and thereby, it needs greater membrane and plate areas to obtain a suitable production rate.

In comparison with the electrolytic reactor for generating hydrogen fuel from black liquor [19], this black liquor processing system has the following innovations:

- the space of the electrolytic reactor was separated into a large anode chamber and a cathode plate by a cationic exchange membrane. As a result, besides hydrogen fuel, sodium hydroxide was generated as the main product in the cathode chamber, and the main lignin product was fully precipitated in the anode chamber;

- the generated hydrogen gas and oxygen gas were generated in the cathode chamber and anode chamber, respectively, and thereby, allowed the generated hydrogen gas be combusted safely;

- a filter was used to separate the lignin precipitates from the acidified mixture, and thereby, allows the system to work continually.

Description of the Treatment Process

First, 200 mL of black liquor was diluted to 900 mL using the recycled acidic filtrate in beaker 1. Secondly, the diluted solution was fed into the electrolytic reactor 2. In the reactor, NaOH and H_2 were generated in cathode chamber 6; an acidic mixture of different organic acids, lignin and some organic polymers was generated in anode chamber 5. This acidified mixture was filtered by membrane filter 3, yielding wet biomass solids (including lignin and other organic precipitates) and an acidic filtrate. Finally, the acidic filtrate was consequently recycled to dilute new 200 mL black liquor. Operating in 240 h, 1000 mL of black liquor was treated completely, generating as products NaOH, H_2, and biomass solids and consuming electrical energy. The images of biomass precipitating in the diluted solution were shown in Figure S1 (the Section 1 of the supplementary materials).

Analytical Methods

The sodium hydroxide concentrations were determined via titration with a 0.1 M H_2SO_4 solution. The low heat value of the organic polymers was determined using a calorimeter. The mass of the residue solids in the acidified filtrate was tried and weighed.

The mass of hydrogen was determined using an infrared analyzer (Gasboard gas analyzer-3100, Sifang Co., Ltd., Wuhan, China) [3]. First, the hydrogen gas generated in the cathode chamber was collected in a tank under water during electrolysis. The collected gas was diluted to 4 L of total volume with air. Third, the diluted gas was conducted into chamber with dry $CaCl_2$ powder and cotton to absorb the water in the moisture. And the drying gas was consequently conducted into the infrared analyzer to determine the hydrogen concentration. The volume of hydrogen was determined by multiplying the concentration and the diluted volume, and the mass of hydrogen was determined by multiplying the hydrogen gas volume and its density.

The ^{1}H-Nuclear Magnetic Resonance (^{1}H-NMR) analysis were carried out on an AVANCE III HD 600 MHz instrument (Bruke, Karlsruhe, Germany), using DMSO and TMS as solvent and reference, respectively. A Fourier Transform Infrared Spectrometer (FTIR) instrument (Nicolet IS 50, Thermo Fisher Scientific, Waltham, MA, USA), was used for analysis. Each powdered sample was mixed with KBr, and the mixture was pressured in a mould to get a sample plate. Then the sample plates were analyzed in the range of 4000–300 cm^{-1}.

RESULTS

Identification of Main Products

As presented in Table 1, treatment of 1000 mL (122 g/L of solids) black liquor from the process described above, led to the generation of 52.10 g of dry biomass solids, 30.70 g of sodium hydroxide and 0.82 g hydrogen gas. There were still 33.63 g of residual solids remaining in the acidified filtrate, and there were other amounts of gaseous and volatile compounds released from the system, including CO_2, O_2, methanol, ethanol, formic acid, *etc.* The numbers in Table 1 were obtained by averaging three trials; the recovery ratios of elemental sodium and biomass are 80.4% and 76%, respectively. The detail calculations were shown in the Section 2 of the supplementary materials, and the reaction of gaseous and volatile compounds generated were shown in Section 3 of the supplementary materials.

Table 1. Masses of products from 1000 mL black liquor

Products	Masses (g)
Sodium hydroxide	30.70
Hydrogen gas	0.82

Biomass solids	52.10
Residue solids	33.63

Energy Balance in Treating 1000 mL of Black Liquor

Figure 2 shows the energy conversions involved in treating 1000 mL of black liquor. In this process, the energy inputs were: 345.6 kJ of electrical energy and 1573.8 kJ of heat of the black liquor; while the outputs were: 1239.49 kJ and 117.2 kJ of the heats of biomass solid and hydrogen gas, respectively, and 526.71 kJ of the residual solids. Determined by a calorimeter, a 23.79 kJ/g of the low heat value of biomass solid was obtained and used in the heat calculation. The detail calculations were shown in the Section 4 of the supplementary materials.

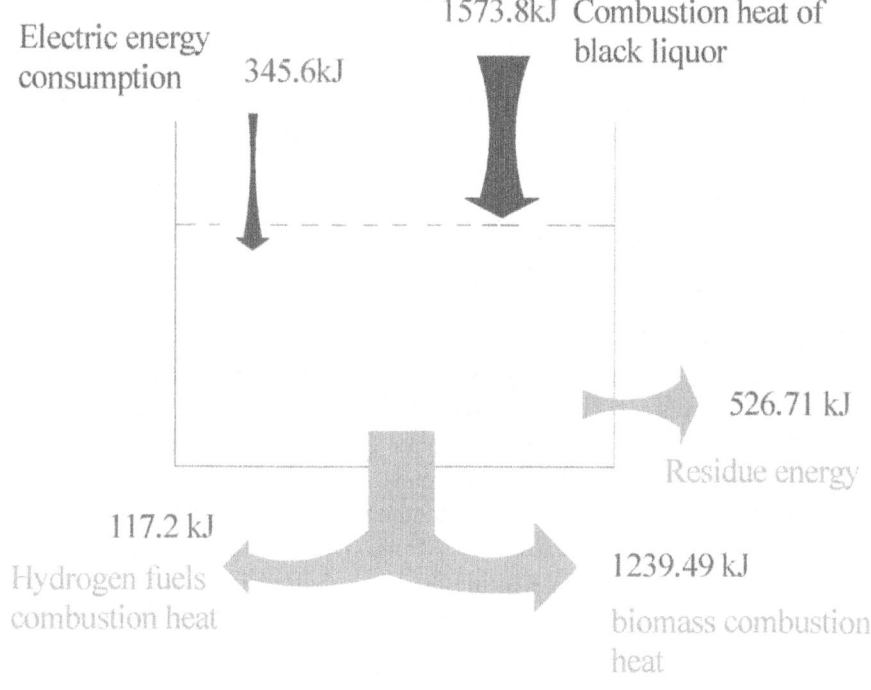

Figure 2. Energy balance involved in treating 1000 mL of black liquor.

The electric energy consumption for hydrogen generation under the studied conditions was 117.07 kWh/(kg H_2), which was significantly higher than 45 kWh/(kg H_2) in the water electrolysis for hydrogen [20]. The higher electric energy consumption was due to the higher working voltage used to obtain a high production rate in this study. Additionally, the ratio between the heat of the generated biomass and the heat of black liquor was 78.8%.

Investigation of the Composition of the Biomass Solids

Generated Biomass Solids

Tested by ashing and weight, the generated biomass solids contained 0.32% of ash. Figure 3 is the FTIR spectrums of biomass solids, lignin and the precipitates from the acidic filtrate. Shown in Figure 3a is the FTIR spectrums of the generated biomass. Where, the character absorption peaks of lignin at 1512.94 cm⁻¹ and 1603.14 cm⁻¹ were presented. Thereby, the generated biomass solids contain lignin can be identified. Besides the character absorption peaks of lignin, there are character absorption peaks at 3396.45 cm⁻¹, 2922.22 cm⁻¹ and 1459.30 cm⁻¹, corresponding to the groups of –OH, –CH and –CH$_2$, respectively.

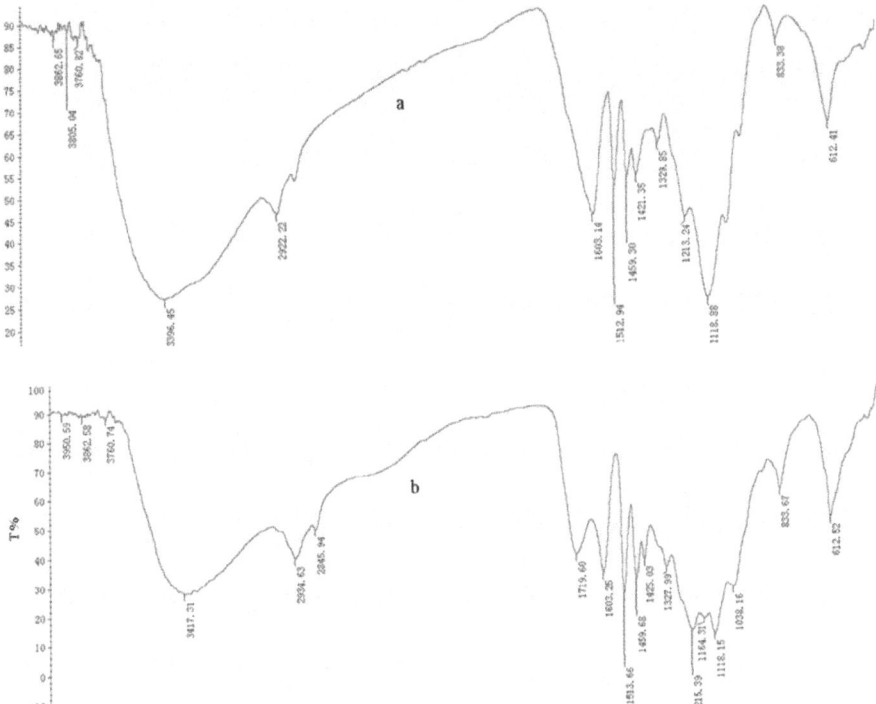

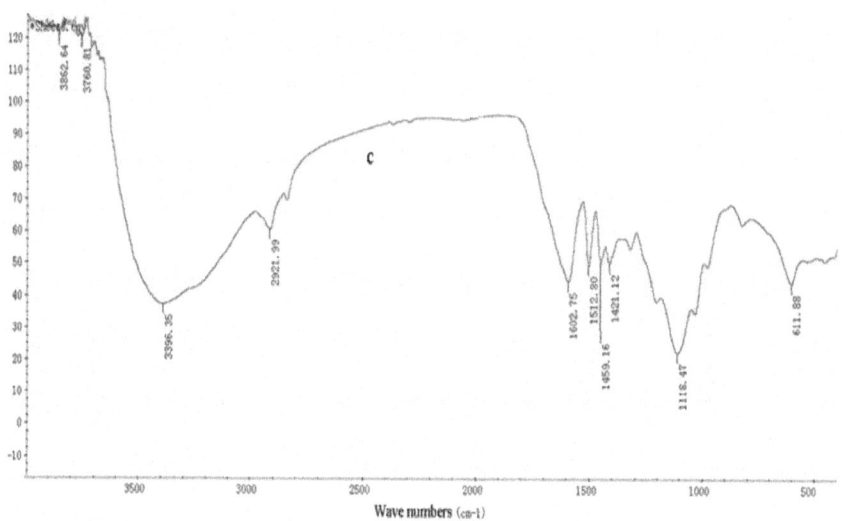

Figure 3. Fourier Transform Infrared Spectrometer (FTIR) spectrums of biomass solids (**a**); lignin (**b**) and the precipitates from the acidic filtrate (**c**).

Figure 4 shows the ¹H-NMR spectra of biomass solids, lignin and the precipitates from the acidic filtrate. Shown in Figure 4a is the ¹H-NMR spectrum of the generated biomass, where, the characteristic absorption peaks of bamboo lignin at 2.51 ppm, 3.31 ppm and 3.75 ppm are present. Thereby, the presence of lignin in the generated biomass solids contain was confirmed. The absorption peaks at 1.27 ppm were assigned to $-CH_3$ groups, and the peaks at 2.51 ppm were assigned to $-CH_2$. The absorption peaks at 3.31 ppm and 3.83 ppm were assigned to $-CH$ groups on the benzene ring. The absorption peaks at 6.4–7.53 ppm, correspond to the $-OH$ groups.

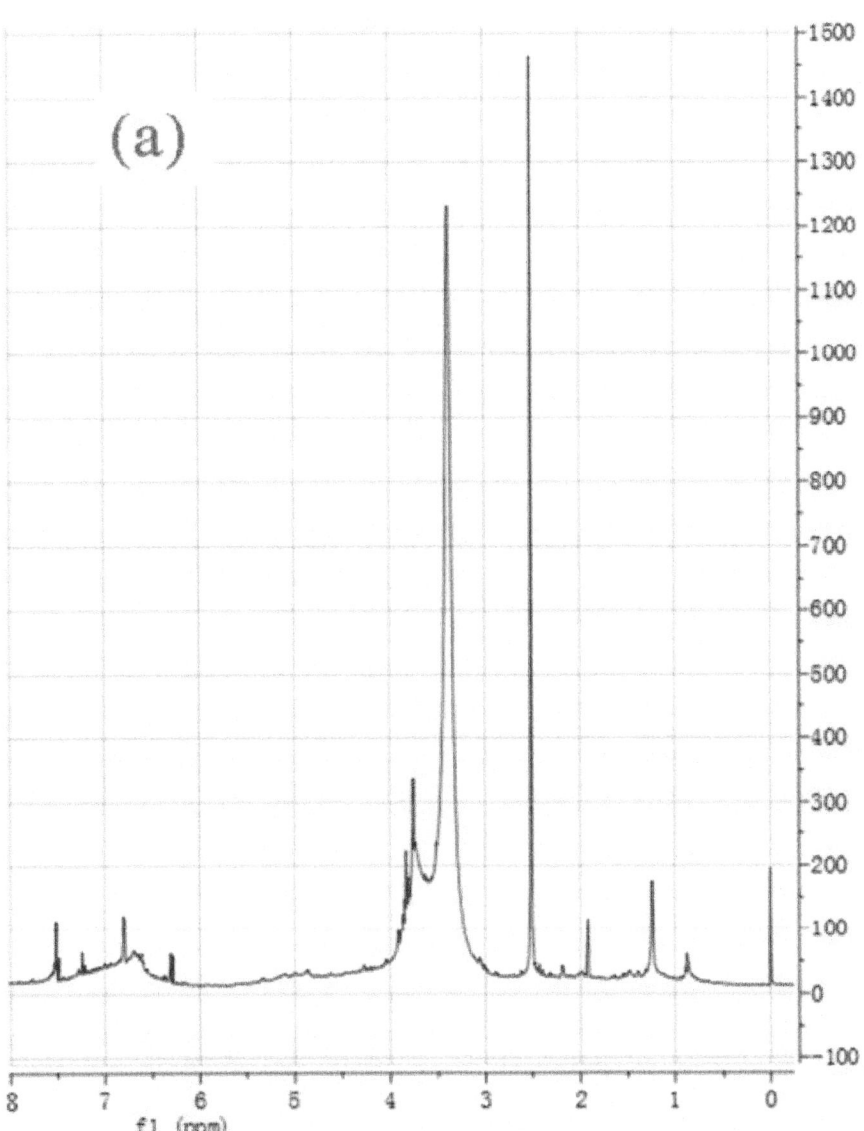

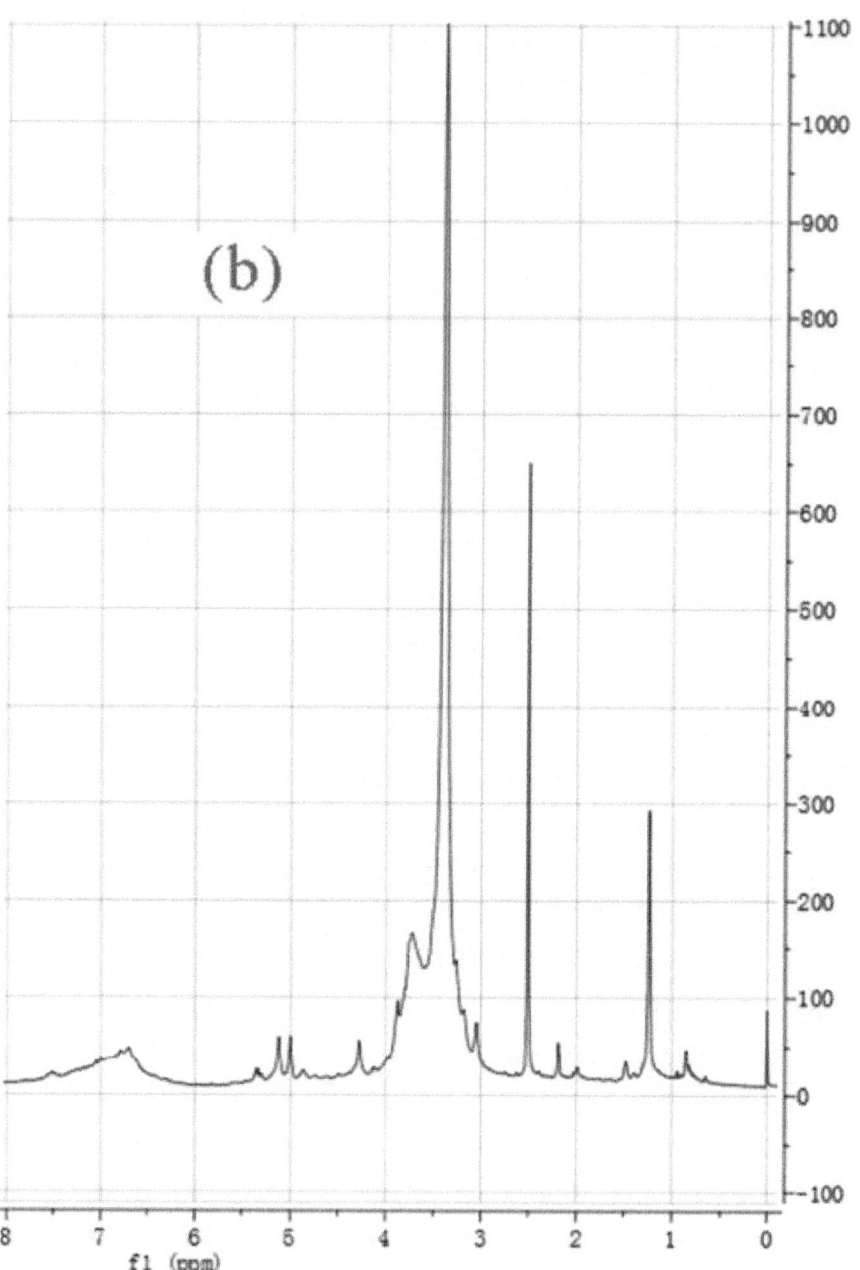

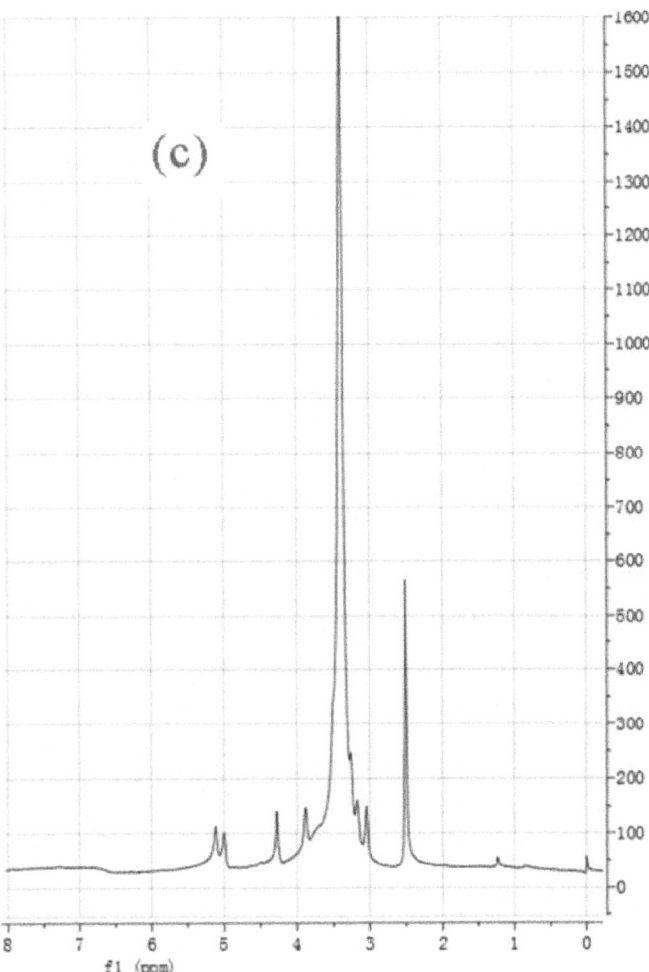

Figure 4. ^{1}H-Nuclear Magnetic Resonance (^{1}H-NMR) spectra of biomass solids (**a**); lignin (**b**) and the precipitates from the acidic filtrate (**c**).

Precipitated Lignin

Additional experiments were performed to further investigate the mass of lignin containing in the generated biomass solids. 1000 mL of the black liquor was diluted to 2000 mL of solution, which was consequently acidified with 0.1 M sulfuric acid, yielding 39.8 g of lignin and 2000 mL of acidic filtrate. Thereby, the biomass solids generated by electrolysis was regarded as containing 39.8 g of lignin. Besides lignin, there was 12.3 g of other organic precipitates contained in the biomass solids. Shown in Figure 3b is the FTIR spectrum of the lignin precipitated by sulfuric acid. Besides the characteristic absorption peaks of lignin at 1513.66 cm^{-1} and 1603.25 cm^{-1}, there are characteristic absorption peaks at 3396.35 cm^{-1}, 2921.99 cm^{-1} and 1459.16 cm^{-1}, corresponding to the –OH, –CH and –CH$_2$ groups, respectively. Shown in Figure 4b is the^{1}H-NMR spectrum of the lignin precipitated by sulfuric acid, presenting the absorption peaks of bamboo lignin at 1.25 ppm, 2.51 ppm, and 3.31 and 3.75 ppm. As mentioned ahead, those peaks were assigned to the –CH$_3$, –CH$_2$, and –CH groups, respectively.

Other Organic Precipitates

Additional experiments were performed to further investigate the generation of the other organic precipitates. After removed lignin by filtration, the remaining 2000 mL of acidic filtrate was fed into the electrolytic reactor and electrolyzed for 96 h. As a result, precipitates were formed, yielding 12.3 g of dry residue solids. Thereby, the generated biomass solids contained 12.3 g of other organic precipitates that can be identified. Shown in Figure 3c is the FTIR spectrum of the organic precipitates generated from the acidic filtrate. There are mainly characteristic peaks at 1512.80 cm^{-1} and 1602.75 cm^{-1}, corresponding to the small units of lignin, thereby, we conclude that the precipitates generated from the acidic filtrate contain small lignin units. Shown in Figure 4c is the ^{1}H-NMR spectrum of the organic precipitates generated from the acidic filtrate. They contained small lignin units, and thereby, its ^{1}H-NMR contained the characteristic absorption peaks of bamboo lignin at 2.50 ppm, 3.34 ppm. However, the peak at 1.25 ppm assigned to –CH$_3$ groups was much smaller and the peak at 3.75 was absent, indicating that the structure of the organic precipitates from the acidic filtrate was somewhat different from that of the lignin precipitated by sulfuric acid and the precipitates obtained by electrolysis. The organic polymerized and precipitated from the acidified filtrate were shown in the Section 5 of the supplementary materials.

\Investigation of the Working Mechanisms of Black Liquor Electrolysis

As shown in Figure 5, the working mechanisms of black liquor electrolysis are described as follows.

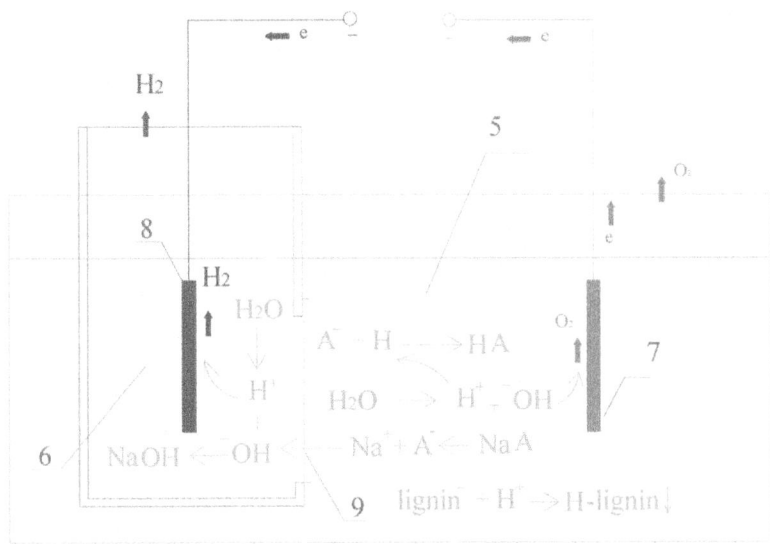

5. Anode chamber; 6. Cathode chamber; 7. Anode plate;
8. Cathode plate; 9. Cationic exchange membrane.

Figure 5. Working mechanisms of black liquor electrolysis.

Black liquor contains sodium lignin and sodium salts of organic acids (NaA), such as sodium form acid and acetate. The sodium lignin dissociates to Na^+ and lignin anion (lignin$^-$), and the sodium salts of organic acids dissociate to Na^+ and organic acidic anions (A$^-$) in solution. Selective blocking by the cationic exchange membrane prevented the organic acidic ions from penetrating the membrane, while the Na^+ can penetrate the membrane and transfer into the cathode chamber [21]. In the cathode chamber, the dissociated H^+ ions accept electrons and become hydrogen gas when they touched the cathode [22]; and the remaining –OH combines with the incoming Na^+ to form sodium hydroxide. In the anode chamber, the dissociated –OH ions lose electrons and become oxygen gas and water upon contacting with the anode [23]; and the remaining H^+ ions combine with the remain inorganic acidic ions (A$^-$) to generate organic acids (HA) [22]. The reactions of sodium hydroxide and lignin generation are expressed as Equations (1)–(8). The generation of organic acid led to the lignin precipitation, which mechanisms are shown as Figure 6.

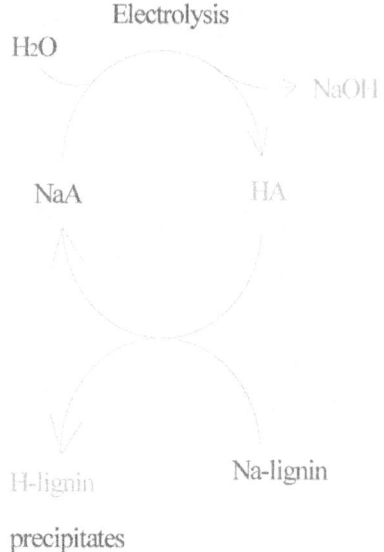

Figure 6. Mechanisms of lignin precipitating in the process of black liquor electrolysis.

The following reactions occur in the cathode chamber:

$$4H_2O \rightarrow 4H^+ + 4^-OH \quad \text{(1) (Water ionization)}$$
$$4^-OH - 4e^- \rightarrow 2H_2O + O_2 \uparrow \quad \text{(2) (Anode reaction)}$$
$$NaA \rightarrow Na^+ + A^- \quad \text{(3) (Ion-dissociation reaction)}$$
$$H^+ + A^- \rightarrow HA \quad \text{(4) (Ion-binding reaction)}$$
$$HA + Na\text{-lignin} \rightarrow H\text{-lignin} \downarrow + NaA \quad \text{(5) (Ion-binding reaction)}$$

The following reactions occur in the cathode chamber:

$$4H_2O \rightarrow 4H^+ + 4^-OH \quad \text{(1) (Water ionization)}$$
$$4H^+ + 4e^- \rightarrow 2H_2 \uparrow \quad \text{(6) (Cathode reaction)}$$
$$4Na^+ + 4^-OH \rightarrow 4NaOH \quad \text{(7) (Ion-binding reaction)}$$

The net reaction for the electrolytic reactor is as follows:

$$Na\text{-lignin} + NaA + 4H_2O \rightarrow 2NaOH + H\text{-lignin} + HA + H_2\uparrow + O_2\uparrow \quad (8)$$

DISCUSSION

Treating black liquor by electrolysis consumes rather large amounts of electric energy, and thereby increases the electric energy consumption cost of a pulp mill. However, compared to combustion [6] and gasification [24], treating black liquor via electrolysis has the following advantages:

- no need to concentrate the thin black liquor, which reduces the enormous amount of energy that is consumed during water evaporation;
- no lime is consumed in the NaOH regeneration process, thereby, eliminating the energy intensive lime cycle, together with the necessary solid residue disposal system;
- reduced emissions of CO_2 gas and other poisonous gases [7];
- large-scale generation of lignin, which has potential to be converted into bioenergy and bio-materials [13,25,26];
- in addition to lignin, other organic components in the solution are recovered via polymerization, thereby reducing the amount of organic pollutants discharged.

Based on the numbers in Figure 2, the electric consumption for treating 1 m³ of black liquor is 96 kWh that costs about $9.6 USD. The generated NaOH is about 30.7 kg that is worth about $10.7 USD. Besides the NaOH, 52.1 kg of biomass solid and 0.82 kg of hydrogen are obtained at the same time; those two products are worth about $11.2 USD. Therefore, although only 80.4% of NaOH was recovered, treating black liquor by electrolysis might be economically feasible. The detail economic calculations were shown in the Section 6 of the supplementary materials.

The black liquor electrolysis process has lots of potential:

- the conventional black liquor recovery process in pulp mills discharges large amounts of solid, liquid and gaseous wastes that have a significant environmental impact on the surrounding areas [27], therefore, it is desirable to develop an alternative technology for processing black liquor;
- the overall energy efficiency of the black liquor electrolysis process is much better than the conventional black liquor process via evaporation, combustion and the lime cycle;
- treating black liquor by electrolysis generates biomass solids including lignin and other organic precipitates that can be converted to biofuels to substitute fossil fuels to reduce the CO_2 emissions, and thereby benefit the environment [28];
- treating black liquor by electrolysis might be fitting to treat the black liquor generated from wheat straw, and the black liquor in a small-scale pulp mill, which get not allow good economic efficiency by installation of general recovery boilers.

CONCLUSIONS

A new black liquor treating system based on the electrolysis concept was studied. The main products were lignin sodium hydroxide, and hydrogen gas. Under the conditions studied, 1000 mL of black liquor (122 g/L solid content) led to the generation of 30.7 g of sodium hydroxide, 52.1 g of biomass solids (containing 39.8 g of lignin and 12.3 g of polymers) and 0.82 g of hydrogen gas, while consuming a total of 345.6 kJ of electrical energy. Therefore, the recovery ratios of elemental sodium and biomass are 80.4% and 76%, respectively. Treating black liquor via electrolysis is an environmentally friendly technology and represents an alternative technology that should be useful in addressing the environmental issues of the future.

SUPPLEMENTARY MATERIALS

Biomass Precipitating from Black Liquor

A 200 mL of the black liquor was diluted to 900 mL water solution, and then the diluted solution was fed into the electrolytic reactor. After operated for 48 h, the diluted solution become acidic with pH = 4–5, and lead to biomass precipitating. And thereby, approximately 900 mL of initial acidic filtrate was obtained by filtering. Figure S1 shows the biomass precipitating in the diluted solution by electrolysis.

Figure S1. Biomass precipitating in the diluted solution by electrolysis.

Detail Calculation on the Recovery Ratios of Sodium Element and Biomass

Recovery Ratio of Sodium Element

The recovery ratio of sodium element (r_{Na}) is the mass of sodium element containing in the recovered sodium hydroxide ($M_{Na, re}$) divided by the mass of sodium element containing in the original black liquor ($M_{Na, BL}$).

$r_{Na} = M_{Na, re}/M_{Na, BL}$

$\quad = (30.4 \times 23/40)/(122 \times 18\%)$

$\quad = 0.804$

$\quad = 80.4\%$

Recovery Ratios of Biomass

The recovery ratio of biomass (r_{bi}) is the mass of biomass recovered ($M_{bi, re}$) divided by the mass of biomass containing in the original black liquor ($M_{bi, BL}$).

$r_{bi} = M_{bi, re}/M_{bi, BL}$

$\quad = 52.1/(122 \times 56.2\%)$

$\quad = 0.760$

$\quad = 76\%$

Where, the 56.2% is the ratio of biomass containing black liquor.

Consideration of Gaseous and Volatile Compounds Release to the Air

There might be some other masses of gaseous and volatile compounds release to the air, including CO_2, O_2, methanol, ethanol and formic acid, etc. Where, the CO_2 was converted from the reactions of Sodium carbonate acidification and anode oxidation, expressed as Equations (1) and (2). The O_2 was converted from water by anode oxidation, expressed as Equations (3) and (4). And the formic acid was converted from the reactions of Sodium formic acid by acidification, expressed as Equation (5). And the methanol and ethanol might be generated from wood in the time cooking and was contained in the original black liquor.

$$NaCO_3 + 2H^+ \rightarrow 2Na^+ + H_2O + CO_2 \uparrow \qquad (1)$$

$$Substrate - 4e^- \rightarrow aH_2O + bCO_2 \uparrow \qquad (2)$$

$$4H_2O \rightarrow 4H^+ + 4^- OH \qquad (3)$$

$$4^- \, OH - 4e^- \rightarrow 2\,H_2O + O_2 \uparrow \qquad \qquad (4)$$

$$HCOONa + H^+ \rightarrow Na^+ + HCOOH \qquad (5)$$

Detail Calculation on Energy Conversions

1. Electrical energy consumption in treating black liquor

   Working in condition of 4 V voltage and average 100 mA for 240 h, the electric energy consumption (E_{ec}) was calculated as following:

   $E_{ec} = V \times I \times t$

   $\quad = 4 \, V \times 0.1 \, A \times 240 \times 3600 \, s$

   $\quad = 345600 \, J$

   $\quad = 345.6 \, kJ$

2. Heat of black liquor

   $H_{BL} = 1000 \, mL \times 1 \, g/mL \times 12.2\% \times 12.9 \, kJ/g$

   $\quad = 1573.8 \, kJ$

3. Heat of biomass solids

   $H_{bs} = 52.1g \times 23.79 \, kJ/g$

   $\quad = 1239.5 \, kJ$

4. Residue heat of reactor

   $H_{residue, \, reactor} = (H_{BL} + E_{ec}) - (H_{bs} + H_{H2})$

   $\quad = (1573.8 + 345.6) - (1239.5 + 117.2)$

   $\quad = 526.7 \, kJ$

5. Heat of hydrogen fuel

   $H_{H2} = 0.82 \, g \times 142.90 \, kJ/g$

   $\quad = 117.2 \, kJ$

6. The energy efficiency of hydrogen generation

   $E_{e, \, H2} = E_{ec}/M_{H2}$

   $\quad = 345.6 \, kJ/0.82 \, g$

   $\quad = 421.46 \, kJ/g$

   $\quad = 117.07 \, kWh/(kg \, H_2)$

7. The ratio between the heats of biomass solid and black liquor

$R = H_{BS}/H_{BL}$

= 1239.49 kJ/1573.8 kJ

= 78.8%

Organic Polymerized and Precipitated from the Acidified Filtrate

Figure S2. Organic compounds polymerized during electrolysis (a: the original acid filtrate;   b: mixture; c: the final acid filtrate separated from the mixture).

The Detail Economic Calculation for Treating 1 m³ Black Liquor

Base on the digits in Table 2, the electric consumption for treating 1 m³ of black liquor is 96 kWh, which is worth about 9.6 dollar, provided the cost of electric energy being 0.1 dollar/kWh. The generated NaOH is about 30.7 kg; which is worth about 10.7 dollar, provided the cost of NaOH being 0.35 dollar/kg. The 52.1 kg of biomass solid is worth 10.4 dollar, provided the cost of   0.2 dollar/kg. And 0.82 kg of hydrogen is worth 0.82 dollar, provided cost of 1 dollar/kg.

ACKNOWLEDGMENTS

The financial support for this project was provided by the Guangxi Natural Science Foundation (Grant #: 2013jjFA20001).

AUTHOR CONTRIBUTIONS

G.N. involved in the new ideal, spectrum analysis and manuscript preparation. S.W. involved in the device preparation. Z.Z. involved in preparation, determination and analysis of products. All authors reviewed the manuscript.

REFERENCES

1. Hamaguchi, M.; Cardoso, M.; Vakkilainen, E. Alternative Technologies for Biofuels Production in Kraft Pulp Mills—Potential and Prospects. *Energies* **2012**, *5*, 2288–2309.

2. Zeng, J.; Tong, Z.; Wang, L.; Zhu, J.Y.; Ingram, L. Isolation and structural characterization of sugarcane bagasse lignin after dilute phosphoric acid plus steam explosion pretreatment and its effect on cellulose hydrolysis. *Bioresour. Technol.* **2014**, *154*, 274–281.

3. Nong, G.; Huang, L.; Mo, H.; Wang, S. Investigate the variability of gas compositions and thermal efficiency of bagasse black liquor gasification. *Energy* **2013**, *49*, 178–181.

4. Saw, W.L.; Nathan, G.J.; Ashman, P.J.; Hupa, M. Influence of droplet size on the release of atomic sodium from a burning black liquor droplet in a flat flame. *Fuel* **2010**, *89*, 1840–1848.

5. Yang, C.Y.; Niu, Y.; Su, H.J.; Wang, Z.; Tao, F.; Wang, X.; Tang, H.; Ma, C.; Xu, P. A novel microbial habitat of alkaline black liquor with very high pollution load: Microbial diversity and the key members in application potentials.*Bioresour. Technol.* **2010**, *101*, 1737–1744.

6. Maček, A. Research on combustion of black-liquor drops. *Prog. Energy Combust. Sci.* **1999**, *25*, 275–304.

7. Carlsson, P.; Wiinikka, H.; Marklund, M.; Grönberg, C.; Pettersson, E.; Lidman, M.; Gebart, R. Experimental investigation of an industrial scale black liquor gasifier. 1. The effect of reactor operation parameters on product gas composition. *Fuel* **2010**, *89*, 4025–4034.

8. Maciel, A.V.; Job, A.E.; Mussel, W.N.; Pasa, V.M.D. Pyrolysis and auto-gasification of black liquor in presence of ZnO: An integrated process for Zn/ZnO nanostructure production and bioenergy generation. *Biomass Bioenergy* **2012**, *46*, 538–545.

9. Naqvi, M.; Yan, J.; Dahlquist, E. Bio-refinery system in a pulp mill for methanol production with comparison of pressurized black liquor gasification and dry gasification using direct causticization. *Appl. Energy* **2012**, *90*, 24–31.

10. García-Trenco, A.; Martínez, A. The influence of zeolite surface-aluminum species on the deactivation of CuZnAl/zeolite hybrid catalysts for the direct DME synthesis. *Catal. Today* **2014**, *227*, 144–153.

11. Eriksson, H.; Harvey, S. Black liquor gasification—Consequences for both industry and society. *Energy* **2004**, *29*, 581–612.

12. Zhang, X.; Wang, T.; Ma, L.; Zhang, Q.; Huang, X.; Yu, Y. Production of cyclohexane from lignin degradation compounds over Ni/ZrO_2–SiO_2 catalysts. *Appl. Energy* **2013**, *112*, 533–538.

13. Yoshikawa, T.; Shinohara, S.; Yagi, T.; Ryumon, N.; Nakasaka, Y.; Tagoa, T.; Masudaa, T. Production of phenols from lignin-derived slurry liquid using iron oxide catalyst. *Appl. Catal. B* **2014**, *146*, 289–297.

14. Welker, C.M.; Balasubramanian, V.K.; Petti, C.; Rai, K.M.; DeBolt, S.; Mendu, V. Engineering plant biomass lignin content and composition for biofuels and bioproducts. *Energies* **2015**, *8*, 7654–7676.

15. Guo, F.; Xiu, Z.; Liang, Z. Synthesis of biodiesel from acidified soybean soapstock using a lignin-derived carbonaceous catalyst. *Appl. Energy* **2012**, *98*, 47–52.

16. Minu, K.; Kurian Jiby, K.; Kishore, V.V.N. Isolation and purification of lignin and silica from the black liquor generated during the production of bioethanol from rice straw. *Biomass Bioenergy* **2012**, *39*, 210–217.

17. Nong, G.; Chen, S.; Xu, Y.; Huang, L.; Zou, Q.; Li, S.; Mo, H.; Zhu, P.; Cen, W.; Wang, S. Artificial photosynthesis of oxalate and oxalate-based polymer by a photovoltaic reactor. *Sci. Rep.* **2014**, *4*.

18. Han, B.; Steen, S.M., III; Mo, J.; Zhang, F. Electrochemical performance modeling of a proton exchange membrane electrolyzer cell for hydrogen energy. *Int. J. Hydrogen Energy* **2015**, *40*, 7006–7016.

19. Ghatak, H.R. Electrolysis of black liquor for hydrogen production: Some initial findings. *Int. J. Hydrogen Energy* **2006**, *31*, 934–938.

20. Chen, Y.X.; Lavacchi, A.; Miller, H.A.; Bevilacqua, M.; Filippi, J.; Innocenti, M.; Marchionni, A.; Oberhauser, W.; Wang, L.; Vizza, F. Nanotechnology makes biomass electrolysis more energy efficient than water electrolysis. *Nat. Commun.* **2014**, *5*.

21. Hong, J.G.; Zhang, B.; Glabman, S.; Uzal, N.; Dou, X.; Zhang, H.; Wei, X.; Chen, Y. Potential ion exchange membranes and system performance

in reverse electrodialysis for power generation: A review. *J. Membr. Sci.* **2015**, *486*, 71–88.

22. Chanda, D.; Hnát, J.; Paidar, M.; Schauer, J.; Bouzek, K. Synthesis and characterization of NiFe$_2$O$_4$ electrocatalyst for the hydrogen evolution reaction in alkaline water electrolysis using different polymer binders. *J. Power Sources* **2015**,*285*, 217–226.

23. Nie, X.; Luo, W.; Janik, M.J.; Asthagiri, A. Reaction mechanisms of CO$_2$ electrochemical reduction on Cu(III) determined with density functional theory. *J. Catal.* **2014**, *312*, 108–122.

24. Sricharoenchaikul, V.; Frederick, W.J., Jr.; Agrawal, P. Carbon distribution in char residue from gasification of Kraft black liquor. *Biomass Bioenergy* **2003**, *25*, 209–220.

25. Lin, X.; Sui, S.; Tan, S.; Pittman, C.U., Jr.; Sun, J.; Zhang, Z. Fast pyrolysis of four lignins from different isolation processes using Py-GC/MS. *Energies* **2015**, *8*, 5107–5121.

26. Wen, J.; Sun, S.; Yuan, T.; Xu, F.; Sun, R. Understanding the chemical and structural transformations of lignin macromolecule during torrefaction. *Appl. Energy* **2014**, *121*, 1–9.

27. Bordado, J.C.M.; Gomes, J.F.P. Atmospheric emissions of Kraft pulp mills. *Chem. Eng. Process.* **2002**, *41*, 667–671.

28. Budzianowski, W.M. Negative carbon intensity of renewable energy technologies involving biomass or carbon dioxide as inputs. *Renew. Sustain. Energy Rev.* **2012**, *16*, 6507–6521.

Chapter 8

CO2 FIXATION BY MEMBRANE SEPARATED NACL ELECTROLYSIS

Hyun Sic Park [1], Ju Sung Lee [1], JunYoung Han [2], Sangwon Park [3], Jinwon Park [1] and Byoung Ryul Min [1]

[1]Department of Chemical and Biomolecular Engineering, Yonsei University, 262 Seongsanno, Seodaemun-gu, Seoul 120-749, Korea

[2]Proton Conductors Section, Department of Energy Conversion and Storage, Technical University of Denmark, Kemitorvet 207, Kgs. Lyngby DK-2800, Denmark

[3]CO_2 Sequestration Department, Korea Institute of Geoscience and Mineral Resources (KIGAM), 124 Gwahak-ro, Yuseong-gu, Daejeon 305-350, Korea

ABSTRACT

Atmospheric concentrations of carbon dioxide (CO_2), a major cause of global warming, have been rising due to industrial development. Carbon capture and storage (CCS), which is regarded as the most effective way to reduce such atmospheric CO_2 concentrations, has several environmental and technical disadvantages. Carbon capture and utilization (CCU), which has been introduced to cover such disadvantages, makes it possible to capture CO_2, recycling byproducts as resources. However, CCU also requires large amounts of energy in order to induce reactions. Among existing CCU technologies, the process for converting CO_2 into $CaCO_3$ requires high temperature and high pressure as reaction conditions. This study proposes a method to fixate $CaCO_3$ stably by using relatively less energy than existing methods. After forming NaOH absorbent solution through electrolysis of NaCl in seawater, $CaCO_3$ was precipitated at room temperature and pressure. Following the experiment, the resulting product $CaCO_3$ was analyzed with Fourier transform infrared spectroscopy (FT-IR); field emission scanning electron microscopy (FE-SEM) image and X-ray diffraction (XRD) patterns were also analyzed. The results showed that the $CaCO_3$ crystal product was high-purity calcite. The study shows a successful method for fixating CO_2 by reducing carbon dioxide released into the atmosphere while forming high-purity $CaCO_3$.

INTRODUCTION

Increases in energy consumption due to population growth and industrial development have had a large effect on global warming by raising carbon dioxide (CO_2) concentrations in the atmosphere [1]. CO_2 is one of the six major gases causing global warming (CO_2, CH_4, N_2O, HFCs, PFCs, SF_6) [2], accounting for an estimated 80% of the greenhouse gases by amount [3]. For this reason, many researchers have focused on studies to reduce atmospheric concentrations of CO_2.

Among the various worldwide attempts to reduce atmospheric CO_2 concentrations, carbon capture and storage (CCS) technology, which captures released CO_2 and storages underground or underwater, is considered the most effective and efficient technology. In CCS technology, CO_2 can be captured before, during, or after coal or gas combustion [4]. CO_2 content in flue gas is reduced by physical, chemical, or biological means, and CCS industries, which process CO_2 after combustion, use chemical solvents [5,6]. From an energy and climate policy point of view, this type of CCS technology is theoretically the most effective method to reduce amount of CO_2 released but also entails the possibility of resistance from the general public due to high investment costs, limitation and uncertainty of potential storage capacity of CO_2. According to Nicholas et al. [7], over 80% of energy used in CCS is consumed in the CO_2 desorption process. In addition, according to Baciocchi et al. [8], CO_2 capture mechanisms require considerable energy and costs. Yu et al. [1] and Gough [9] have noted possible safety problems (due to earthquake or volcanic activity) from long-term CO_2 storage, while Holloway [10] reported on its risks, citing the example of a CO_2 leak case in Cameroon's Nyos Lake. Damen et al. [11] reported on five types of risks for underground carbon dioxide storage (CO_2 and CH_4 leakage, seismicity, ground movement, displacement of brine) [12]. In addition to these problems, according to Mazzoldi et al. [12], there is a possibility of CO_2 leakage caused by corrosion or external damage of pipeline of high-pressure transportation system, which was reported in the oil industry literature [13].

Because of these problems, studies related to the carbon capture and utilization (CCU) technology that recycles carbon dioxide as a resource has attracted attention. Existing CCU technologies for the production of calcium carbonate can be divided into direct and indirect reactions between solid chemicals and CO_2 gas, and aqueous system methods. Solid calcium carbonate production processes use relatively greater energy than liquid production processes. For example, according to research by Lackner et al. [14], Stasiulaitiene et al. [15] and Fagerlund et al. [16,17,18], the carbonation process for magnesium hydroxide obtained from serpentine requires high temperature (over 500 °C) and high-pressure (over 20 bar) conditions. In contrast, the liquid method according to Gerdemann *et al.* [19] and Khoo *et al.* [20] requires relatively lower carbonation temperature and pressure conditions of 185 °C and more than 40 bar, and 170 °C and 1 bar, respectively. However, the method which uses an aqueous system to produce carbonate also requires a reaction condition of high temperature and pressure. As with the solid calcium carbonate production method, it clearly requires great amounts of energy for processes such as cool-down in high-temperature and high-pressure conditions, compared to CCS separation methods [21]. For these reasons, existing methods cannot be considered as optimal alternatives for reducing atmospheric CO_2 concentrations. For optimal CO_2 reduction, methods that

- satisfy environmental considerations by addressing possible CO_2 leaks, and

- reduce energy consumption compared to existing carbonation processes conducted at high-temperature and in high-pressure conditions must be devised.

This study proposes a carbonation process which uses relatively less energy than traditional process, through an electrolysis technology using ceramic membrane. CO_2 usually reacts with alkaline solutions as absorbent, with NaOH (Sodium Hydroxide) and NH_4 (Ammonium) solution, MEA (Mono-ethanol Amine), DEA (Di-ethanol Amine), and MDEA (N-methyl Diethanolamine) being the most common examples [22]. Previous studies use magnesium hydroxide ($Mg(OH)_2$) in order to absorb CO_2. Existing processes such as the one by Nduagu *et al.* [23] require the use of ammonium magnesium salt heated to over 500 °C in order to extract magnesium from serpentine or other magnesium silicates. They also require high temperature reaction conditions for the separation process of magnesium oxide (MgO) and silicon dioxide (SiO_2) from serpentine, and the hydration process after separation. Accordingly, this study uses NaOH (Sodium Hydroxide) as absorbent for CO_2 through electrolysis of NaCl in seawater. The proposed method enables formation of NaOH using low voltage of 1–4 amperes.

This technology was proven to produce alkaline solution at low voltages by CALERA Corporation of the US [24]. Furthermore, the electrolysis process using a ceramic membrane requires no additional pressure or temperature. Additionally, the carbonation process does not require a separate process for the separation of the absorbed CO_2. This enables it to reduce energy consumed in the process of CO_2 desorption, and to significantly reduce energy consumption through reaction at room pressure and room temperature. Moreover, the chemical conversion solution produced after the carbonation reaction can be recycled as feed solution for the electrolysis to produce NaOH. Accordingly, this study may find its significance in overcoming the problems of existing CO_2 reduction technologies (CCS and CCU technology) by stably fixating CO_2 at room temperature and room pressure with relatively less energy, while producing a metal carbonate to generate income.

EXPERIMENTAL SECTION

Materials and Electrolysis Device

Sodium chloride (NaCl), 99.5% (Mn = 58.43 g/mol) used as feed for electrolytic reaction was purchased from Samchun chemical (Gyeonggi-do, Korea). Calcium chloride ($CaCl_2$), 95.9% (Mn = 110.98 g/mol) was purchased from Kanto (Tokyo, Japan) to produce the carbonate, carbon dioxide (CO_2) gas was purchased from Samheung (Gyeonggi-do, Korea). Calcium carbonate ($CaCO_3$), ≥99.0% (Mn = 100.09 g/mol) used to determine formation of final product purity of $CaCO_3$ was purchased from Sigma-Aldrich (St Louis, MO, USA), and de-ionized water was used as solvent in the experiment.

Figure 1 is a schematic diagram of the electrolysis device designed for the experiments. The device was built with 10 mm acrylic material. In general, ion-exchanged membrane or ceramic membrane is used for electrolysis device. Ferro et al. [25] reported that an ion-conducting ceramic membrane around the graphite anode is necessary in electrolytic cells where reactive metals, such as calcium, magnesium and sodium, are produced to minimize the possibility of back reactions. Therefore, in this study ceramic membrane (Korea Material scientific, Changwon, Korea) which has less than 0.2 μm pore size of alumina material, was used. The ceramic membrane used in this experiment has excellent physical properties and can be used semi-permanently, compared with MF membrane. Electrolytes in the acrylic water tank are divided into cathodes and anodes by a ceramic membrane, a negative electrode was used as the stainless steel, a positive electrode was used as the graphite. To increase the contact area between CO_2 bubbles and absorbent produced by electrolysis, a high-pressure air stone (DAE Yang Air Stone Ind., Co, Busan, Korea) was used, which generates fine CO_2 bubbles.

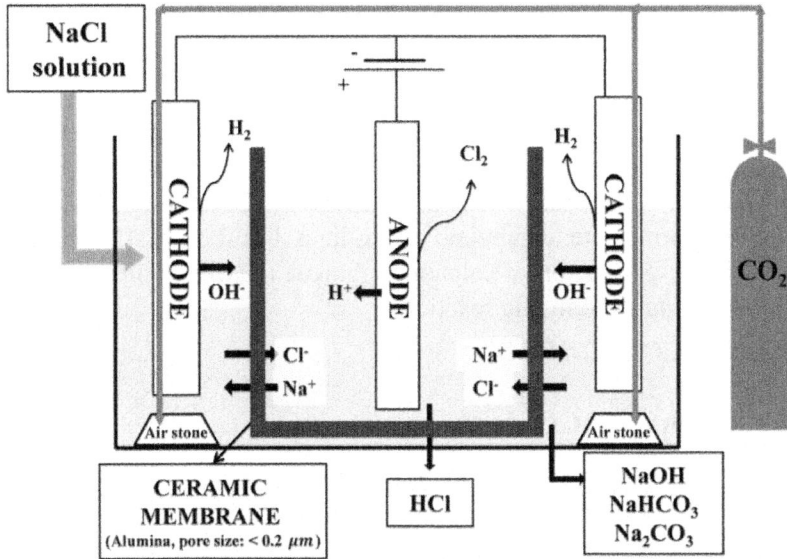

Figure 1. Schematic diagrams of the electrolysis device for the electrolysis of NaCl solutions.

Electrolysis of NaCl Solution

The method used in this study is based on chemical absorption and conversion method in CCS technology. Among the alkali solutions used as CO_2 absorbents, amine-family absorbents were excluded due to toxicity, while magnesium hydroxide ($Mg(OH)_2$) used in existing processes was excluded due to high temperature and high pressure reaction conditions. For these reasons, NaOH, which does not require high temperature and high pressure reaction conditions, was obtained through electrolysis of seawater and used as an absorbent. In the case of the NaCl solution used as feed for electrolysis, various concentration ranges were used based on concentrations (NaCl 2%–6%, 5L volume) in seawater and seawater concentrate. The NaCl solutions were prepared in the electrolysis device as in Figure 1, changes in pH over time were checked using pH electrodes (ORION 3-STAR Benchtop pH/ISE meter, Thermo Scientific Korea, Seoul, Korea), and the electrolysis was conducted using even an current of 1–4 amperes. Electrolysis was conducted for 10–15 min at a temperature of around 25 °C and atmospheric pressure, and the underlying reaction is as follows:

$$2Cl^- \rightarrow Cl_2 + 2e^- \quad \text{(anode, oxidation reaction)} \tag{1a}$$

$$2H_2O + 2e^- \rightarrow H_2 + 2OH^- \quad \text{(cathode, reduction reaction)} \tag{1b}$$

$$2NaCl + 2H_2O \rightarrow 2NaOH + H_2 + Cl_2 \qquad (1c)$$

CO_2 *Gas Capture Using Sodium Hydroxide*

NaOH is produced at the "negative" electrode as CO_2 absorbent. CO_2 gas was injected at an even flow rate into the NaOH feed solution through the CO_2 generator. The weak acidic CO_2 gas combined with alkaline NaOH to form the intermediate compound of sodium bicarbonate ($NaHCO_3$) form needed for the generation of calcium carbonate ($CaCO_3$). This is represented by the following neutralizing reaction:

$$NaOH + CO_2 \rightarrow NaHCO_3 \qquad (2)$$

Precipitation of $CaCO_3$ *by the Titration*

$CaCO_3$ was generated through precipitation by adding $CaCl_2$ feed solution to sodium bicarbonate ($NaHCO_3$) produced in this study.

According to Henry's law and the solubility constant, the pH value affects the distribution of ionic CO_2; in order to increase the conversion rate to $CaCO_3$, bicarbonate should be turned into carbonate by adjusting pH level. Figure 2 shows changes of carbonic acid (H_2CO_3), bicarbonate (HCO_3^-), and carbonate (CO_3^{2-}) compositions in accordance with pH at 25 °C. Sodium bicarbonate ($NaHCO_3$) is converted into sodium carbonate (Na_2CO_3) when pH level is higher than 12. Converted Na_2CO_3 is precipitated as $CaCO_3$ by adding $CaCl_2$ solution. The reaction by which the highly reactive Ca^{2+} and Na^+ form $CaCO_3$ through substitution reaction is as follows:

$$Na_2CO_3 + CaCl_2 \rightarrow 2NaCl + CaCO_3 \qquad (3)$$

Reaction was conducted by the titration method, and the Ca^{2+} ion selective electrode (orion ionplus calcium electrode, Thermo Scientific Orion, Beverly, MA, USA) was used in order to confirm the end of the carbonate reaction. Changes in the concentration of the Ca^{2+} ion selective electrode were observed by adding $CaCl_2$ solution in prepared concentrations. $CaCl_2$ feed solution added by the titration method is fully consumed through substitution reaction, so the concentration at the Ca^{2+} ion selective electrode was close to zero as the reaction was in progress. When the CO_3^{2-} was fully reacted, Ca^{2+} ion selective electrode concentration also changed, and the reaction was deemed as completed and terminated. The precipitate after the reaction was filtered using a GF/C film (Whatman® Glass microfiber filters, Whatman, Clifton, NJ, USA). It was dehydrated in a vacuum oven for 24 h at room temperature. After dehydrating, the output was a white powder form.

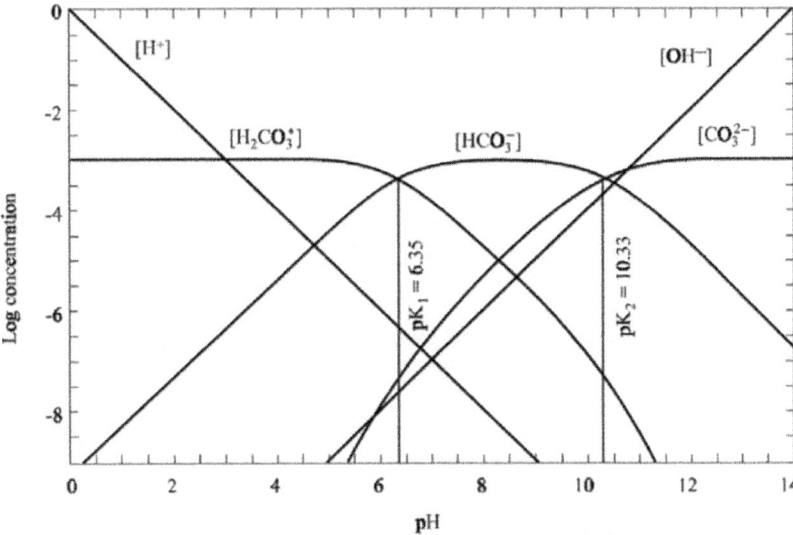

Figure 2. Carbonate-bicarbonate speciation in water v. pH at 25 °C (Reproduced with permission from [26]. Copyright 2012, John Wiley & Sons).

Characterization of CaCO₃

Fourier transform infrared spectroscopy (Spectrum 100 FT-IR Spectrometer, PerkinElmer, Norwalk, CT, USA) was used to confirm molecular structure of $CaCO_3$, the final product of the experiment. Each sample spectrum was measured in the 4000–600 cm^{-1} wavenumber ranges after drying for over 24 h in a vacuum oven. The particles of the product in the form of powder was applied on carbon tape, and then coated with platinum for several minutes. Shape and size of the coated particles was confirmed by using field emission scanning electron microscopy (FE-SEM, JEOL-6701F, JEOL, Tokyo, Japan). Measurements of confirmed particles were done through X-ray diffraction (XRD, Ultima IV, RIGAKU, Tokyo, Japan) pattern analysis at 40 kV/30 mA output.

RESULTS AND DISCUSSION

The Metal Carbonate Produced by the CO₂ Fixation.

In Figure 1, as electrolysis is conducted in the electrolysis device, Na^+ and OH^- ions, and H^+ and Cl^- ions are separated into the "negative" and "positive" electrodes, respectively, to form alkaline NaOH and acidic HCl. Figure 3 shows the changes in pH over time measured with a pH installed in the "negative" electrode. Feed solution of electrolysis using a NaCl solution (2%–6% concentration) of five liters. Figure 3 shows that regardless of concentration, NaCl solution approaches a pH of 12 over time at 1–4 A conditions (8–20 V voltage, approx. 25 °C and atmospheric pressure). The time to end the electrolysis was determined when the pH value was no longer increased. Electrolysis was completed over 10–15 min in all cases, and reaction time decreased as ampere was increased. It was shown that sodium bicarbonate, a CO_2 absorbent, could be produced without high temperature, high pressure conditions. The NaOH produced through electrolysis satisfied the reaction conditions through pH control, after which reaction took place over 2–3 h with $CaCl_2$ in feed solution form, representing a substantial reduction in reaction time over naturally-occurring $CaCO_3$ composite reaction.

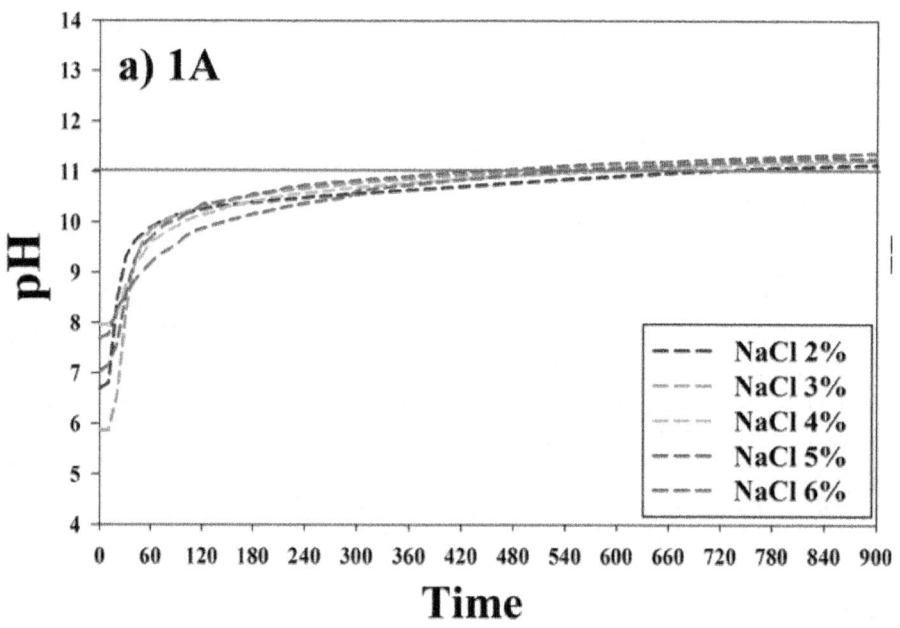

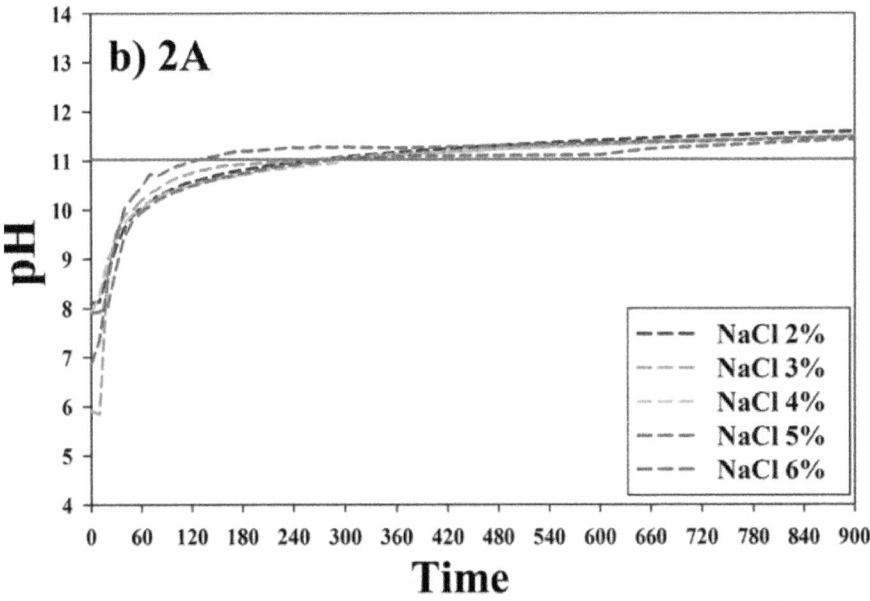

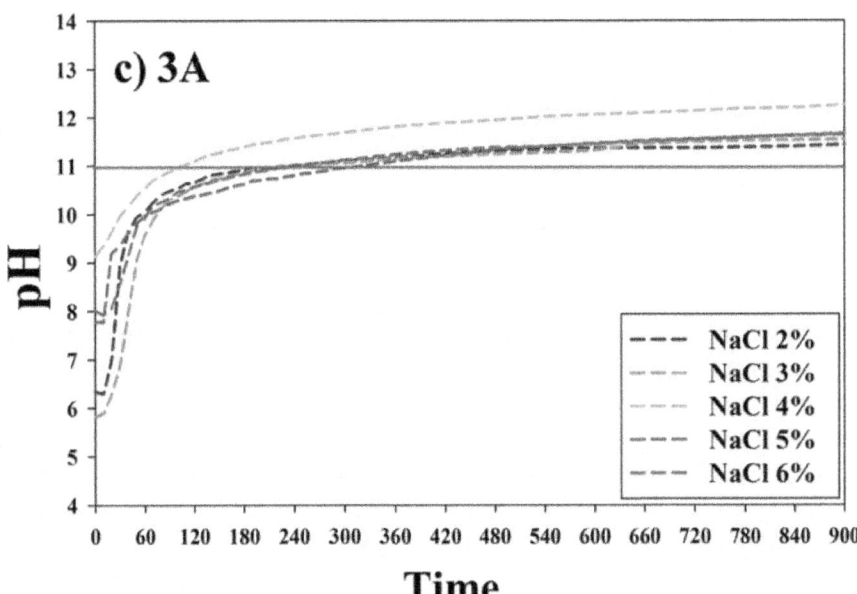

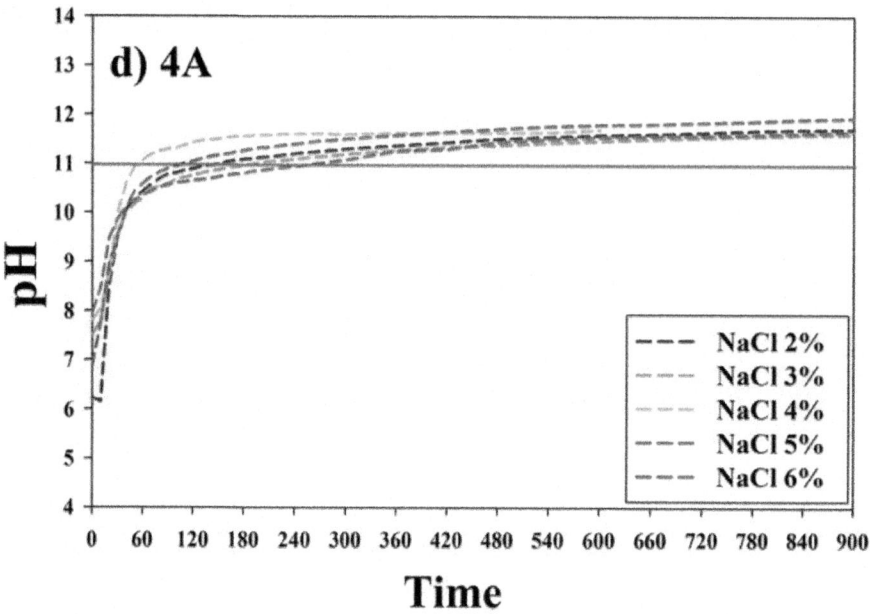

Figure 3. Result of the electrolysis of NaCl solution. (**a**) 1A condition; (**b**) 2A condition; (**c**) 3A condition; (**d**) 4A condition.

Characteristics of the Metal Carbonate: FT-IR Analysis

Fourier transform infrared spectroscopy (FT-IR) was used to confirm the molecular structure of the white-powder form precipitate obtained through the experiment. Figure 4 shows a comparison between FT-IR spectra of pristine $CaCO_3$ and those of experiment-derived $CaCO_3$ particles. Figure 4 shows that the characteristic peaks of the pristine $CaCO_3$ and those of experiment-derived $CaCO_3$ particles are consistent. Table 1 shows the FT-IR peak values of $CaCO_3$ particles, which confirms that they are consistent with characteristic peaks values of 1418 cm^{-1} (C–O function group, stretching mode in calcite), 876 cm^{-1} (C–O function group, In-plan bending), and 713 cm^{-1} (C–O function group, Out-plan bending in calcite) for $CaCO_3$ in calcite form. This is also consistent with research by Tlili *et al.* [27] and Wu *et al.* [28] on FT-IR spectra for $CaCO_3$.

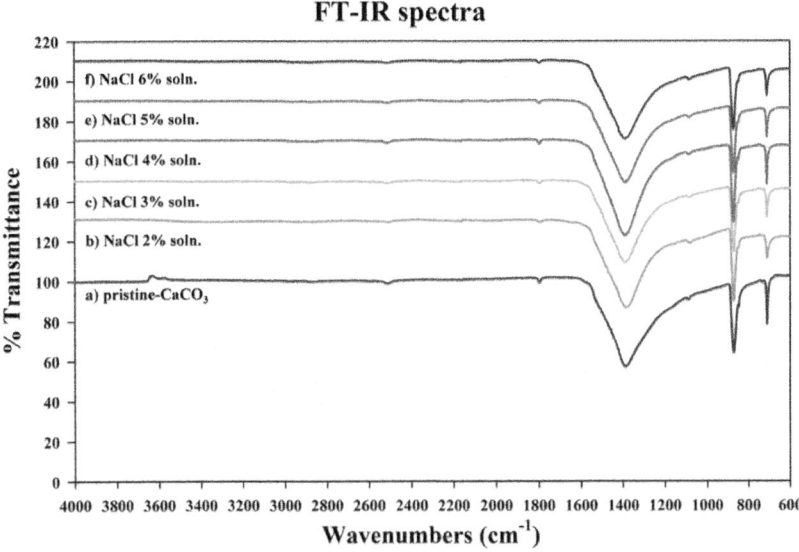

Figure 4. FT-IR spectrum of (**a**) pristine-CaCO$_3$; (**b**) NaCl 2% soln; (**c**) NaCl 3% soln; (**d**) NaCl 4% soln; (**e**) NaCl 5% soln; (**f**) NaCl 6% soln.

Table 1. Assignment of peaks for CaCO$_3$

Wavenumber (cm^{-1})	Function Groups	Vibration
2513	Combination	*1080 cm^{-1} and 1440 cm^{-1}*
1796	Combination	*1080 cm^{-1} and 713 cm^{-1}*
1418	C–O	Stretching mode (1490 cm^{-1}, 1420 cm^{-1} *in vaterite; 1418 cm^{-1} in calcite; 1465 cm^{-1} in aragonite*)
1080	O–C–O	Stretching mode
876	C–O	In-plane bending
848	C–O	In-plane bending (only appeared in vaterite and aragonite)
713	C–O	Out-plane bending (750 cm^{-1} *in* vaterite; 713 cm^{-1} *in calcite; 707 cm^{-1}, 692 cm^{-1} in aragonite*)

Characteristics of the Metal Carbonate: FE-SEM Image Analysis

In general, CaCO$_3$ is known to exist in three crystals forms—cubic (calcite), spherical (aragonite), and columnar (vaterite) [29,30,31]. Through comparative analysis with FT-IR spectra of pristine $CaCO_3$, the molecular structure of the precipitate obtained through the experiment was confirmed as being that of calcite. However, when compared with values in Table 1, it was confirmed as having the vibration of $CaCO_3$ in cubic form. Accordingly, field emission scanning electron microscopy (FE-SEM) images were taken in order to get a more clear confirmation of the crystal form. The FE-SEM images in Figure 5show that the $CaCO_3$ obtained through the experiment were observed to be of cubic form. Other crystal forms of calcite were not identified, and changes in crystal form did not vary with increases in feed solution concentrations.

Figure 5. SEM images of the sample precipitated by the carbonation. (×10,000) (**a**) Calcite-CaCO$_3$; (**b**) NaCl 2% soln; (**c**) NaCl 3% soln; (**d**) NaCl 4% soln; (**e**) NaCl 5% soln; (**f**) NaCl 6% soln.

Characteristics of the Metal Carbonate: The XRD Pattern Analysis

While FT-IR and SEM images confirmed that the precipitate calcite was $CaCO_3$, X-ray diffraction (XRD) pattern analysis was conducted on the precipitate to confirm the crystal structure. Figure 6 shows the XRD patterns of pristine $CaCO_3$ and experiment-derived samples. The XRD patterns of the selected $CaCO_3$ samples display sharp reflection lines between 20° and 70° at the characteristic 2-θ positions. A comparison with the JCPDS values in Figure 6 shows that the 2-θ values of the $CaCO_3$ obtained through the experiment are consistent with the known 2-θ values of calcite of 29.32° (d-value 3.04 Å), 39.34° (d-value 2.29 Å), and 43.08° (d-value 2.10 Å) [32]. No other 2-θ values were detected, other than those for characteristics peaks in calcite form. The XRD pattern analysis shows that the $CaCO_3$ (white-powder form precipitate) obtained through the experiment is calcite of high purity.

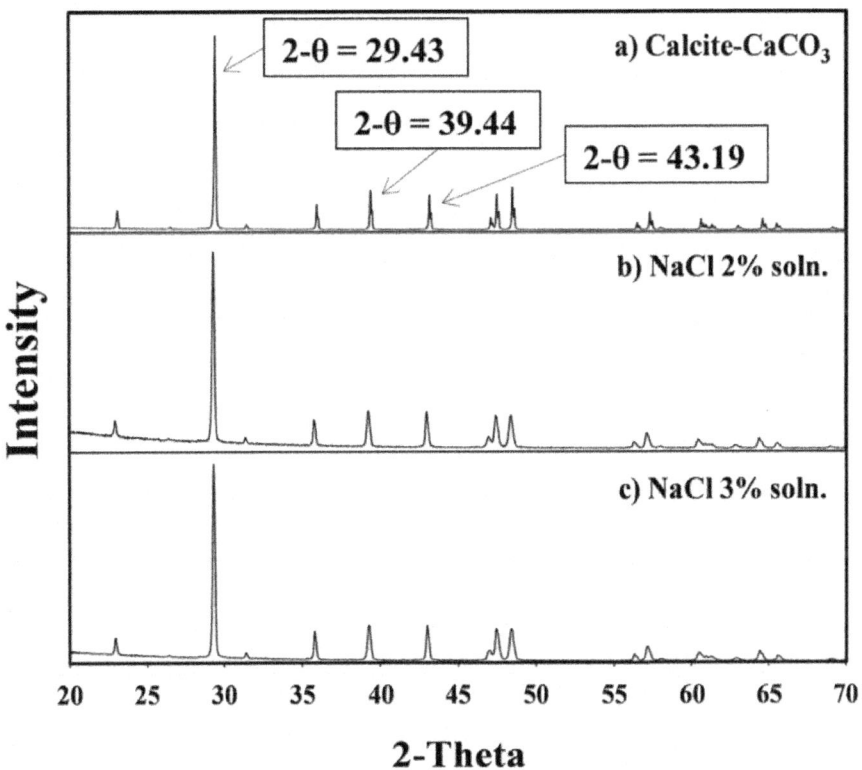

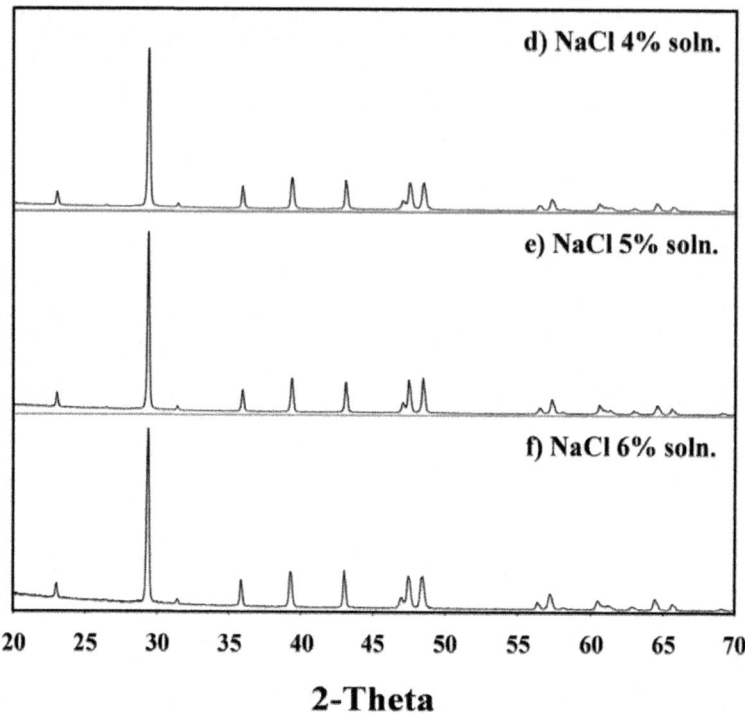

Figure 6. XRD patterns of the sample precipitated by the carbonation. (**a**) Calcite-$CaCO_3$; (**b**) NaCl 2% soln; (**c**) NaCl 3% soln; (**d**) NaCl 4% soln; (**e**) NaCl 5% soln; (**f**) NaCl 6% soln.

CONCLUSIONS

This study proved that $CaCO_3$ could be stably formed by absorbing CO_2 at normal room temperature and pressure conditions. Among absorbents used in existing CCS technology, a series of toxic amine was excluded, and a more environment-friendly method for CO_2 fixation was realized by using NaOH as absorbent. In addition, an electrolysis method which uses ceramic membrane at 25 °C, and atmospheric conditions with voltage of 8–24 V for 10–15 min, was used for the formation of NaOH. Through this method, NaOH was produced more effectively. In the case of the carbonation process, the precipitate was stably formed through a substitution reaction between Ca^{2+} and Na^+, by adding $CaCl_2$ at room temperature and atmospheric conditions. It was also confirmed as being calcite through FT-IR spectra measurements. Furthermore, the contents of the chemical solvent resulted from $CaCO_3$ precipitate formation consisted of sodium chloride (NaCl) and sodium hydroxide (NaOH), which can be reused as feed solution for electrolysis via an additional concentration process. Through SEM image and XRD pattern analyses, it was revealed that the precipitate $CaCO_3$ powder was high-purity calcite-$CaCO_3$ of cubic crystal form. $CaCO_3$ exists in calcite of cubic form, aragonite of spherical form, and vaterite of columnar form, of which cubic calcite is most stable [33]. Aragonite transforms into calcite at 380–470 °C, and vaterite is the most unstable [34,35].

Natural formation of $CaCO_3$ requires extensive time and entails numerous reciprocal interactions. The method presented not only does not require high temperature reaction conditions but also makes it possible to form $CaCO_3$ in a short time. Moreover, the $CaCO_3$ produced is calcite of high purity which does not require an additional refinement process. According to Siefert and Litster [7], the most of energy used in CCS is consumed in the desorption process. However, the presented method does not use the desorption process, so the energy required to remove CO_2 does not need to be considered. According to Svensson et al. [36], in the transport process, in the range from 1 €/ton up to 18 €/ton of CO_2, costs arise. However, the presented method does not require a transport process of CO_2, so it can reduce the cost as much as 18 €/ton of CO_2. In order to overcome disadvantages of existing methods using serpentine or other magnesium silicate at high temperatures (over 500 °C) and high pressure (over 20 bar) conditions to produce $Mg(OH)_2$, the method which uses electrolysis at room temperature and pressure is used to produce absorbent. High temperature (over 500 °C) and high pressure (over 20 bar) conditions incurs costs, so this study provides economic benefits.

As demonstrated through this experiment, this study proposes a more stable carbonation method for more environment-friendly and economical CO_2 fixation and recycling of chemical ingredients. On the basis of the aforementioned considerations, we would like to be able to reduce the CO_2 emissions through recycling the CO_2 with the method that we have proposed.

ACKNOWLEDGMENTS

This work was supported by the Human Resources Program in Energy Technology of the Korea Institute of Energy Technology Evaluation and Planning (KETEP), granted financial resource from the Ministry of Trade, Industry & Energy, Republic of Korea. (No. 20154010200810).

AUTHOR CONTRIBUTIONS

Hyun Sic Park performed the experiments, analyzed the data, and wrote first draft; Ju Sung Lee, JunYoung Han, Sangwon Park analyzed the data; Byoung Ryul Min and Jinwon Park revised the manuscript.

REFERENCES

1. Yu, K.M.K.; Curcic, I.; Gabriel, J.; Tsang, S.C.E. Recent advances in CO_2 capture and utilization. *ChemSusChem* **2008**, *1*, 893–899.

2. Khoo, H.H.; Tan, R.B. Environmental impact evaluation of conventional fossil fuel production (oil and natural gas) and enhanced resource recovery with potential CO_2 sequestration. *Energy Fuels* **2006**, *20*, 1914–1924.

3. West, T.O.; Peña, N. Determining thresholds for mandatory reporting of greenhouse gas emissions. *Environ. Sci. Technol.* **2003**, *37*, 1057–1060.

4. Zeebe, R.E.; Zachos, J.C.; Caldeira, K.; Tyrrell, T. Carbon emissions and acidification. *Sci. N. Y. Wash.* **2008**, *321*, 51.

5. Abbott, T.M.; Buchanan, G.W.; Kruus, P.; Lee, K.C. 13c nuclear magnetic resonance and Raman investigations of aqueous carbon dioxide systems. *Can. J. Chem.* **1982**, *60*, 1000–1006.

6. Inoue, R.; Ueda, S.; Wakuta, K.; Sasaki, K.; Ariyama, T. Thermodynamic consideration on the absorption properties of carbon dioxide to basic oxide. *ISIJ Int.* **2010**, *50*, 1532–1538.

7. Siefert, N.S.; Litster, S. Exergy and economic analyses of advanced IGCC-CCS and IGFC-CCS power plants. *Appl. Energy* **2013**, *107*, 315–328.

8. Baciocchi, R.; Corti, A.; Costa, G.; Lombardi, L.; Zingaretti, D. Storage of carbon dioxide captured in a pilot-scale biogas upgrading plant by accelerated carbonation of industrial residues. *Energy Procedia* **2011**, *4*, 4985–4992.

9. Gough, C. State of the art in carbon dioxide capture and storage in the UK: An experts' review. *Int.J. Greenh. Gas Control* **2008**, *2*, 155–168.

10. Holloway, S. Underground sequestration of carbon dioxide a viable greenhouse gas mitigation option. *Energy* **2005**,*30*, 2318–2333.

11. Damen, K.; Faaij, A.; Turkenburg, W. Health, safety and environmental risks of underground CO_2 storage—Overview of mechanisms and current knowledge. *Clim. Chang.* **2006**, *74*, 289–318.

12. Mazzoldi, A.; Hill, T.; Colls, J.J. CFD and gaussian atmospheric dispersion models: A comparison for leak from carbon dioxide transportation and storage facilities. *Atmos. Environ.* **2008**, *42*, 8046–8054.

13. Burgherr, P.; Hirschberg, S. *Comparative Assessment of Natural Gas Accident Risks*; Paul Scherrer Institute: Villigen PSI, Switzerland, 2005.

14. Lackner, K.S.; Butt, D.P.; Wendt, C.H. Progress on binding CO_2 in mineral substrates. *Energy Convers. Manag.* **1997**, *38*, S259–S264.

15. Stasiulaitiene, I.; Fagerlund, J.; Nduagu, E.; Denafas, G.; Zevenhoven, R. Carbonation of serpentinite rock from Lithuania and Finland. *Energy Procedia* **2011**, *4*, 2963–2970.

16. Fagerlund, J.; Nduagu, E.; Romão, I.; Zevenhoven, R. CO_2 fixation using magnesium silicate minerals part 1: Process description and performance. *Energy* **2012**, *41*, 184–191.

17. Fagerlund, J.; Nduagu, E.; Zevenhoven, R. Recent developments in the carbonation of serpentinite derived $Mg(OH)_2$ using a pressurized fluidized bed. *Energy Procedia* **2011**, *4*, 4993–5000.

18. Fagerlund, J.; Teir, S.; Nduagu, E.; Zevenhoven, R. Carbonation of magnesium silicate mineral using a pressurised gas/solid process. *Energy Procedia* **2009**, *1*, 4907–4914.

19. Gerdemann, S.J.; O'Connor, W.K.; Dahlin, D.C.; Penner, L.R.; Rush, H. Ex situ aqueous mineral carbonation. *Environ. Sci. Technol.* **2007**, *41*, 2587–2593.

20. Khoo, H.; Bu, J.; Wong, R.; Kuan, S.; Sharratt, P. Carbon capture and utilization: Preliminary life cycle CO_2, energy, and cost results of potential mineral carbonation. *Energy Procedia* **2011**, *4*, 2494–2501.

21. Dou, B.; Song, Y.; Liu, Y.; Feng, C. High temperature CO_2 capture using calcium oxide sorbent in a fixed-bed reactor. *J. Hazard. Mater.* **2010**, *183*, 759–765.

22. Song, H.-J.; Lee, S.; Park, K.; Lee, J.; Chand Spah, D.; Park, J.-W.; Filburn, T.P. Simplified estimation of regeneration energy of 30 wt.% sodium glycinate solution for carbon dioxide absorption. *Ind. Eng. Chem. Res.* **2008**, *47*, 9925–9930.

23. Nduagu, E. Mineral Carbonation: Preparation of Magnesium Hydroxide [$Mg(OH)_2$] from Serpentinite Rock. Master Thesis, The Åbo Akademi University, Åbo, Finland, 2008.

24. Gilliam, R.J.; Decker, V.; Seeker, W.R.; Boggs, B.; Jalani, N.; Albrecht, T.A.; Smith, M. Low-Voltage Alkaline Production from Brines. U.S. Patent 12/703,605, 10 February 2010.

25. Ferro, P.; Mishra, B.; Olson, D.; Averill, W. Application of ceramic membrane in molten salt electrolysis of $CaO\text{-}CaCl_2$. *Waste Manag.* **1998**, *17*, 451–461.

26. Stumm, W.; Morgan, J.J. *Aquatic Chemistry: Chemical Equilibria and Rates in Natural Waters*; John Wiley & Sons: Hoboken, NY, USA, 2012; Volume 126.

27. Tlili, M.; Amor, M.B.; Gabrielli, C.; Joiret, S.; Maurin, G.; Rousseau, P. Characterization of $CaCO_3$ hydrates by micro- raman spectroscopy. *J. Raman Spectrosc.* **2002**, *33*, 10–16.

28. Wu, G.; Wang, Y.; Zhu, S.; Wang, J. Preparation of ultrafine calcium carbonate particles with micropore dispersion method. *Powder Technol.* **2007**, *172*, 82–88.

29. Lyu, S.-G.; Sur, G.-S.; Kang, S.-H. A study of crystal shape of the precipitated calcium carbonate formed in the emulsion state. *J. Korean Ins. Chem. Eng.* **1997**, *35*, 186–191.

30. Saylor, C.H. Calcite and aragonite. *J. Phys. Chem.* **1928**, *32*, 1441–1460.

31. Kontoyannis, C.G.; Vagenas, N.V. Calcium carbonate phase analysis using XRD and FT-Raman spectroscopy. *Analyst* **2000**, *125*, 251–255.

32. McMurdie, H.F.; Morris, M.C.; Evans, E.H.; Paretzkin, B.; Wong-Ng, W.; Ettlinger, L.; Hubbard, C.R. Standard X-ray diffraction powder patterns from the JCPDS research associateship. *Powder Diffr.* **1986**, *1*, 64–77.

33. Yoshioka, S.; Kitano, Y. Transformation of aragonite to calcite through heating. *Geochem. J.* **1985**, *19*, 245–249.

34. Rao, M.S. Kinetics and mechanism of the transformation of vaterite to calcite. *Bull. Chem. Soc. Jpn.* **1973**, *46*, 1414–1417.

35. Koga, N.; Kasahara, D.; Kimura, T. Aragonite crystal growth and solid-state aragonite-calcite transformation: A physico-geometrical relationship via thermal dehydration of included water. *Cryst. Growth Des.* **2013**, *13*, 2238–2246.

36. Svensson, R.; Odenberger, M.; Johnsson, F.; Strömberg, L. Transportation systems for CO_2 application to carbon capture and storage. *Energy Convers. Manag.* **2004**, *45*, 2343–2353.

Chapter 9

A DEMONSTRATION OF CARBON-ASSISTED WATER ELECTROLYSIS

Bruce C.R. Ewan [1] and Olalekan D. Adeniyi [2]

[1]Chemical & Biological Engineering Department, University of Sheffield, Mappin Street, Sheffield S1 3JD, UK

[2]School of Engineering and Engineering Technology, Federal University of Technology, Gidan Kwano Campus PMB 65, Minna, Niger State, Nigeria

ABSTRACT

It is shown that carbon fuel cell technology can be combined with that of high temperature steam electrolysis by the incorporation of carbon fuel at the cell anode, with the resulting reduction of the required electrolysis voltage by around 1 V. The behaviour of the cell current density and applied voltage are shown to be connected with the threshold of electrolysis and the main features are compared with theoretical results from the literature. The advantage arises from the avoidance of efficiency losses associated with electricity generation using thermal cycles, as well as the natural separation of the carbon dioxide product stream for subsequent processing.

INTRODUCTION

High temperature electrolysis of water has been the subject of a number of studies in recent years [1,2,3] and the benefits derive from the positive entropy change and hence reducing free energy change with temperature. Economic analyses [4] have shown that a significant portion of the cost of hydrogen production by this route derives from the price of the electricity, and the modest electricity consumption savings at higher temperature are an attempt to reduce this element. Although sustainable electricity generation technologies are increasing (e.g., wind, solar), it has been noted [5] that the major proportion of electricity is still generated from fossil fuel sources, and electricity generation costs from these are inevitably associated with efficiencies of around 40%, which are typical of thermal generation cycles.

In parallel with developments in high temperature electrolyser technology, there have been significant improvements and developments in fuel cell technologies, and high temperature electrolysis has benefited from these through the use of high temperature solid oxide materials such as zirconia [2]. The fuels used for fuel cells cover a range including hydrogen, simple hydrocarbons, alcohols and carbons arising from a range of sources. Given the similarity in the high temperature technologies associated with fuel cells and electrolysis cells, it is natural to consider a combination of these devices into a single unit in which the electricity generated from the fuel cell, rather than the power station, is used directly in the electrolysis process. The work of Pham *et al.* [5] and Martinez-Frias *et al.* [6] have demonstrated this principle through the use of natural gas fed to the anode (fuel electrode) of a solid oxide electrolysis cell operating at 900 °C, where a reduction of around 1 V was achieved in the required electrolysis voltage.

For the reaction:

$$CH_4(g) + 2O_2(g) = CO_2(g) + 2H_2O(g) \qquad\qquad \Delta G° = -800 \text{ kJ/mol}$$
(1)

this is consistent with a fuel cell EMF, $E° = -\Delta G°/nF = 1.04$ V (n = number of electrons, F = Faraday constant, 96,487 C/mol).

The advantage in such an approach to electrolysis is seen to be the avoidance of the efficiency losses associated with burning the fuel in a thermal cycle for electricity production, but also through the natural separation of CO_2 product gas from other non-condensable gases in the air supply, which greatly reduces carbon capture costs.

The direct carbon fuel cell (DCFC) is a special kind of high temperature fuel cell that directly uses carbon as fuel supplied to the anode (fuel electrode) and has the potential to reduce the complexities of reforming hydrocarbon raw materials to simpler fuels such as hydrogen. The DCFC has been shown to offer significantly higher thermal efficiencies [7] for electrical power generation compared to combustion routes and even for other fuel cell types using different fuels. The raw materials for powering a DCFC are solid, carbon-rich fuels, and much of the effort in recent years has been devoted to fossil fuel carbon sources, such as coal and petroleum coke.

For the solid oxide fuel cell, since transport is taken to be that of the O^{2-} ion, the cell reactions are:

cathode $O_2 + 4e^- \rightarrow 2O^{2-}$ (2)

anode $C + 2O^{2-} \rightarrow CO_2 + 4e^-$ (3)

A further development in this approach of using a carbon fuel to aid the water electrolysis process was examined theoretically by Gopalan et al. [8] by the use of coal fuel particles dispersed in a molten silver anode connected to a solid oxide electrolyte. The analysis was aimed at efficiency evaluations under a range of cell operating conditions and including any waste heat recovery. Importantly, the contribution of the carbon oxidation free energy was included, based on the cell operating temperature of 1300 K and the oxidation to carbon monoxide:

$$C + 0.5O_2(g) = CO(g) \qquad\qquad \Delta G° = -226.5 \text{ kJ/mol} \qquad (4)$$

In this case the predicted reduction in the applied cell voltage becomes 1.17 V.

The relationship between current density i, the required cell voltage E_c, and the area specific resistances (ASR) for the cell (R_i), and due to concentration polarisation (R_p), was given by:

$$-i = \frac{E_c}{(R_i + R_p)} = \frac{(E_{app} + E_{Nernst})}{(R_i + R_p)} \qquad (5)$$

where $E_c = E_{app} + E_{Nernst}$ and E_{Nernst} is the standard cell potential (E°) for the overall reaction, plus a voltage contribution due to the non-standard partial pressures of reactants and products.

For an electrolysis process based on the anode Equation (4), the overall reaction is:

$$C + H_2O(g) = CO(g) + H_2(g)$$

and at 1300 K, $\Delta G° = -50.5$ kJ/mol and E° = 0.262 V. This implies that the overall process should proceed spontaneously, although there are energy losses due to the current resistances in the circuit and a nett applied voltage (E_{app}) is required to drive the process. The authors made estimates of expected efficiencies and E_{app} values based on typical values for R_i and R_p. These are discussed further below. The possible use of carbon to reduce the electrolysis voltage has a particular value due to the wide range of sources available, allowing advantage to be taken of the lowest cost sources. In general, this approach provides an additional technology option, either for the generation of hydrogen for use in the chemical industry, avoiding the use of premium hydrocarbon fuels, or for the gasification of solid carbons for use in electricity generation using gas turbines.

Gasification processes are of renewed interest for the processing of solid fuels due to the benefits of using gas turbine (GT) technology combined with steam cycles, the so-called combined cycle routes. Such combined cycle processes are capable of achieving 50%+ in energy efficiencies, and gasification of the solid fuel is required in order to take advantage of GT technology.

The present work demonstrates the initial experimental observations in using carbon to assist the electrolysis process and compares these with some of the predictions from the above work.

EXPERIMENTAL SECTION

Results from three separate configurations are presented below, based around a solid oxide supported electrolyte cell geometry and consisting of a direct carbon fuel cell, an electrolysis cell and a combination using a carbon assisted electrolysis cell.

Solid Oxide Cell Construction and Operation

The solid oxide cell used in the experiments consisted of a button cell of stabilized zirconia, where the overall cell containment used is shown in Figure 1. The electrode assembly shown in Figure 2 was held between two open alumina tubes of 24 mm internal diameter, 3 mm wall, and oriented vertically. The end of each tube was held within a closed metal chamber of stainless steel, which included built-in flanges and allowed steel, spring loaded tie bars to be fixed between each end of the system to hold the electrode assemblies in place. The overall length of the tubular assembly was 280 mm.

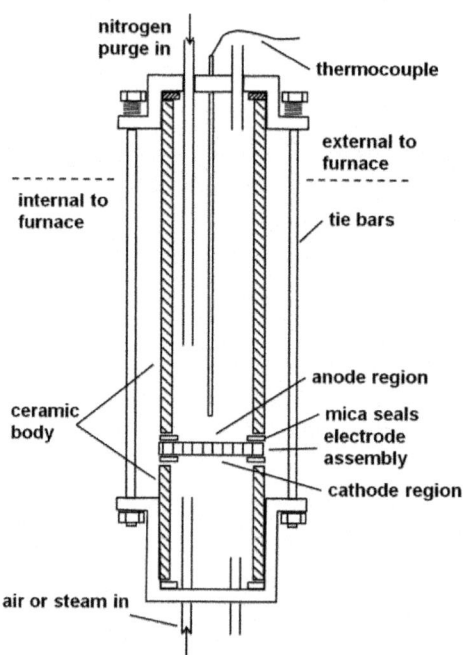

Figure 1. Cell geometry used for insertion into furnace.

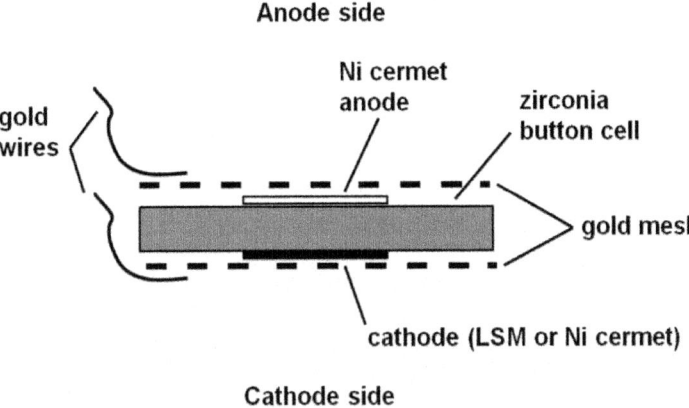

Figure 2. Electrode assembly used in fuel cell and electrolysis cell.

The lower 80% of the tubular system could then be located within a furnace and the cell operating temperature was monitored locally using a sheathed K-type thermocouple which entered through the upper metal chamber.

Inlet and outlet tubes were provided through the upper and lower metal chambers to allow some purging on the anode side and the required reactant gas on the cathode side (steam side).

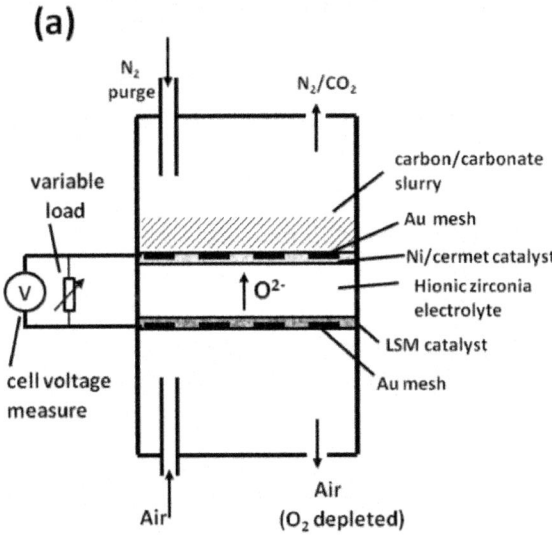

anode reaction: $C + 2O^{2-} \rightarrow CO_2 + 4e^-$

cathode reaction: $O_2 + 4e^- \rightarrow 2O^{2-}$

ΔG overall at 800 °C = −396 kJ/mol

(b)

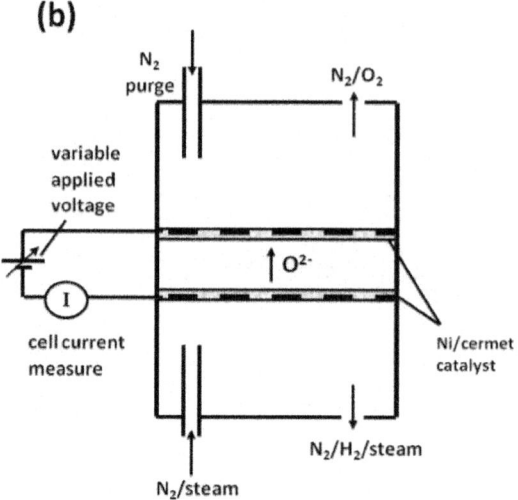

anode reaction: $2O^{2-} \rightarrow O_2 + 4e^-$

cathode reaction: $2H_2O + 4e^- \rightarrow 2O^{2-} + 2H_2$

ΔG overall at 800 °C = 188.7 kJ/mol H_2O

(c)

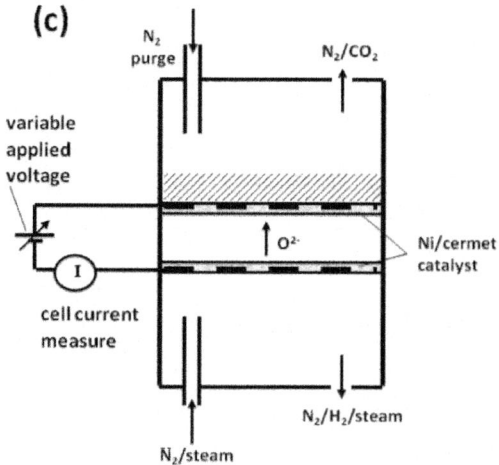

anode reaction: $C + 2O^{2-} \rightarrow CO_2 + 4e^-$

cathode reaction: $2H_2O + 4e^- \rightarrow 2O^{2-} + 2H_2$

ΔG overall at 800 °C = −9.3 kJ/mol H_2O

Figure 3. Overall schematic of the operation of the three cell geometries used. Each cell body is immersed in a furnace environment. (**a**) carbon fuel cell; (**b**) electrolysis cell (EC); (**c**) carbon assisted a electrolysis cell (CAEC).

The central button cell element was of 25 mm diameter (supplied by Fuel Cell Materials, Lewis Center, OH, USA) and consisted of a Hionic electrolyte (stabilized zirconia, 150 μm in thickness) with a central 12.5 mm diameter active layer (50 μm thickness) on each side. The choice of these active layers depended on the process being investigated as described below.

An overall schematic of the operation of the three cell geometries used is shown in Figure 3a–c. This identifies the inlet and outlet gases, location of fuel and electrolyte/catalyst combinations employed.

Carbon Fuel and Preparation

Based on experience in working with biomass materials in carbon fuel cells, the carbon chosen for the demonstration was the char from pyrolysed miscanthus (*M. giganteus*). The biomass samples of particle size of 0.5 to 1.0 mm, were dried at 100 °C before pyrolysing in a cylindrical, electrically heated furnace at a heating rate of 7 °C/min and up to the operating temperature of 800 °C, where samples were held for 30 min. [9]. Nitrogen gas was used to purge the system during and after the pyrolysis process until it cooled down to 200 °C. The proportion of char produced in each case was in the range 22–24 wt %. Although these chars show some graphitic structure there is also a high degree of amorphous carbon, similar to that observed in many coal samples.

The biomass char was ground by ball milling and size analysis on the samples using laser diffraction sizing showed that 50% of the weight fell within the particle size range 2.2–8.1 μm. At the operating temperature of the fuel cell, the fuel is presented as a slurry of 15 wt %, dispersed in a molten carbonate mixture of Li_2CO_3/K_2CO_3 mixed in the ratio of 46.6 wt % Li_2CO_3 and 53.4 wt % K_2CO_3, giving a melting point close to 500 °C [10].

Carbon Fuel Cell Operation

In common with several other reported works using SOFCs, the anode side of the cell was chosen for its catalytic behavior toward oxidation and consisted of a 50 μm bilayer of Ni-yttria stabilized zirconia/Ni-gadolinium doped ceria (Ni-YSZ/Ni-GDC).

The cathode layer consisted of a 50 μm bilayer of lanthanum strontium manganite/lanthanum strontium manganite-gadolinium doped ceria (LSM/LSM-GDC) which was suited to the oxygen reduction step. Cells with these layers already attached are provided by the manufacturer.

As supplied, the Ni on the anode side is in the oxidized form and this was subjected to a reducing atmosphere consisting of 5 vol% hydrogen in nitrogen for 1 h at 900 °C prior to use. Any subsequent processing steps which required elevated temperatures ensured that the same protective reducing atmosphere was provided. This cell element was further processed with the addition of gold mesh current collectors on anode and cathode sides, which were attached using silver ink to the edges and a further heat cycle at 900 °C for 20 min. The silver ink consists of a suspension of fine silver particles in a terpene oil and is available from Fuel Cell Materials. In order to ensure good continuity between the electrode surfaces and their respective gold meshes, nickel ink was then applied sparingly to the anode mesh surface, with any excess removed and LSM ink to the cathode mesh in the same way. The electrode assembly was then dried ready for use. Gold wires (0.4 mm diameter) lead the current from the cells. These wires had flattened ends and were held in contact with the gold mesh elements by the compressive force of the external springs.

The 15 wt % of the carbon fuel in carbonate mixture was supplied to the anode side of the cell at a fixed mass of 3.0 g for all experiments. Air was supplied to the cathode side at a rate of 2 L/min (room temperature and atmospheric pressure) and a light flow of nitrogen gas was connected to the inlet at the anode (to purge the CO_2 produced from the system).

Electrolysis Cell Operation

The anode oxidation process taking place in conventional hydrogen fuelled cells to produce water vapour is reversible and therefore, for the electrolysis process involving proton reduction, the same nickel/zirconia cermet electrode material is an appropriate choice for the active layer of the cathode side of the cell (steam side). For the anode side evolving oxygen, the active layer was LSM, being the reverse reaction involved in oxygen gas transformation during fuel cell operation.

Steam was supplied to the cathode side of the cell by means of nitrogen carrier gas at a concentration of 30 vol% of steam in a nitrogen flow rate of 2 L/min. Trace heating of supply lines was provided to avoid steam condensation.

Carbon Assisted Electrolysis Cell Operation

The operation of the carbon assisted electrolysis cell was achieved by a combination of the two separate processes, electrolysis and carbon fuel cell. Since the electrode processes involve hydrogen production on the cathode side and carbon oxidation on the anode side, both sides of the zirconia electrolyte are coated with a 50 μm layer of Ni/zirconia cermet. The materials supplied were identical to those used in the separate cell processes, *i.e.*, 30 vol% of steam in a nitrogen flow rate of 2 L/min supplied to the cathode and 3.0 g of char fuel mixture to the anode.

RESULTS AND DISCUSSION

The performance of the carbon fuel cell (DCFC) at 800 °C is shown in Figure 4 in terms of cell voltage developed and power density, where the external load was varied from open circuit to 1.2 Ω. The current-voltage curve shows behaviour which is consistent with that of most direct-carbon fuel cells, *i.e.*, a voltage close to the predicted open circuit value at zero current, falling approximately linearly with increase in current density due to ohmic losses and followed by a more rapid fall at high current densities due to mass transfer polarisation at one or both of the electrodes. The open circuit voltage for this system is found to be in the range 1.15–1.2 V, which is consistent with other recently reported values for carbons, e.g., 1.2 V with petcoke [11] and coal [12] as carbonate/fuel slurries. The current density range shown is typical for this cell geometry and the results with several fabricated cells have shown closely similar behavior.

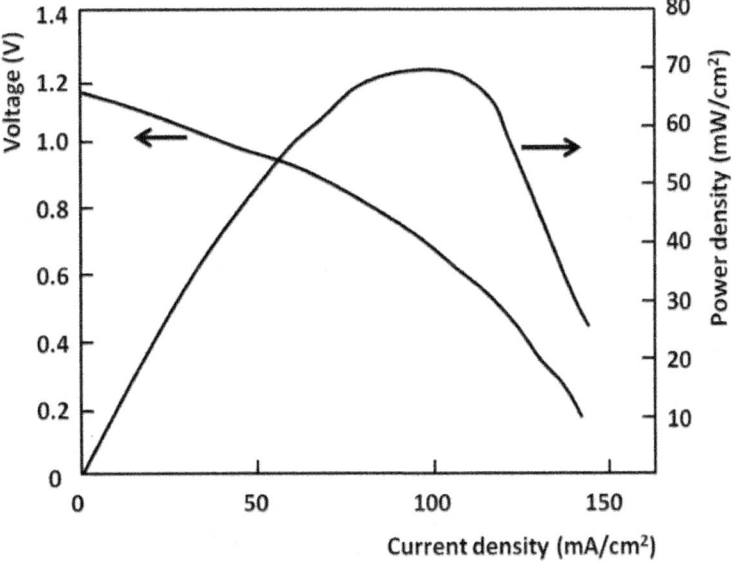

Figure 4. Fuel cell voltage and power *vs.* current density for cell at 800 °C using miscanthus char.

Figure 5 shows the combined results for the electrolysis cell (EC) and the carbon assisted electrolysis cell (CAEC). The EC results are shown at three temperatures of operation 700, 750 and 800 °C (curves A, B and C) and these show the variation of applied cell voltage and measured current density.

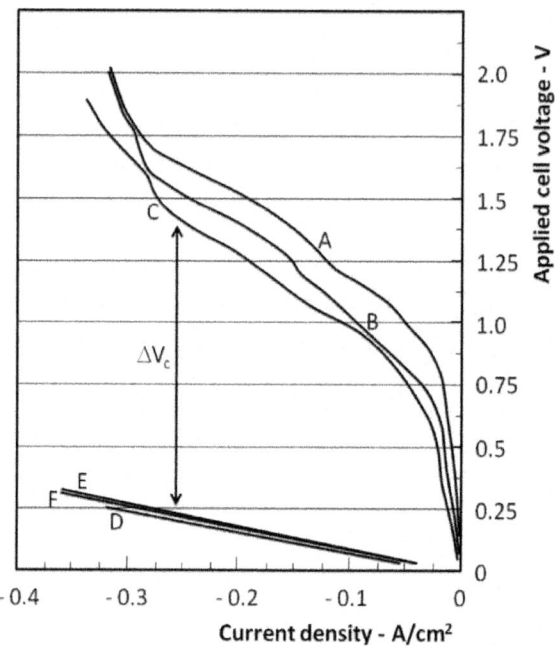

Figure 5. Current density *vs.* applied voltage during electrolysis. A, B and C correspond to conventional electrolysis at 700, 750 and 800 °C respectively. D, E and F involve electrolysis with carbon at the anode, at 700, 750 and 800 °C respectively.

At voltages up to around 0.7 V, the current density is low and increases linearly due to zirconia electrical conductivity, while in the region of the 0.7–0.9 V threshold, the gradient of the curves increases indicating the onset of electrolysis. This is consistent with the expected electrolysis threshold behaviour at elevated temperature and is similar to the behaviour reported by others using the same SOFC materials [13]. At higher current densities, there is evidence of non-linear behaviour, which is attributed to mass transfer effects.

The operation of the CAEC configuration is represented by the curves D, E and F. Curve D was obtained at 700 °C by gradually increasing the applied voltage from 0 V. In some cases during this start up process, the current/voltage relationship would show some unstable behaviour in the region of 0–0.5 V applied to the cell. When the process stabilised, within a 1 min timescale, the current-voltage behaviour would then follow the curve D shown. It was then found that any further changes in voltage would result in current densities which fell on the curve D.

This behaviour is attributed to the onset of electrolysis and the resultant feedback contribution of the carbon oxidation to the driving force of the cell as oxide ion begins to arrive at the anode of the cell. The contribution of the carbon to the electrolysis cell voltage is indicated in Figure 5 as ΔV_c. This is in the range 0.8–1.2 V and is consistent with the expected fuel cell voltage contribution associated with the carbon fuel cell. It is noted however, that this ΔV_c value is larger than might be expected from the carbon fuel cell at the CAEC current densities used. For example, at $0.1 A/cm^2$, the fuel cell voltage is around 0.7 V and is expected to be lower in the $0.2–0.3 A/cm^2$ range seen in Figure 5. An important difference with the CAEC system is the removal of the oxygen transformation step at the cathode of the cell ($O_2 + 4e^- \rightarrow 2O^{2-}$), where it is replaced by a water reduction step ($H_2O + 2e^- \rightarrow H_2 + O^{2-}$). The oxygen transformation step is known to contribute to significant overvoltage losses (0.4–0.6V), depending on catalyst surface used, whereas the water or H^+ reduction process carries a much lower voltage loss (less than 0.1 V). This may account for the fact that much of the open circuit voltage of the carbon fuel cell is available to transfer to the reduction in the electrolysis voltage at the higher current densities used. This may also provide an explanation for the negligible effect of temperature on the curves D–F, which suggest low activation energy processes are involved at both electrodes.

Curves E and F correspond to operation of the CAEC at 750 °C and 800 °C respectively and were obtained by reducing the applied cell voltage and corresponding current to $-35 mA/cm^2$ with the system at 700 °C, and then raising the furnace temperature to the new target value. Raising the applied voltage levels at the target temperature resulted in the strongly linear behaviour shown by curves E and F, which were closely similar to that of curve D obtained following start up. This reduction of the required electrolysis voltage in the presence of carbon is similar to the reported reduction by 1 V using methane as the reducing fuel [6], as depicted in Equation (1).

The behaviour of the EC and CAEC systems can be compared with the predictions of the work of reference [8]. The cell processes in the present work are carried out between 973 and 1073 K compared to 1300 K for the work of reference [8], and the overall standard free energy changes in the lower temperature range are derived from:

$$C + H_2O(g) = CO(g) + H_2(g) \qquad \Delta G° = -7.5 \text{ kJ/mol} \qquad (6)$$

$$C + 2H_2O(g) = CO_2(g) + 2H_2(g) \qquad \Delta G° = -10.4 \text{ kJ/mol} \qquad (7)$$

Based on the input steam pressure on the cathode side, and making an estimate for the average values of steam (0.25 atm), hydrogen produced (0.075 atm) and CO or CO_2 (1 atm) on the anode side, an estimate can be made for the E_{Nernst} values for Equation (6) (0.08 V) and Equation (7) (0.05 V) for use with Equation (5).

For the EC operation shown in Figure 5, based on the approximately linear region of the curves, the average value of the ASR for the curves A, B and C is 2.5 Ωcm^2, measured between the voltage values of 1.5 and 1.0 and the corresponding current values of -0.25 A and -0.05 A, whereas for the CAEC it is close to 0.83 Ωcm^2.

The introduction of carbon on the anode side appears not only to reduce the required electrolysis voltage by around 1 V as expected, but also to significantly reduce the ASR value of the cell, since the cathode side of the cell is the same in both cases. It is noted that the measured ASR values are those for the overall cell and that differences in the electrode processes may have an effect on the resulting values. For the EC, the electrodes employ a Ni/cermet on the hydrogen side and LSM on the oxygen release side, whereas for the CAEC, Ni/cermet is used on both sides. Other studies [13,14] have reported similar ASR values as the present work for the overall cells and have also shown that high temperature electrolysis cells may show significantly different ASR values depending on whether they operate in electrolysis mode or fuel cell mode. The optimisation of the catalyst surfaces for the specific duties required will remain an active area of development for these systems.

Referring to Figure 5 it can be seen that the curves D, E and F, when extrapolated, intercept the voltage axis on the negative side. This corresponds to the open circuit condition ($-i = 0$), where $E_{app} = -E_{Nernst}$ and is consistent with the small and negative $\Delta G°$ values for Equations (6) and (7) and their corresponding $-E°$ values at the temperature of operation. Due to their similar values it is not clear from the curve positions which of these dominates the anode reaction and will be the subject for further study.

CARBON ENERGY REQUIREMENTS

The impact of the use of a CAEC system is likely to arise through the energy savings of the primary fossil fuel source and therefore a comparison of the conventional high temperature EC and CAEC systems from this point of view is a useful one to make. The electrolysis systems cannot operate at the minimum thermodynamic voltage since this corresponds to almost zero current and the practical operation voltages can be identified from the many works published in the area of solid oxide electrolysis cells. Examples of such reported work [13,14] indicate that, at 800 °C, typical current densities are 0.2 A/cm^2 and operational voltages are 1.3 V. Based on the present work, the direct contribution of the carbon in the cell means that the corresponding operational voltage can be set at 0.4 V. These values are sufficient to enable a comparison of the carbon energy (input to the power station and the CAEC) required to achieve a fixed volume of hydrogen, and Table 1 collects together the key energy parameters for the production of 1 m³ of hydrogen at standard conditions.

Table 1. Comparison of the primary carbon fuel required to produce 1 m³ of hydrogen at standard conditions for the EC and CAEC systems operating at 800 °C. Carbon fuel energy is taken as 33 MJ/kg

Operating parameters	EC system	CAEC system	Units
Moles of H_2 /m³	44.62		Moles
Operating current	0.20		A
Applied voltage	1.30	0.40	V
Direct electrical power	0.26	0.08	W
Electrical energy/m³ H_2	3.11	0.94	kWh
Source carbon energy at 40% electrical production efficiency	7.77	2.39	kWh
Additional cell carbon	-	0.27	kg
Cell carbon energy/m³ H_2	-	2.45	kWh
Total carbon energy used/m³ H_2	**7.77**	**4.84**	kWh

It can be seen from the table that the CAEC system uses only 62% of the primary carbon fuel compared to the conventional EC system, which will have cost and CO_2 emissions implications.

Reference has been made to the analysis provided by Manage *et al.* [4], which provides a comprehensive review of the relationship between solid oxide electrolyser operating parameters and the cost of hydrogen production by steam electrolysis. A key conclusion from this work is the importance of the cost contributions to hydrogen of natural gas in the case of steam reforming (SMR) and electricity in the case of electrolysis. For SMR, the cost of hydrogen is US$2.50/kg which is equivalent to US¢7.5/kWh, whereas the lowest cost of electricity for electrolysis is around US¢4/kWh.

When high temperature electrolysis is used, the energy cost savings associated with the use of a CAEC versus an EC system can be calculated, based on the data from reference [4]. For the EC system the electrical energy cost/m³ of hydrogen is US¢12.44/m³ (3.11 × US¢4). Based on coal costs of US$65/tonne for low sulphur coal (30 MJ/kg, 8.33kWh/kg), then for the CAEC system, the electricity and carbon energy costs are calculated as US¢5.52/m³ (0.94 × US¢4 + 0.27 × US¢6.5). This cost saving reflects both the reduced amount of electricity used and the lower cost of the substitute carbon fuel used directly in the cell.

CONCLUSIONS

Carbon-assisted electrolysis has an equivalence to an electrochemical gasification process and has some advantages over conventional thermal gasification processes, in particular the automatic separation of products, which is particularly useful given the continued interest in CCS technologies, in which existing CO_2 separation stages carry a severe energy penalty.

The demonstration has shown that existing solid oxide technologies are very able to handle carbon char within an EC configuration, providing the full thermodynamic advantage associated with the chemical energy of the carbon, and avoiding Carnot type losses associated with thermal routes to electricity generation.

From other reported works on carbon fuel cells, a wide range of carbons also become potential sources for gasification by this route, enabling a wider carbon 'catchment area' for the conversion of solid to gaseous fuels for use in gas turbines and combined cycles, where hydrogen is increasingly being used in combination with other gaseous fuels. Some tailoring of the products may also be possible based on the temperature of operation, since, at lower temperatures (below 1000 K) the H_2/CO_2 couple is favored while at higher temperatures the H_2/CO pair is favored, offering a route to syngas. Although the large scale solid oxide technology is not yet available to take advantage of this cost saving, it is clear that further work in this area has potential benefits for the hydrogen and syngas markets.

REFERENCES

1.   Brisse, A.; Schefold, J.; Zahid, M. High temperature water electrolysis in solid oxide cells. *Int. J. Hydrogen Energy* **2008**,*33*, 5375–5382.

2.   Ni, M.; Leung, M.K.H.; Leung, D.Y.C. Technological development of hydrogen production by solid oxide elecrolyzer cell (SOEC). *Int. J. Hydrogen Energy* **2008**, *33*, 2337–2354.

3.   Mingyi, L.; Bo, Y.; Jingming, X.; Jing, C. Thermodynamic analysis of the efficiency of high-temperature steam electrolysis for hydrogen production. *J. Power Sources* **2008**, *177*, 493–499.

4.   Manage, M.N.; Hodgson, D.; Milligan, N.; Simons, S.J.R.; Brett, D.J.L. A techno-economic appraisal of hydrogen generation and the case for solid oxide electrolyser cells. *Int. J. Hydrogen Energy* **2011**, *36*, 5782–5796.

5.   Pham, A.Q.; Wallman, H.; Glass, R.S. Natural Gas-Assisted Steam Electrolyzer. US Patent No. 6051125, April 2000.

6.   Martinez-Frias, J.; Pham, A.-Q.; Aceves, S.M. A natural gas-assisted steam electrolyzer for high-efficiency production of hydrogen. *Int. J. Hydrogen Energy* **2003**, *28*, 483–490.

7.   Cao, D.; Sun, Y.; Wang, G. Direct carbon fuel cell: Fundamentals and recent developments. *J. Power Sources* **2007**, *167*, 250–257.

8.   Gopalan, S.; Ye, G.; Pal, U.B. Regenerative, coal-based solid oxide fuel cell-electrolysers. *J. Power Sources* **2006**, *162*, 74–80.

9.   Sensöz, S. Slow pyrolysis of wood barks from *Pinus brutia* Ten. and product composition. *Bioresour. Technol.* **2003**, *89*, 307–311. [PubMed]

10. Cherepy, N.J.; Krueger, R.; Fiet, K.J.; Jankowski, A.F.; Cooper, J.F. Direct conversion of carbon fuels in a molten carbonate fuel cell. *J. Electrochem. Soc.* **2005**, *152*, A80–A87.

11. Kouchachvili, L.; Ikura, M. Performance of a direct carbon fuel cell. *Int. J. Hydrogen Energy* **2011**, *36*, 10263–10268.

12. Li, X.; Zhu, Z.; de Marco, R.; Bradley, J.; Dicks, A. Evaluation of raw coals as fuel for direct carbon fuel cells. *J. Power Sources* **2010**, *195*, 4051–4058.

13. Yu, B.; Zhang, W.; Xu, J.; Chen, J. Status and research of highly efficient hydrogen production through high temperature steam electrolysis at INET. *Int. J. Hydrogen Energy* **2010**, *35*, 2829–2835.

14. Stoots, C.M.; O'Brien, J.E.; Condie, K.G.; Hartvigsen, J.J. High-temperature electrolysis for large-scale hydrogen production from nuclear energy—Experimental investigations. *Int. J. Hydrogen Energy* **2010**, *35*, 4861–4870.

Chapter 10

CATHODIC USING OF ZRB2-αSIC AND TIB2-αSIC FOR PEM ELECTROLYSIS AND WATER ELECTROLYSIS AT LOW TEMPERATURE

Kafoumba Bamba, Nahossé Ziao

Laboratoire de Thermodynamique et de PHysico-Chimie du Milieu, UFR-SFA, Université Nangui Abrogoua, Abidjan, Côte d'Ivoire

ABSTRACT

39 mol% SiC of ceramic pellets ZrB_2-αSiC and TiB_2-αSiC were synthesized by the reactive hot pressure RHP process at 1850°C under 40 Mpa in vacuum. The XR diffraction displays the absence of other reagents apart from ZrB_2, SiC and TiB_2 confirming the purity of the pellets. The cathodic exploitation of both of them through electrochemical study shows that TiB_2-αSiC is the most active for Hydrogen Evolution Reaction (HER) and Hydrogen Oxidation Reaction (HOR) in 0.5 M of H_2SO_4 solution at room temperature. Moreover, the kinetic exploitation shows that for both pellets the system is controlled by mass transport when they are used as HER. However, in the case of HOR, the system is controlled by the electron transfer.

INTRODUCTION

Proton Exchange Membrane Fuel Cell PEMFC remains one of the competitive methods to produce a renewable energy up today. The combustible hydrogen provides a protected environmental clean energy; it is abundant and lightweight. The product from its oxidation that is water is environmentally benign [1] . The chemical energy per mass of hydrogen (142 MJ/Kg) is three times that of other chemical fuels like hydrocarbons (47 MJ/Kg). So far, its best production comes from the use of water as electrolyte according to the absence of monoxide carbon which is liable to depreciate the performance of the PEMFC [2] [3] . Therefore, the cleanest way for its production is to use sunlight in combination with photovoltaic cells and water electrolysis [4] .

Considering these advantages, setting PEMFC requires however several attentions. We have the hydrogen storage, the efficiency of both electrodes, the tolerability of electrolyte toward electrodes and the efficiency of ion exchange membrane used. At the cathode of the cell, and because of their high activity for the hydrogen evolution reaction (HER), platinum/palladium black and carbon supported platinum or palladium nano-particles are most of the time used [5] -[7] . However, considering that the platinum remains the most expensive catalyst, doing without that precious material in the processes and replacing it by an active and cheap efficient electrode are a great desire to reduce the cost associated with the production of energy by electrochemical methods. Since then, many types of materials have also been used to reduce proton in hydrogen or to oxidize hydrogen. As the reaction is made in acidic area, those materials are confronted with corrosion [8] . Recently, Lonné et al. [9] experienced ceramic electrode based on ZrB_2-αSiC and coated with a proton conducting SiO_2-rich glass layer; they found a tremendous result with resistance and lack of corrosion in H_2SO_4 electrolyte but its polarization behavior toward reduction of hydrogen displays a high over-potential at about -0.5 V. Before, Monticelli et al. studied the corrosion asset of the naked ZrB_2-αSiC in aqueous solution. They found it a better candidate as a cathode electrode for hydrogen reduction and they emphasized on the anticorrosion asset of the ceramic that depends on the amount of SiC [10] . In this study, we test for the first time both ceramic components (61 mol% ZrB_2-αSiC and 61 mol% TiB_2-αSiC) as cathode rotating disk electrode RED for hydrogen evolution reaction HER and hydrogen oxidation reaction HOR in PEMFC.

Conductivity characteristic of ZrB_2-αSiC has been developed in previous articles emphasizing its resistance in corrosion even at high temperature (2000°C), its high conductivity depending on the amount of α-SiC in the sample [11] -[14] . Whereas TiB_2-αSiC was described in other articles as exhibiting hardness, elastic modulus, high melting point and good corrosion resistance to chemical and oxidation resistance at temperature up to 1100°C [15] -[17] . Therefore, the electrical conductivity of TiB_2-αSiC increases if the percentage of TiB_2 is more than that of α-SiC in the sample and decreases when temperature increases [18] . This value that is about 6.1 to $6.5 \times 10^5 S \cdot m^{-1}$ for a temperature between 25°C and 927°C is higher than that of ZrB_2-αSiC (2 $S \cdot m^{-1}$ at 80°C).

EXPERIMENTAL

Sintering of ZrB$_2$-αSiC and TiB$_2$-αSiC

The starting reagents were TiB$_2$, ZrB$_2$ and α-SiC with respectively an average size of 1.2 - 2 μm, 4.7 - 5.3 μm and 0.7 - 1.0 μm. They were used as received with high purity about 98%+. Both ZrB$_2$-αSiC and TiB$_2$-αSiC were prepared according to the same protocol thanks to the route described explicitly in ref.9. Nevertheless, the sintering process has been slightly modified to well improve the density of the pellets. In fact, the sintering was made following a program.

First, during 1 h, the sample is heated with a slope of 15°C/min from home temperature until 1000°C. The High pressure 40 MPa equivalent of 1282 kg is then applied on the sample. The temperature dwelling at 1000°C for 15 min, the heating is then pursued for 1 h always with the slope of 15°C/min until 1850°C. Then, this temperature is maintained during 2 hours whence the high pressure is stopped and the cooling starts with the slope of 30°C/min. This final operation last out 1 h. Figure 1 displays the detail program.

After cooling the samples, they are cleaned out of BN that was coated on the graphite die and weighted. Density of 98.7% for ZrB$_2$-αSiC and 99% for TiB$_2$-αSiC were obtained.

X-Ray Diffraction

After polishing the pellets, a D5000 diffractometer equipped with a Cu as the anticathode and a back mono- chromator was used for characterization. The wavelength used was λ = 1.5406 Å. The diffraction data were collected at a constant rate of 0.02° min^{-1} over an angle range of 2θ = 10° - 60°.

Electrochemical Measurements

Cyclic voltammetric and polarization were carried out at room temperature in a standard three-electrode cell over a model 362 scanning potentiostat. The scanning was controlled by the software "potentiostat". The solution was 0.5 M H$_2$SO$_4$ aqueous solution (Aldrich and Millipore MiliQ+ water). It was deoxygenated by bubbling pure nitrogen or argon to chase all trace of oxygen molecule. The reference electrode was RHE and the counter electrode was glassy carbon. The working electrode was the ceramic ZrB$_2$-αSiC and TiB$_2$-αSiC with geometry area of 18.46 mm^2, mounted on a rotating disc electrode RDE (CTV 101) provided by radiometer analytical.

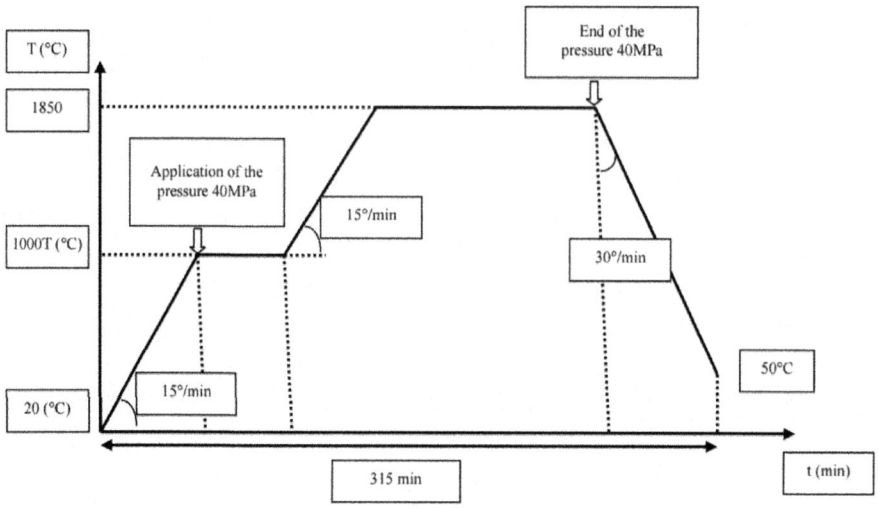

Figure 1. Sintering program of ZrB_2-αSiC and TiB_2-αSiC.

First the time, the pellets are surrounded by non conducting sheath (polyethylene, PTFE) constructed so that the faces of the electrode and the sheath are flush and only the face of disc electrode is in contact with the electrolyte solution. The speed of RDE for HER and HOR varies from 0 to 3000 rpm. The reactions on both ceramic catalysts were performed at quasi stationary conditions with sweep rate at 2 mV/s. The hydrogen was provided in the solution with a mass flow controller "Brooks 5850TR" (15 ml/min). Before recording the polarization, voltammery of bar platinum was carried out to notice the purity of the electrolyte with the sweep rate of 50 mV/s. The platinum used was a small disk with 0.15 cm² as geometric area.

RESULTS AND DISCUSSION

Electrodes Constituents

After sintering and polishing the ceramic pellets, they were submitted to XRD diffraction. Figure 2displays their patterns.

Both pellets show that the α-SiC is present with a small amount. It is also shown that no compound apart from the reagents was detected. The patterns also confirm the resistance of reagents at the temperature higher than 1800°C during high pressure HP sintering. This result has been confirmed by Zhao et al. [19] .

Electrochemical Analyses

Ceramics pellets have been exploited to perform an electrochemical test. So, the RDE (2 mv·s^{-1}, 0.5 M H$_2$SO$_4$, room temperature) measurements were performed to evaluate their electrocatalytic properties. The tests were undertaken at different rate between 0 to 3000 rounds per minute (rpm). They were also made both for hydrogen evolution reaction (HER) and for hydrogen oxidation reaction HOR.

Hydrogen Evolution Reaction

Many authors have been interested of the HER over different types of metal used as indicative electrode. So far, a consensus has not been reached on the predominant reaction mechanism for the electrochemical formation of hydrogen molecule resumed by the following reaction:

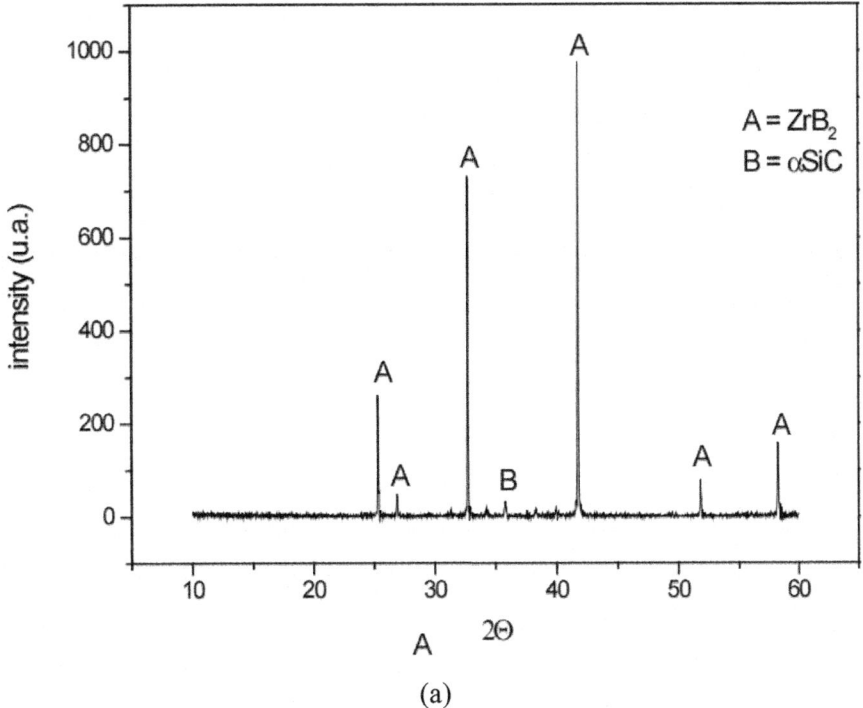

(a)

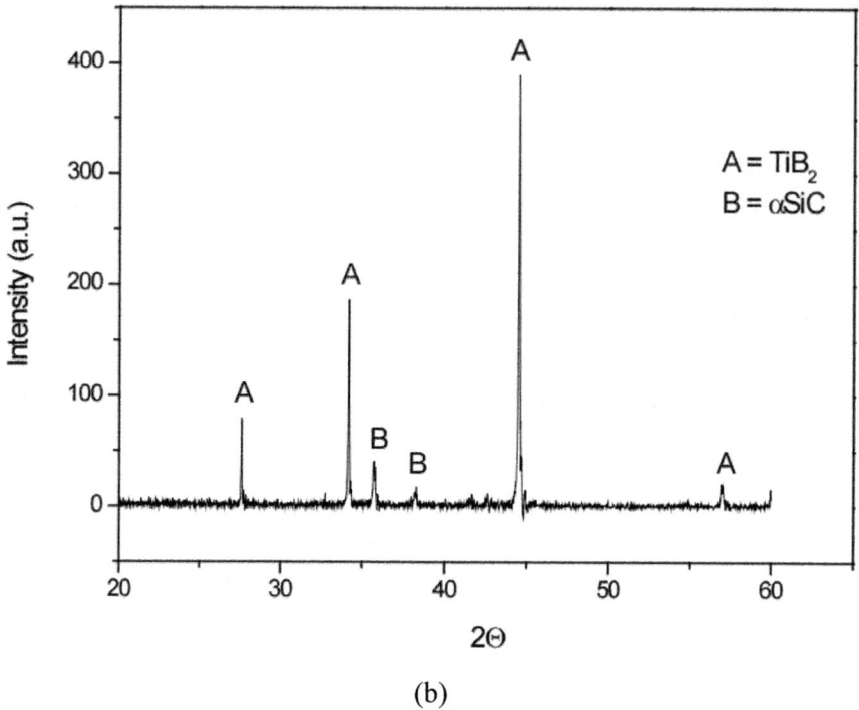

(b)

Figure 2. XRD of 61 mol% of (a) ZrB_2-αSiC; (b) TiB_2-αSiC.

$$2H^+ + 2e^- \rightarrow H_2 \text{ [20] - [23]} \tag{1}$$

This reaction is divided up through three electron transfer steps:

$$H^+ + e^- \rightarrow H_{ads} \text{(Volmer)} \tag{2}$$

$$H_{ads} + H^+ + e^- \rightarrow H_2 \text{ [24] (Heyrovsky) (3)}$$

$$2H_{ads} \rightarrow H_2 \text{ (Tafel)} \tag{4}$$

The Volmer step (Equation (2)) that is the initial adsorption of the proton is admitted to be the fastest. And the subsequent step relies on two possible routes: the Tafel reaction (Equation (4)) is the homolytic step and the Heyrovsky (Equation (3)) that corresponds to the heterolytic step [23] . Nevertheless, the Tafel level is most of the time neglected assuming that the hydrogen is less adsorbed on the electrode surface for electrode made of Ni, Bi, C, etc. Therefore, the most common reactions held up are Volmer-Heyrovsky route [25] [26] .

Figure 3 displays the polarization curves of ZrB$_2$-αSiC and TiB$_2$-αSiC as RDE at room temperature in acidic middle as explained above. The cathodic potential was comprised between 0 V to −1 V/RHE. Nevertheless, the potential was limited to −0.8 V/RHE for rotation speed less than 3000 rpm due to the existence of hydrogen at the electrode surface.

It shows that the hydrogen is produced at high overpotential when scanning in negative side. At 0 rpm regarding the steady-state, the potential is −0.541 V/RHE and −0.413 V/RHE over ZrB$_2$-αSiC (a) and TiB$_2$-αSiC (b) respectively. These values go down when the disk speed is high. They reach −0.417 V and −0.334 V/RHE at the rotation speed of 3000 rpm where the curves are well linear comparing to those at low speeds. This is caused by a lack of hydrogen bubbles formed by HER on the electrode surface that is responsible for the ohmic drop.

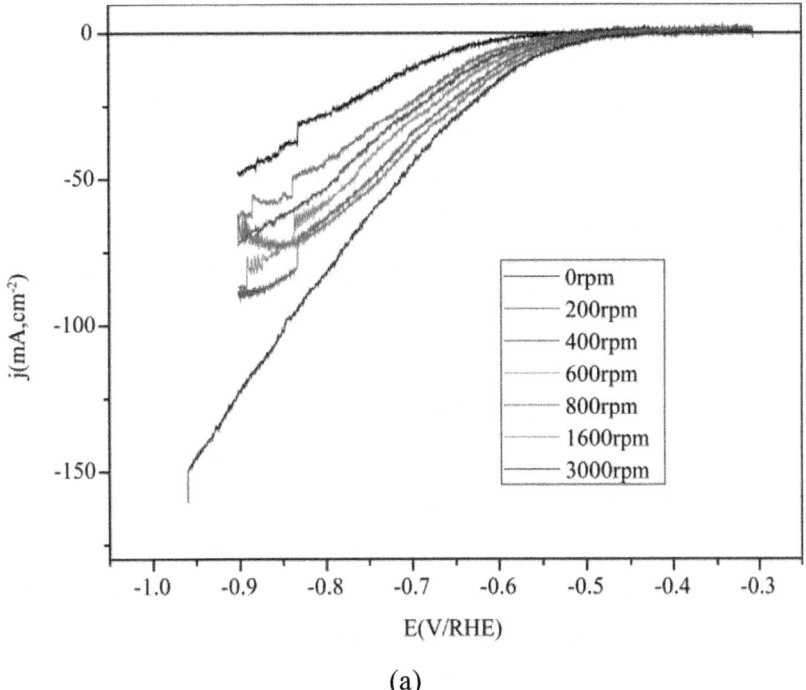

(a)

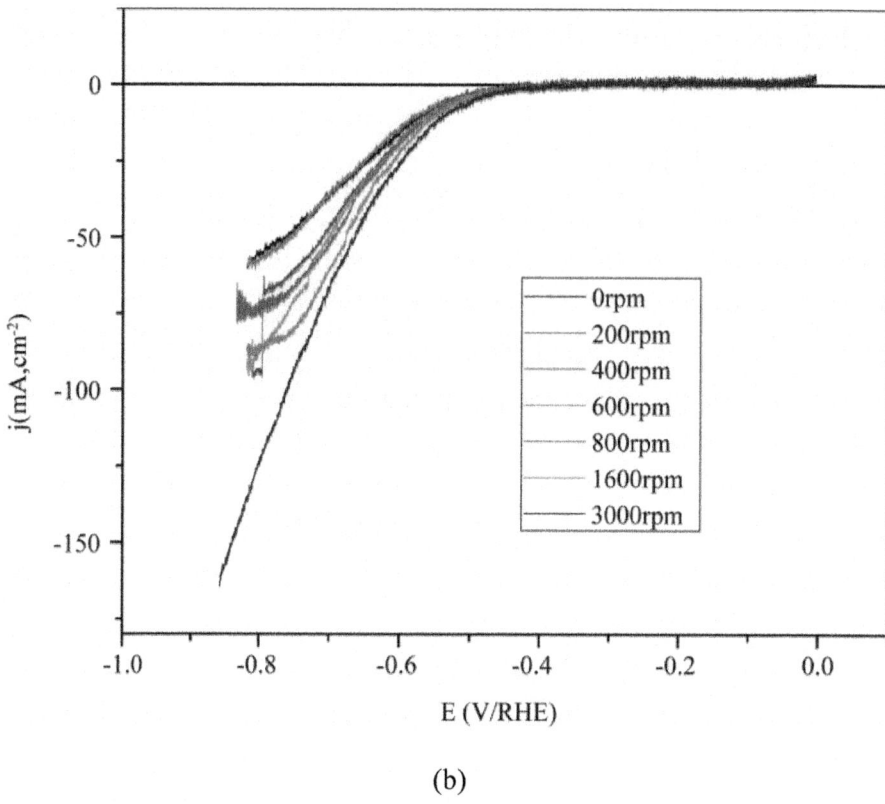

(b)

Figure 3. Polarization curves or linear voltammetry of (a) ZrB_2-αSiC and (b) TiB_2-αSiC mounted on RDE with various speeds at room temperature in H_2SO_4 0.5 M. These data were collected to construct the Koutechy-Lewich plot.

Moreover, the comparison of both pellets shows that TiB_2-αSiC is the most active since its activity requires low overpotential. Always at the speed of 0 rpm, comparison of both electrodes has been made with the linear voltammetry performed on bare platinum as displayed on Figure 4. It shows up that the bare platinum remains the most active electrode as sustained in literature [27] -[29] . Its overpotential for HER is −0.035 V and hydrogen production continues for −0.8 V without ohmic drop of the current. It means that the active sites of the platinum are not soon blocked up by hydrogen bubbles. Moreover, ZrB_2-αSiC displays not merely the highest overpotential but also its active sites are soon affected by hydrogen produced at its surface (−0.764 V) bringing out the ohmic drop.

The analysis of the curves was carried out using Koutecky-Levich equation:

$$\frac{1}{|j|} = \frac{1}{Bw^{1/2}} + \frac{1}{j_k} \qquad (5)$$

where $B = 0.62nFD_0^{2/3}v^{-1/6}C_0^*$ is the Levich constant and the slope of the equation, w is the RDE rate; j_k and j are respectively the kinetic (absence of mass transfer effect) and total current density at a given electrode potential. v is the kinematic viscosity of the solution. Its value in aqueous middle is 10^{-2} cm²·s⁻¹ [30] ; D_0 is the diffusion coefficient and C_0^* is the concentration of the solution. In fact, the current defined by Koutecky-Levich is the contribution of both diffusion and charge transfer currents.

$$\frac{1}{|j|} = \frac{1}{j_{dif}} + \frac{1}{j_k} \quad \text{with} \quad j_{dif} = Bw^{1/2} \qquad (6)$$

Figure 5 and Figure 6 display the Koutecky plots of both ceramic electrodes. The linear plots obtained with the software Origin Pro 8 permit to determine for each potential the constant B and the current density j_k corresponding to the system when controlled by the electron transfer. From these plots, the total number of electron n can be calculated.

The Table 1 displays the j_k, B and D_0 values coming of the Koutecky plots of both RDE electrodes for each potential.

Assuming that n is the total number of electron transferred during the hydrogen evolution, and the kinematic viscosity of the solution admitted to be 10^{-2} cm²·s⁻¹ for aqueous solution, we can calculate the diffusion coefficient for each value of B given in Table 1 through the equation

$$D_0 = \left(\frac{B}{0.62nFAC_0^*v^{-1/6}} \right)^{3/2} . \qquad (7)$$

The electrode geometric surface is 0.18 cm² for the pellets diameter of 4.85 mm.

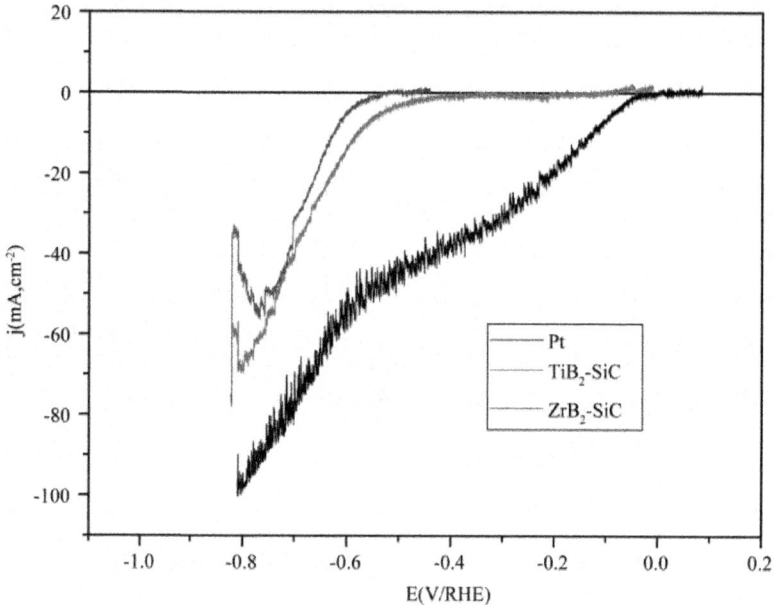

Figure 4. Linear voltammetry of Pt, TiB$_2$-αSiC and ZrB$_2$-αSiC in steady-state in H$_2$SO$_4$ 0.5 M at room temperature (2 mV/s, 0 rpm).

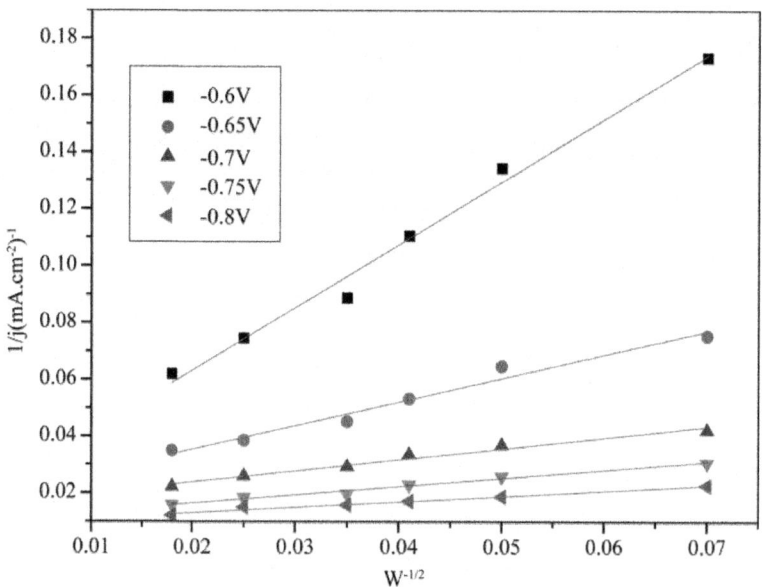

Figure 5. Koutecky-Levich plots 1/j vs. w$^{-1/2}$ for proton reduction at ZrB$_2$-αSiC rotating disk electrode (H$_2$SO$_4$ 0.5 M, room temperature, 2 mV·s^{-1}).

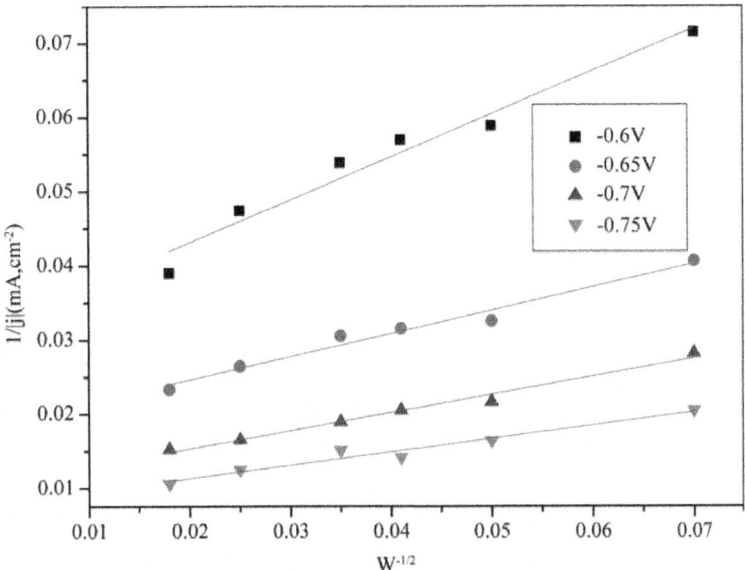

Figure 6. Koutecky-Levich plots $1/j$ vs. $w^{-1/2}$ for proton reduction at TiB$_2$-αSiC rotating disk electrode (H$_2$SO$_4$ 0.5 M, room temperature, 2 mV·s^{-1}).

Table 1. Kinetic parameters recorded with ZrB$_2$-αSiC and TiB$_2$-αSiC RDE for each overpotential

Electrode	Parameters	Potential (V/RHE)				
		−0.60	−0.65	−0.70	−0.75	−0.80
ZrB$_2$-αSiC	$|j_k|$ (mA·cm^{-2})	52.30	52.50	69.40	92.00	106.20
	B	0.45	1.20	2.56	3.47	5.26
	D_0 (10^{-6} cm^2·s^{-1})	0.85	3.70	12.00	18.00	34.00
TiB$_2$-αSiC	$|j_k|$ (mA·cm^{-2})	31.80	54.60	95.40	128.50	
	B	1.72	3.19	4.10	5.58	
	D_0 (10^{-6} cm^2·s^{-1})	6.40	16.00	23.00	37.00	

Hydrogen Oxidation Reaction HOR

It is performed by bubbling and dissolving hydrogen gas through the electrolyte solution (0.5 M H$_2$SO$_4$). For each rate of the RDE, linear voltamogram was carried out from 0 to 3000 rpm. The limited potentials were −0.8 and 0 V/RHE with identical parameters during the HER. The potential sweep was 2 mV/s. The general equation concerning HOR in acidic solution is:

$$H_2 \leftrightharpoons 2H^+ + 2e^-.$$
(8)

This reaction takes place at the interface through three recognized kinetic steps.

$H_{2elec} \leftrightharpoons 2H_{ads}$ is the Tafel step
(9)

$H_{2elec} \leftrightharpoons H_{ads} + H_{aq}^+ + e^-$ is the Heyrovsky step
(10)

And the Volmer step is $H_{ads} \leftrightharpoons H_{aq}^+ + e^-$.
(11)

Nevertheless, we can note the diffusion step of molecular hydrogen from bulk solution to the electrode surface $H_{2sol} \leftrightharpoons H_{2elec}$. As there is no layer formed at ceramic electrodes surface during HOR, Tafel step will be neglected for weakly adsorption as well as in the HER reaction [16].

The Figure 7 below displays the polarization curves of ZrB$_2$-αSiC. Before, a voltammetry of bare platinum was recorded in presence of argon atmosphere to confirm the purity of the bulk solution. The anodic potential was comprised between −0.9 V and 0 V/RHE. However, the potential was limited at 0 V because since −0.43 V/RHE, the limited current density was already reached with 8.4 mA·cm^{-2} regardless the rotation speed.

Figure 7 shows that apart from the steady sate, the curves are almost the same for each RDE speed. It shows that the HOR curves don't start with the same value of current as RDE speed increases and all the curves are limited by the same current density (8.4 mA·cm^{-2}) from −0.508 V to 0.0 V/RHE although the rate of the rotating disk increases. This remark shows that the curves don't depend on the RDE rotation rate. Therefore, the system is not controlled by diffusion. Thus, the concentration of the dissolved hydrogen is the same in the bulk solution and at the electrode surface, regardless the electrode reaction.

Figure 8 displays the polarization curves of RDE TiB$_2$-αSiC. It has been recorded from −0.6 V to −0.2 V for electrode speeds comprised between 0 to 3000 rpm. It is observed as well as in Figure 7that there is no real difference between the curves when the speed increases. Therefore, we can conclude that the system doesn't depend on diffusion step.

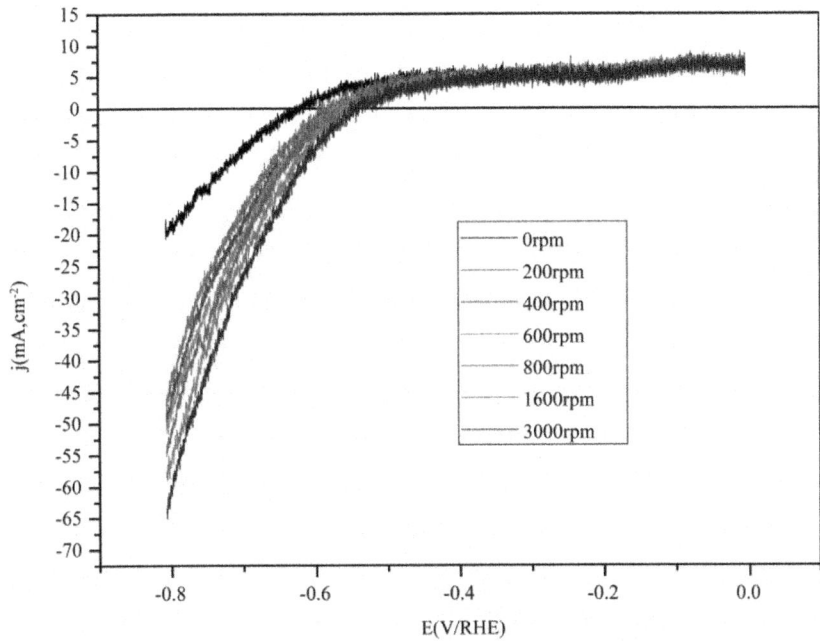

Figure 7. Hydrogen oxidation polarization at RDE ZrB$_2$-αSiC in 0.5 M H$_2$SO$_4$ at 2 mV/s and room temperature. H$_2$ saturated was dissolved in the electrolyte. The rotating speed varied between 0 rpm to 3000 rpm.

The limit current recorded for each RDE rate is 4.91 mA·cm^{-2}. In this case, it is impossible to determine the Levich plots. Therefore, there is no proportionality between the limit current and the speed of the RDE. Then, to determine the kinetic parameters for ceramic electrodes during HOR, Figure 9 displays the linear voltammo- grams of both electrodes at 0 rpm. It confirms that TiB$_2$-SiC is more active than ZrB$_2$-SiC with the limiting current reached respectively at −0.25 V/RHE and −0.0053 V/RHE.

CONCLUSIONS

In this study, non-oxide ceramic pellets ZrB$_2$-αSiC and TiB$_2$-αSiC with 39 mol% SiC were synthesized by hot pressure sintering methods at 1850°C. The DRX performed on them shows the presence of only all raw materials used as ZrB$_2$, SiC and TiB$_2$ confirming their non-destruction and their purity. The electrocatalytic activity of both pellets was determined as rotating disk electrode for hydrogen evolution reaction (HER) and hydrogen oxidation reaction (HOR) for rotating speed comprised between 0 and 3000 rpm.

For HER characterization, we discover that the TiB$_2$-αSiC electrode is more active than ZrB$_2$-αSiC with production of hydrogen for an overpotential of −0.413 V/RHE and the current drop occurs by 0.8 V/RHE at a steady state in H$_2$SO$_4$ 0.5 M solution. Moreover, the comparison with bare platinum with the same parameters reveals that the platinum remains the most active electrode with −0.035 V/RHE of overpotential.

Regarding HOR reaction, we found out that with both electrodes, the rotating disk is useless as the limit current observed remains the same regardless of the disk speed. It means that the determining step is not governed by diffusion.

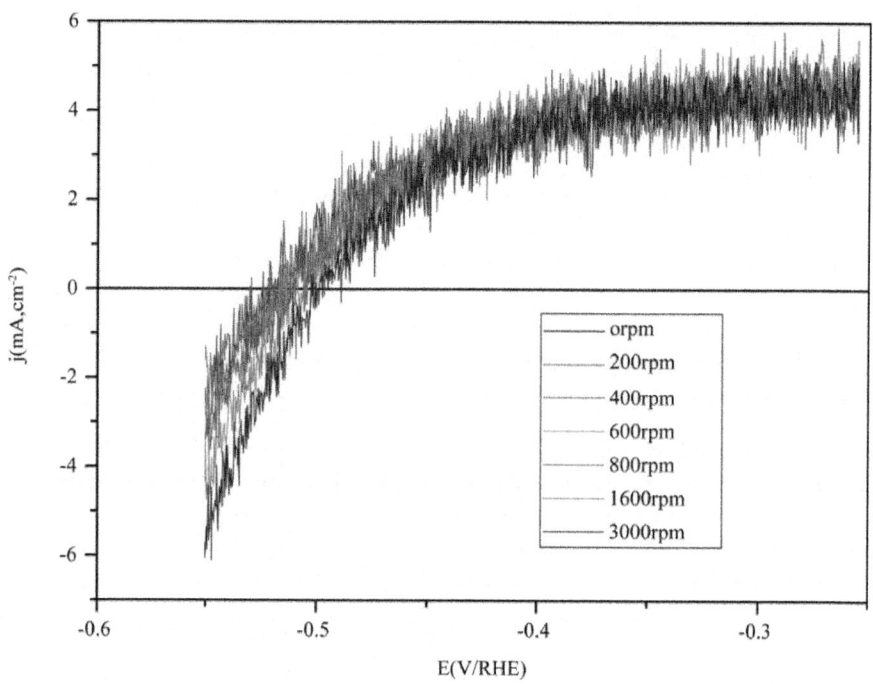

Figure 8. Hydrogen oxidation polarization at TiB$_2$-αSiC RDE in 0.5 M H$_2$SO$_4$ at 2 mV/s and room temperature. H$_2$ saturated was dissolved in the electrolyte. The rotating speed varied between 0 rpm to 3000 rpm.

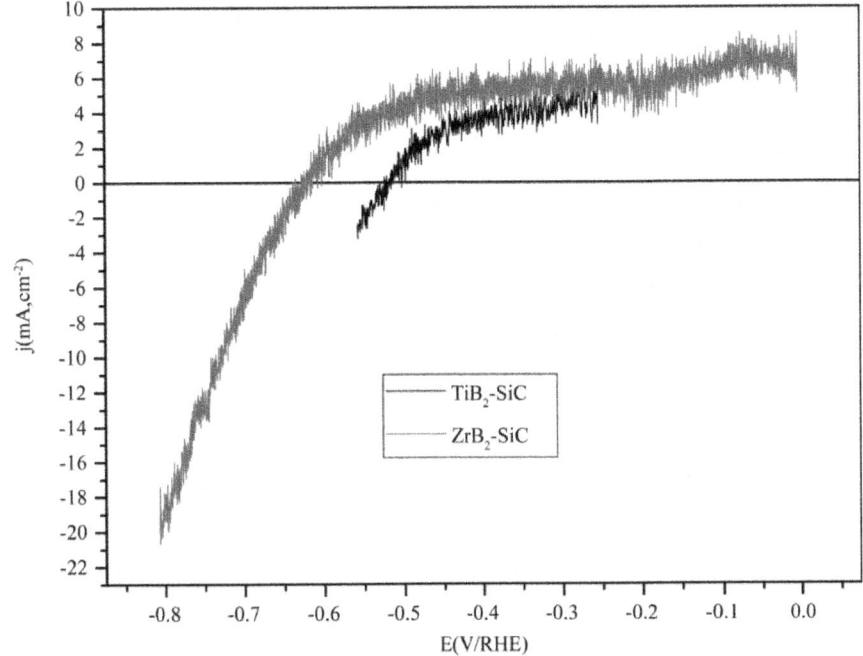

Figure 9. Linear voltammetry of TiB$_2$-αSiC and ZrB$_2$-αSiC in steady-state in H$_2$SO$_4$ 0.5 M at room temperature (2 mV/s, 0 rpm) for HOR reaction.

On behalf of the resistance of both pellets in acidic middle, we need to vary the amount of αSiC in the compounds to increase the electrical conductivity. This will certainly reduce the overpotential for hydrogen evolution and make these ceramic electrodes better candidates as PEMFC electrodes than Pt electrode.

REFERENCES

1. Schlapbach, L. and Züttel, A. (2001) Hydrogen-Storage Materials for Mobile Applications. Nature, 414, 353-358. http://dx.doi. org/10.1038/35104634

2. Maillard, F., Peyrelade, E., Soldo-Olivier, Y., Chatenet, M., Chainet, E. and Faure, R. (2007) Is Carbon-Supported Pt-WOx Composite a CO-Tolerant Material? Electrochimica Acta, 52, 1958-1967. http://dx.doi. org/10.1016/j.electacta.2006.08.024

3. Micoud, F., Maillard, F., Gourgaud, A. and Chatenet, M. (2009) Unique CO-Tolerance of Pt-WOx Materials. Electrochemistry Communications, 11, 651-654. http://dx.doi.org/10.1016/j.elecom.2009.01.007

4. Gratzel, M. (2001) Photoelectrochemical Cells. Nature, 414, 338-344. http://dx.doi.org/10.1038/35104607

5. Matsumoto, F. (2012) Ethanol and Methanol Oxidation Activity of PtPb, PtBi, and PtBi2 Intermetallic Compounds in Alkaline Media. Electrochemistry, 80, 132-138.

6. Mayousse, E., Maillard, F., Foudo-Onana, F., Sicardy, O. and Guillet, N. (2011) Synthesis and Characterization of Electrocatalysts for the Oxygen Evolution in PEM Water Electrolysis. International Journal of Hydrogen Energy, 36, 10474-10481. http://dx.doi.org/10.1016/j.ijhydene.2011.05.139

7. Grigoriev, S.A., Millet, P. and Fateen, V.N. (2008) Evaluation of Carbon-Supported Pt and Pd Nanoparticles for the Hydrogen Evolution Reaction in PEM Water Electrolysers. Journal of Power Sources, 177, 281-285. http://dx.doi.org/10.1016/j.jpowsour.2007.11.072

8. Siroma, Z., Tanaka, M., Yasuda, K., Inaba, K. and Tasaka, A. (2006) Electrochemical Corrosion of Carbon Materials in an Aqueous Acid Solution. Electrochemistry, 75, 258-260.

9. Lonné, Q., Glandut, N., Labbe, J.-C. and Lefort, P. (2011) Fabrication and Characterization of ZrB2-SiC ceramic Electrodes Coated with a Proton Conducting, SiO2-Rich Glass Layer. Electrochimica Acta, 56, 7212-7219. http://dx.doi.org/10.1016/j.electacta.2011.04.027

10. Monticelli, C., Zucchi, F., Pagnoni, A. and Dal Colle, M. (2005) Corrosion of a Zirconium Diboride/Silicon Carbide Composite in Aqueous Solutions. Electrochimica Acta, 50, 3461-3469. http://dx.doi.org/10.1016/j.electacta.2004.12.023

11. Levine, S.R., Opila, E.J., Halbig, M.C., Kiser, J.D., Singh, M. and Salem, J.A. (2002) Evaluation of Ultra-High Temperature Ceramics Foraeropropulsion Use. Journal of the European Ceramic Society, 22, 2757-2767. http://dx.doi.org/10.1016/S0955-2219(02)00140-1

12. Yan, Y.J., Huang, Z.R., Dong, S.M. and Jiang, D.L. (2006) Pressureless Sintering of High-Density ZrB2-SiC Ceramic Composites. Journal of the American Ceramic Society, 89, 3589-3592. http://dx.doi.org/10.1111/j.1551-2916.2006.01270.x

13. Rezaie, A., Fahrenholtz, W.G. and Hilmas, G.E. (2007) Effect of Hot Pressing Time and Temperature on the Microstructure and Mechanical Properties of ZrB2-SiC. Journal of Materials Science, 42, 2735-2744. http://dx.doi.org/10.1007/s10853-006-1274-2

14. Han, J.C., Hu, P., Zhang, X.H., Meng, S.H. and Han, W.B. (2008) Oxidation-Resistant ZrB2-SiC Composites at 2200 °C. Composites

Science and Technology, 68, 799-806. http://dx.doi.org/10.1016/j.compscitech.2007.08.017

15. Tian, W., Kita, H., Hyuga, H., Kondo, N. and Nagaoka, T. (2010) Reaction Joining of SiC Ceramics Using TiB2-Based Composites. Journal of the European Ceramic Society, 30, 3203-3208. http://dx.doi.org/10.1016/j.jeurceramsoc.2010.07.017

16. Blanc, C., Thevenot, F. and Goeuriot, D. (1999) Microstructural and Mechanical Characterization of SiC-Submicron TiB2 Composites. Journal of the European Ceramic Society, 19, 561-569. http://dx.doi.org/10.1016/S0955-2219(98)00227-1

17. Zhao, G.L., Huang, G.Z., Liu, H.L., Zou, B., Zhu, H.T. and Wang, J. (2014) Microstructure and Mechanical Properties of TiB2-SiC Ceramic Composites by Reactive Hot Pressing. International Journal of Refractory Metals & Hard Materials, 42, 36-41. http://dx.doi.org/10.1016/j.ijrmhm.2013.10.007

18. Li, W.J., Rong, T. and Goto, T. (2005) Preparation of TiB2-SiC Eutectic Composite by an Arc-Melted Method and Its Characterization. Materials Transactions, 46, 2504-2508. http://dx.doi.org/10.2320/matertrans.46.2504

19. Zhao, G.L., Huang, C.Z., Liu, H.L., Zoua, B., Zhu, H.T. and Wang, J. (2014) A Study on In-Situ Synthesis of TiB2-SiC Ceramic Composites by Reactive Hot Pressing. Ceramics International, 40, 2305-2313. http://dx.doi.org/10.1016/j.ceramint.2013.07.152

20. Markovic, N.M. and Ross Jr., P.N. (2002) Surface Science Studies of Model Fuel Cell Electrocatalysts. Surface Science Reports, 45, 117-229. http://dx.doi.org/10.1016/S0167-5729(01)00022-X

21. Kunimatsu, K., Senzaki, T., Tsushima, M. and Osawa, M. (2005) A Combined Surface-Enhanced Infrared and Electrochemical Kinetics Study of Hydrogen Adsorption and Evolution on a Pt Electrode. Chemical Physics Letters, 401, 451-454. http://dx.doi.org/10.1016/j.cplett.2004.11.100

22. Mukerjee, S., Srinivasan, S., Soriaga, M.P. and McBreen, J. (1995) Role of Structural and Electronic Properties of Pt and Pt Alloys on Electrocatalysis of Oxygen Reduction: An in Situ XANES and EXAFS Investigation. Journal of the Electrochemical Society, 142, 1409-1422. http://dx.doi.org/10.1149/1.2048590

23. Skulason, E., Karlberg, G.S., Rossmeisl, J., Bligaard, T., Greeley, J., Jonsson, H. and Norskov, J.K. (2007) Density Functional Theory Calculations for the Hydrogen Evolution Reaction in an Electrochemical

Double Layer on the Pt(111) Electrode. Physical Chemistry Chemical Physics, 9, 3241-3250. http://dx.doi.org/10.1039/b700099e

24. Mello, R.M.Q. and Ticianelli, E.A. (1997) Kinetic Study of the Hydrogen Oxidation Reaction on Platinum and Nafion? Covered Platinum Electrodes. Electrochimica Acta, 42, 1031-1039. http://dx.doi.org/10.1016/S0013-4686(96)00282-4

25. Lasia, A. (2002) Applications of Electrochemical Impedance Spectroscopy to Hydrogen Adsorption, Evolution and Absorption into Metals. In: Conway, B.E. and White, R.E., Eds., Modern Aspect of Electrochemistry, Vol. 35, Chap. 1, Kluwer Academic Publishers, New York, 1-49. http://dx.doi.org/10.1007/0-306-47604-5_1

26. . Lasia, A. and Rami, A. (1990) Kinetics of Hydrogen Evolution on Nickel Electrodes. Journal of Electroanalytical Chemistry and Interfacial Electrochemistry, 294, 123-141. http://dx.doi.org/10.1016/0022-0728(90)87140-F

27. Croissant, M.J., Napporn, T., Léger, J.-M. and Lamy, C. (1998) Electrocatalytic Oxidation of Hydrogen at Platinum-Modified Polyaniline Electrodes. Electrochimica Acta, 16, 2447-2457. http://dx.doi.org/10.1016/S0013-4686(97)10157-8

28. Rau, M.S., Marozzi, C.A., Gennero de Chialvo, M.R. and Chialvo, A.C. (2010) Kinetic Study of the Hydrogen Oxidation Reaction on Membrane Coated Electrodes. Part II: Applications. The Open Electrochemistry Journal, 2, 1-5. http://dx.doi.org/10.2174/1876505X01002010001

29. Marozzi, C.A., Gennero-Chialvo, M.R. and Chialvo, A.C. (2009) Kinetic Study of the Hydrogen Oxidation Reaction on Membrane Coated Electrodes. Part I: Theoretical Aspects. The Open Electrochemistry Journal, 1, 49-55. http://dx.doi.org/10.2174/1876505x00901010049

30. Fatisson, J. (2005) Elaboration de nouveaux matériaux d'électrodes obtenus par autoassemblage de polyélectrolytes, nanoparticules et biomolécules: Etudes physico-chimiques et applications. Ph.D., University Joseph Fourier, Grenoble, 32.

Chapter 11

HYDROGEN PRODUCTION BY WATER ELECTROLYSIS EFFECTS OF THE ELECTRODES MATERIALS NATURE ON THE SOLAR WATER ELECTROLYSIS PERFORMANCES

Romdhane Ben Slama

Unit of Research: Environment, Catalysis & Processes Analysis, National School of Engineers of Gabes, University of Gabes, Gabes, Tunisia.

ABSTRACT

Our contribution in the production of hydrogen, vector of energy, consists in testing the water electrolysis by photovoltaic solar energy. The realization of some electrolysers whose electrodes are various materials, showed a clear difference from the point of view produced hydrogen flow, conversion efficiency, energy specific consumption and the electrodes lifespan. This made it possible to classify materials, by performances descending order, as follows: copper, lead, bronzes, aluminum, stainless, graphite and steel. However lead has a too low flow and aluminum corrodes quickly. Steel admits poor yield and lifespan. Then, we retain primarily copper like anode metal. To increase the hydrogen produced flow by electrolysis, the electrolysers parallel assembly choice is essential. According to the hour of the day, the evolution of the parameters such as consumed current, efficiency, and specific energy differs from a material with another, which can be explained by the variation of solar energy during the day.

INTRODUCTION

The importance of Hydrogen as an energy vector is not any more to show that is to supply the fuel cells or the internal combustion engines, while respecting the environment.

Various ways exist to produce hydrogen. Let us quote the reforming using the natural gas, cracking at high temperature etc.

The hydrogen solar production by brackish water electrolysis can be profitable because solar energy is free, abundant and clean, our Tunisian climate lends itself to it.

Thus, among the multiple hydrogen production ways, we chose the solar process of water electrolysis because we have a considerable potential of solar radiation. The electrolysers are locally manufactured and the process does not require high temperatures, not easily controllable and requiring solar continuation systems and a high investment [1-5].

Works are published on the water electrolysis but do not deal with the electrodes materials nature and their corrosion [6-10]. However some authors treat anode corrosion [11-14] and prefer to use platinum.

In the present experimental study, our initial goal is to produce hydrogen by water electrolysis, using photovoltaic solar energy [3-8,15]. However, we are confronted with the anode corrosion problem, therefore, seven different materials were tested; It is about copper, the mild steel, the stainless steel, bronze, graphite, aluminum and lead. During the tests, current soup and produced hydrogen flow are measured, and are deduced then the efficiency and energy necessary.

EXPERIMENTAL PROTOCOL

Parameters of Calculation

Hydrogen production flow rate: $Qv = V/t$ (m³/s) with:

- Absorptive power by the electrolyser:

- $Pa = U \cdot I$ (W).

- Useful power of the electrolyser: $Pu = PCI. \, Q \cdot \rho$ (W) with PCI: lower thermal value of hydrogen (119.9×10^6 J/Kg) ρ: density of hydrogen (0.09 Kg/m³).

- Consumed electric power: $W = Pa \cdot t$ (J).

- Useful efficiency: $\eta = PCI \cdot (V/(Pa \cdot t)) \cdot \rho$ (−).

- Consumed electric power per unit of volume: $W/V = Pa \cdot t/V$ (J/cm³).

Photovoltaic Module and Electrolysers

A photovoltaic model with its panel, electrolysers and their electrodes made in various materials, are represented by the following photograph (**Figure 1**).

ROLE OF THE ELECTRODES MATERIALS ON THE ELECTROLYSIS AND THE ANODE CORROSION

It is that the electrodes materials nature has an influence, not only over their lifespan, but also on the hydrogen production and the energy consumption.

Corrosion Phenomenon

Corrosion is the result of the phenomenon which occurs when a metal is in contact with a gas reagent or liquid in wet underworld. It results from the interaction of the material surface with the surrounding medium. It can occur dry if the temperature is high; it is hot oxidation, or in aqueous underworld with the lower part of 100°C, it is the wet corrosion which appears by a dissolution of metal following an electrochemical phenomenon and the formation of corrosion products.

It was noticed during electrolysis that the electrode anode corrodes until its rupture. Thus, we made a systematic study according to the material nature (steel, copper, stainless steel, graphite, aluminum, bronze and lead) to observe this phenomenon and to determine the operation life until total rupture of the electrodes. The influence of material nature on the performances was shown by plotting the load curves, efficiency and the consumed specific electric power.

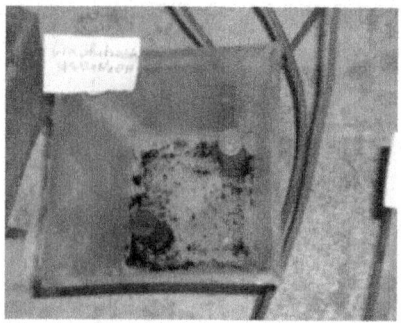

Figure 1. Photographs of the photovoltaic model and the electrolysers with their corroded electrodes.

UniInfluence of the Electrodes Nature on the Electrolysis Performances

The materials have varied electrochemical potentials, thus it is normal that they do not have the same performances hydrogen production: flow rate, efficiency, consumed specific energy. In our tests, seven electrolysers are assembled in parallel connexion (May 28, 2009).

Hydrogen Flow Rate Produced According to Time

To make comparisons between materials under identical conditions, the electrolysers are assembled in parallel with the photovoltaic module, and of measurements (current, voltage and time of filling of the test tube) are taken.

We deduce the produced hydrogen flow rate, the energetic efficiency and the consumed specific electric power.

It is noted that according to the hour, the produced hydrogen flow is not constant (**Figure 2**), because the terminal voltage of the photovoltaic module is not constant, contrary to the laboratory electric generators.

All the curves do not have the same appearance, a priori because of the materials conductivity (resistivity) variation; it is the case between copper and lead.

Hydrogen Production Efficiency According to Time

This efficiency is deduced following the measurements taken on the consumption by the electrolyser and the flow from produced hydrogen (**Figure 3**).

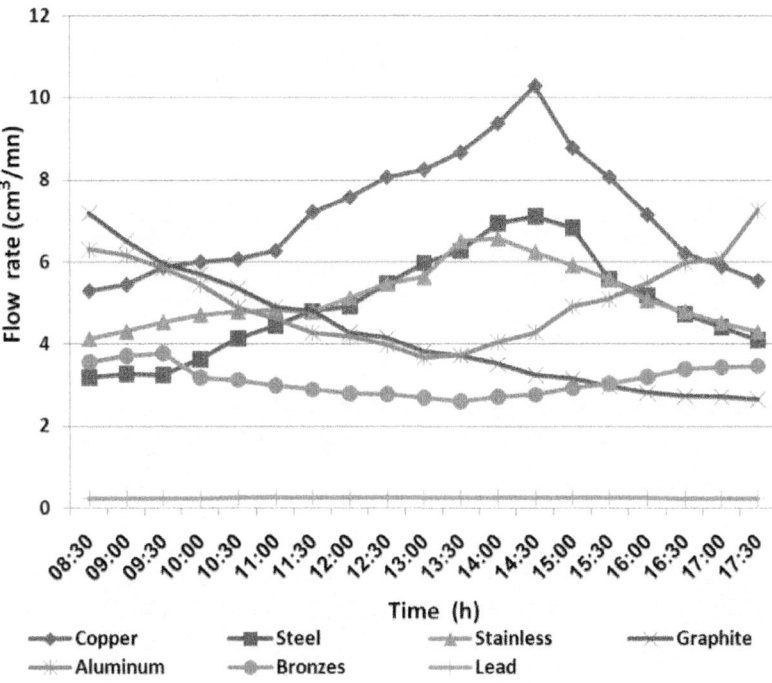

Figure 2. Produced hydrogen flow according to time and electrodes materials.

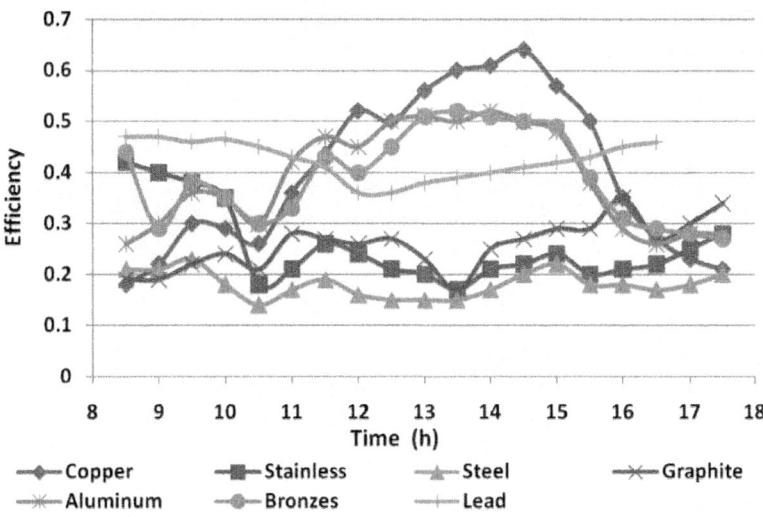

Figure 3. Useful efficiency according to time and electrodes materials (May 28, 2009).

The efficiency evolves according to the hour of the day for this parallel assembly of the electrolysers. For some materials (copper, bronzes and aluminum), this efficiency is maximum at the after midday beginning, for the others (lead, graphite, stainless and steel), the variation is more reduced and the efficiency is weaker (50% and 20% respectively).

Power Consumption per Unit of Produced Hydrogen Volume

For a material, the variation of the power consumption is weak throughout the day (**Figure 4**). The materials which had the best efficiency admit relatively weak power consumption here, and conversely.

Comparison, by the Crossed Sorting Method, between Electrodes Materials and Classification from the Productivity Point of View and Corrosion Resistance

The cross sorting is a hierarchisation method. In these case of comparison between various materials, the numbers corresponding to the difference of flow, efficiency etc between two materials given, are allotted.

Thereafter, a total of numbers is determined and a classification of the materials is granted.

Classification According to the Produced Hydrogen Flow

In **Table 1** of crossed sorting, the indicated number represents the produced hydrogen flow difference in cm^3/mn.

For a given material, one counts these numbers in line and column.

Among materials tested, copper is classified the first from the point of view of produced hydrogen flow, come then in the preferably descending order: stainless, aluminum, steel, graphite, bronze and lead in the last.

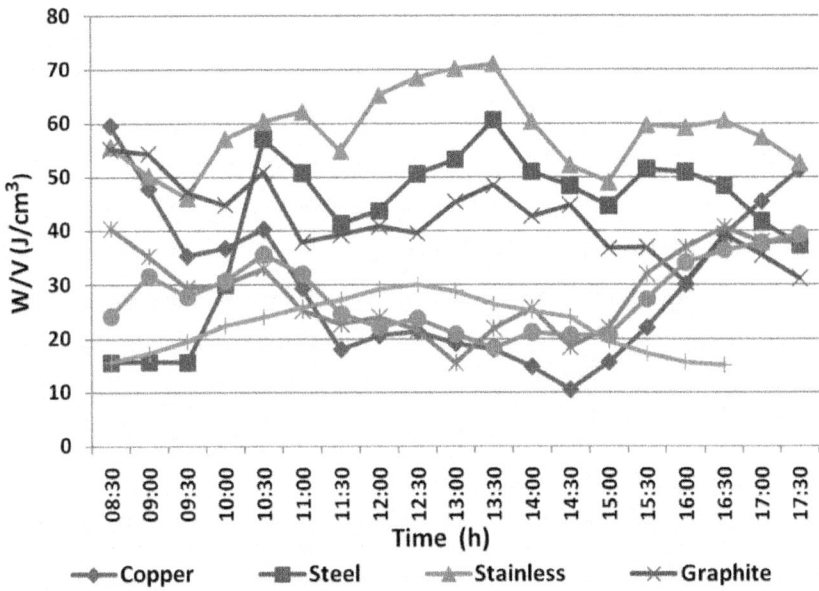

Figure 4. Specific power consumption according to time and electrodes materials.

Classification According to the Energetic Efficiency η

By applying the crossed sorting, **Table 2** makes it possible to classify materials by preferably order from the energetic efficiency point of view.

As for the flow, copper is classified the first from the point of view energetic efficiency, come then in the order descending preferably bronze, lead, aluminum, stainless and graphite, and in the last steel.

Classification According to the Energy Specific Consumption W/V (J/cm³)

By applying the crossed sorting, **Table 3** makes it possible to classify materials by preferably order from the specific consumed energy point of view for the hydrogen production.

As for the flow and the efficiency, copper is always classified the first from specific energy consumption point of view in exico with the aluminum and lead, come then in the preferably descending order bronze, graphite, aluminum, stainless and lastly steel.

Lifespan of Electrodes

Table 4 makes it possible to classify materials according to their lifespan.

Here, copper is classified second. The first rank is in favor of lead, and then come in preferably descending order: bronze, graphite, stainless, steel, and aluminum in last.

Total Classification

The nature of materials influences thus all the performances of electrolysis: flow of produced hydrogen, energetic efficiency, consumption of energy and lifespan of the electrodes.

The total classification (**Table 5**) puts copper in first; it was already for three classifications out of four. Let us note that in another field, copper is also usually used for the water conduits, strongly subjected with corrosion.

Table 1. Application of the sorting crossed for the produced hydrogen flow

	Steel	Stainless	Graphite	Aluminum	Bronzes	Lead	Total	Row
Copper	Copper 3	Copper 3	Copper 3	Copper 3	Copper 4.5	Copper 8	24.5	1st
	Steel	0	Steel 1	0	Steel 2	Steel 5	8	4th
		Stainless	Stainless 2	0	Stainless 3	Stainless 5	10	2nd
			Graphite	Alumin 1	Graph 2	Graph 4	6	5th
				Alumin	Alumin 3	Alumin 5	9	3rd
					Bronzes	Bronzes 3	3	6th
						Lead	0	7th

Table 2. Application of the sorting crossed for the hydrogen production efficiency

	Steel	Stainless	Graphite	Aluminum	Bronzes	Lead	Total	Row
Copper	Copper 2	Copper 2	Copper 2.5	Copper 1	0	0	7.5	1st
	Steel	Stainless 1	Graphite 1	Alumin 2	Bronzes 2.5	Lead 2.2	0	7th
		Stainless	0	Alumin 2	Bronzes 2	Lead 2	1	5th
			Graphite	Alumin 2	Bronzes 2	Lead 2	1	5th
				Alumin	0	0	6	4th
					Bronzes	0	6.5	2nd
						Lead	6.2	3rd

Table 3. Application of the sorting crossed for consumed specific energy

	Steel	Stainless	Graphite	Aluminum	Bronzes	Lead	Total	Row
Copper	Stainless 30	Stainless 20	Graph 20	0	0	0	0	1st
	Steel	Steel 15	Steel 10	Steel 30	Steel 30	Steel 30	115	7th
		Stainless	Stainless 5	Stainless 20	Stainless 20	Stainless 20	85	6th
			Graphite	Graph 20	Graph 20	Graph 22	82	5th
				Alumin	0	0	0	1st
					Bronzes	Bronzes 5	5	4th
						Lead	0	1st

Table 4. Application of the sorting crossed for the lifespan of the electrodes

	Steel	Stainless	Graphite	Aluminum	Bronzes	Lead	Total	Row
Copper	Copper 112	Copper 91	Copper 80	Copper 128	Copper 16	Lead 364	427	2nd
	Steel	Stainless 19	Graph 32	Steel 16	Bronzes 96	Lead 476	16	6th
		Stainless	Graph 11	Stainless 37	Bronzes 75	Lead 455	37	5th
			Graphite	Graph 48	Bronzes 64	Lead 444	91	4th
				Aluminium	Bronzes 112	Lead 492	0	7th
					Bronzes	Lead 380	347	3rd
						Lead	2611	1st

Table 5. Total classification of materials

Row	Flow Qv	Efficiency	W/V	Duration	Somme	Total Row
Copper	1	1	1	2	5	1st
Steel	4	7	7	6	24	7th
Stainless	2	5	6	5	18	5th
Graphite	5	5	5	4	19	6th
Aluminum	3	4	1	7	15	3rd
Bronzes	6	2	4	3	15	3rd
Lead	7	3	1	1	12	2nd

The second rank is in favor of lead, however this one is not favorite for us because of its too weak hydrogen flow. Let us note however that lead is usually used in the electric battery perhaps because of its durability. After, come bronze and aluminum. This last is to be excluded because of its too weak lifespan, itself due to its too low electrochemical potential. The last three materials are graphite, stainless and the steel which is also in efficiency and consumption of energy. If we eliminate the materials which are classified at least once at the last row i.e. steel, lead, and aluminum. The remaining materials are in the preferably descending order: copper, bronze, stainless and graphite.

Some Common Mistakes

By having different electrodes, it is that the performances increase. One of the electrodes remained out of copper, the other out of iron or aluminum.

Electric Current Consumed According to the Terminal Voltage of the Electrolyser

These experiments are carried out for the same conditions such as the temperature with 25°C and the atmospheric pressure.

One takes the water of tap with the addition of NaCl like electrolyte for all the tests with the various electrodes types.

The variation of the consumed current follows one of the two paces, according to the electrodes metals nature (**Figure 5**):

- For the homogeneous electrodes (copper used in our case), the consumed current remains weakest;

- However for the heterogeneous electrodes, the current is more raised and increases definitely more quickly, already even with the weak voltages applied. The current also changes when we permute the power supply between two metals of the couple of electrodes, anodes and cathode.

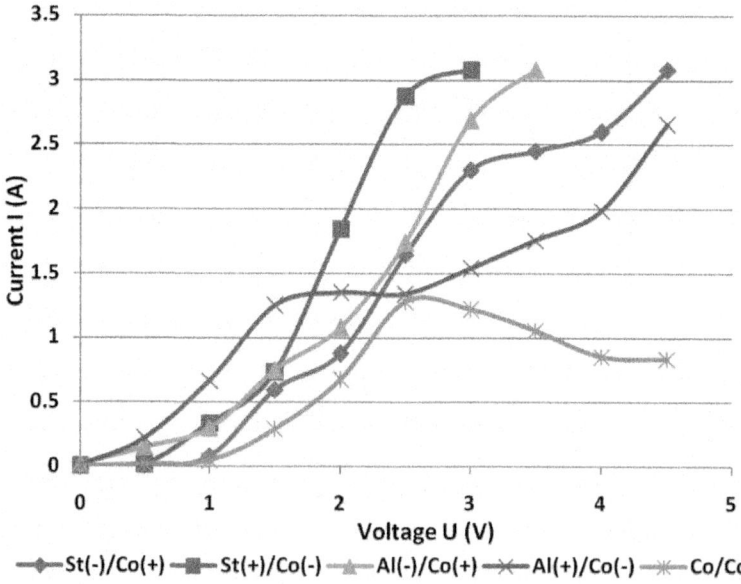

Figure 5. Current intensity variation according to the terminal voltage of the electrolyser.

Released Hydrogen Flow

All these practice works are made with the ambient air, a temperature of approximately 25°C and under the atmospheric pressure. The hydrogen release flow is calculated by knowing the filling time of a test tube, volume 27 cm^3.

The histogram of **Figure 6** makes it possible to emit a certain number of remarks:

Firstly, it is noticed that all the electrolysers have remarkable flows.

Secondly, the lowest flow is in the case where the anode and cathode are out of copper. The homogeneous choice of the two copper electrodes, like usually makes, generates a hydrogen flow lower than if the electrodes are heterogeneous.

Thirdly, the best produced hydrogen flows are for the electrolysers whose their electrodes steel/copper (the anode is out of steel and cathode is out of copper) and aluminum/copper (the anode is out of copper and cathode is out of aluminum).

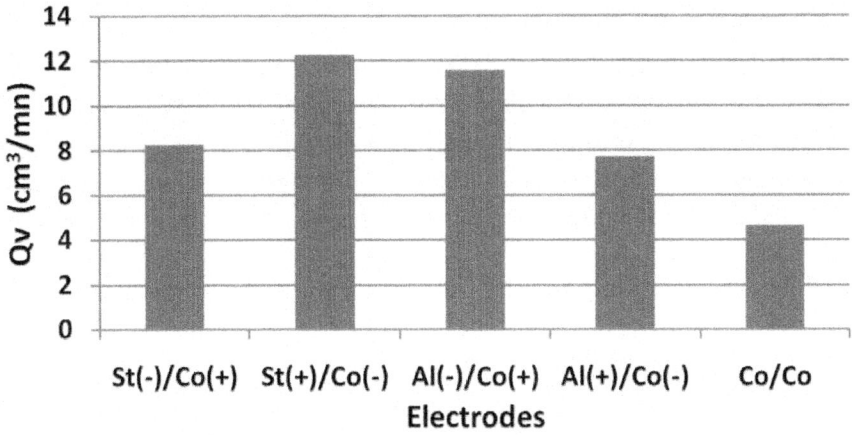

Figure 6. Produced hydrogen flow according to the metal couple of the electrolyser electrodes. Salted tap water (200 g/l).

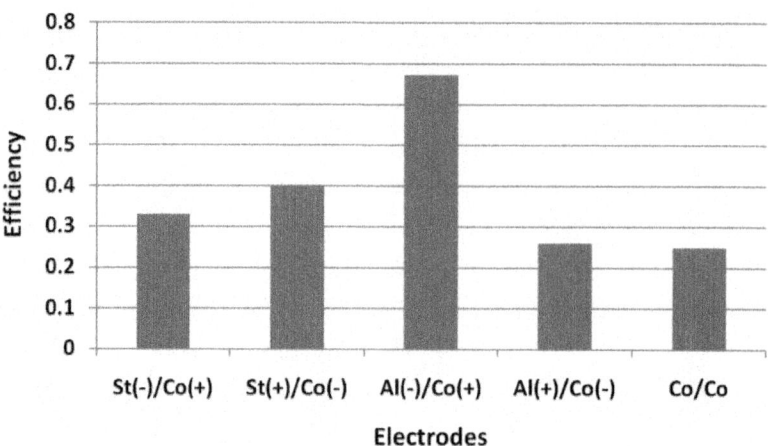

Figure 7. Electrolysis efficiency according to the electrodes metal couple. Salted tap water (200 g/l).

Electrolyser Efficiency

The efficiency characterizes the released hydrogen quantity, on the level of cathode, by the consumed electric power throughout one given time.

The histogram of **Figure 7** shows, that for the electrolyser whose metals of electrodes are homogeneous, until now used, the output is in fact weakest, as for the case of the flow. The couple aluminum/copper, with copper anode, makes it possible to reach a considerable efficiency. This can be explained by the difference of electropositivity between copper and aluminum.

Consumed Electric Power

Contrary to the hydrogen flow and the efficiency which one seeks to maximize, here, the consumed electric power is rather to minimize (**Figure 8**). There still, it is the electrodes couple of aluminum/copper, with copper anode, which is the best; its advantage is accentuated. Even if the electrodes couple of copper/copper consumes low power, it is not favorite because it generates a low flow and an efficiency compared to the couple aluminum/ copper.

Consumed Specific Electric Power

The consumed electric power per produced hydrogen volume unit is the best parameter of comparison between the electrodes metal couples. Theoretical energy necessary for the decomposition of the water molecule in hydrogen and oxygen is of 286 kJ/mol, that is to say 12.76 J/cm^3.

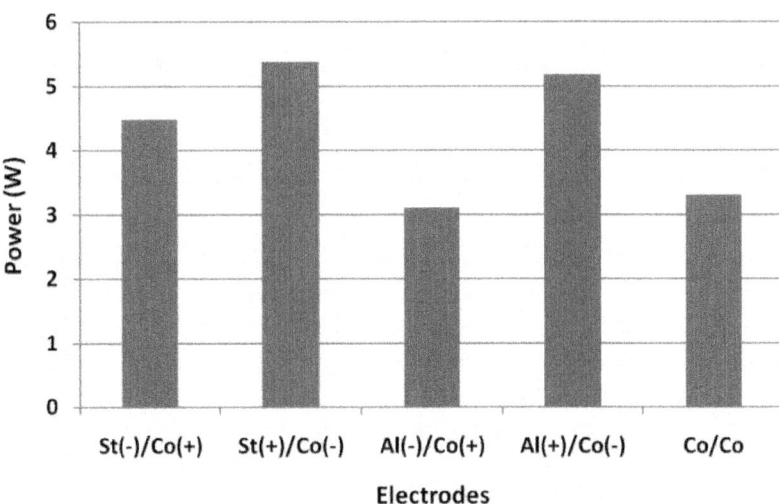

Figure 8. Consumed electric power during electrolysis according to the metal couple of the electrodes. Salted tap water (200 g/l).

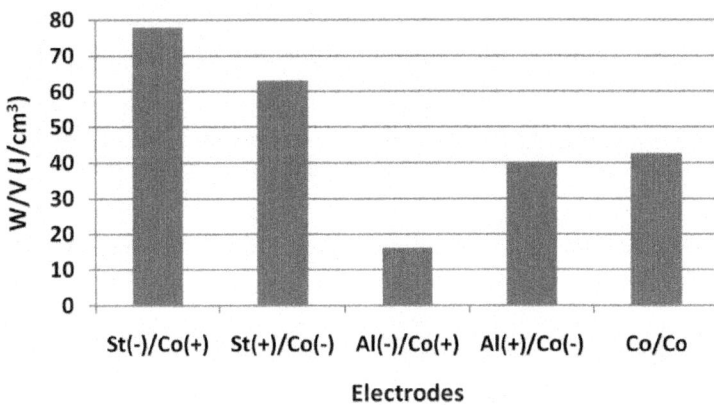

Figure 9. Electric power consumed by the eletrolyser according to the metal of electrode. Tap water (200 g/l).

The results of **Figure 9** prove once again that the electrodes couple of Aluminum/copper, with copper anode, is better and even much better than the other couples of studied materials, the performance point of view, and even from the point of view lifespan of the copper anode as that was shown during previous studies.

INTERPRETATION

The influence of the electrodes materials nature of electrolysers was shown that it is on the hydrogen production flow, the energetic efficiency, consumed specific energy or their lifespan before deterioration.

Classification according to all these comparison criterions makes it possible to order objectively materials.

However, with heterogeneous electrodes the performances can be still improved; it is the case for the electrodes aluminum/copper, with copper anode to resist at the corrosion. The flow and the efficiency more than doubled. On the other hand, the specific power consumption is in contrary well reduced.

CONCLUSIONS AND PROSPECTS

In the case of electrolysers including electrodes made out of homogeneous materials, the performances concerning the produced hydrogen flow, energetic efficiency and power consumption are better for copper and bronze among studied materials. Lead appears to be interesting, however, the corresponding hydrogen flow is too low. Aluminum is in the same way interesting but its lifespan is too weak. The other tested materials: steel, stainless and graphite are to be drawn aside for their weak performances so much the hydrogen flow, the efficiency and the power consumption.

The tests carried out with heterogeneous materials electrodes in the same way related to the determination of these same performances.

Like the electrodes couple of Aluminum/copper gives the best performances (high flow and efficiency, weak power consumption), we envisages to do better by replacing aluminum by magnesium; this last has indeed an electrochemical potential lower than the aluminum and thus the potential difference with copper is increased.

NOMENCLATURES

I: Electrical current (A);

U: Voltage (V);

P: Power (W);

PCI: Lower heating value (J/kg);

Qv: Flow rate (m^3/s);

V: Volume of the test tube (m^3);

t: Tube filling time (s);

W: Electrical energy (J);

ρ: Density of hydrogen (Kg/m^3);

Al: Aluminum;

Co: Copper;

St: Steel.

Indices:

a: Absorbed;

u: Useful;

nom: Nominal;

ab: Absorbed.

REFERENCES

1. E. Bilgen, "Solar Hydrogen from photovoltaic Eletrolizer Systems," Energy Conversion and Management, Vol. 42, No. 9, 2001, pp. 1047-1057. doi:10.1016/S0196-8904(00)00131-X

2. S. H. Jensen, et al., "Hydrogen and Synthetic Fuel Production from Renewable Energy Sources," International Journal of Hydrogen Energy, Vol. 32, No. 15, 2007, pp. 3253-3257.

3. R. Ben Slama, "Production of Hydrogen by Electrolyse of Water and Photovoltaic Energy," Proceeding of the 3rd International congress on Renewable Energies and Environment CERE, Mahdia, 6-8 November 2006.

4. R. Ben Slama, "Tests on the Solar Hydrogen Production by Water Electrolysis," Proceeding of the JITH, Albi, 28- 30 August 2007.

5. R. Ben Slama, "Solar Hydrogen Generation by Water Electrolysis," Proceeding of the 1st Francophone Conference on Hydrogen: Energy Vector, Sousse, 9-11 May 2008, pp. 7-13.

6. R. Ben Slama, "Génération d'Hydrogène par Electrolyse Solaire de l'Eau," Proceeding des Journées Annuelles 2008 Société Française de Métallurgie et de Matériaux, Paris, 4-6 June 2008.

7. R. Ben Slama, "Influence of Material of the Electrodes on the Solar Water Electrolysers for Production of Hydrogen," 2nd International Conference on Hydrogen Energy: ICHE10, Hammamet, 9-11 May 2010.

8. R. Ben Slama, "Comparaison Entre les Matériaux d'Electrodes d'Electrolyseur Pour la Production d'Hydrogène Solaire," 1ère

Conférence Maghrébine sur les Matériaux et l'Energie, Gafsa, 26-28 Mai 2010.

9. F. Jomarda, J. P. Ferauda and J. P. Caire, "Numerical Modeling for Preliminary Design of the Hydrogen Production Electrolyzer in the Westinghouse Hybrid Cycle," International Journal of Hydrogen Energy, Vol. 33, No. 4, 2008, pp. 1142-1152

10. S. A. Grioriev, et al., "Pure Hydrogen Production by PEM Electrolysis for Hydrogen Energy," International Journal of Hydrogen Energy, Vol. 31, No. 2, 2006, pp. 171-175.doi:10.1016/j.ijhydene.2005.04.038

11. P. H. Floch, et al., "On the Production of Hydrogen via Alkaline Electrolysis during Off-Peak Periods," International Journal of Hydrogen Energy, Vol. 32, No. 18, 2007, pp. 4641-4647. doi:10.1016/j.ijhydene.2007.07.033

12. F. Jomard, et al., "Numerical Modelling for Peliminary Design of the Hydrogen Production Elecrolyzer in the Westinghouse Hybrid Cycle," International Journal of Hydrogen Energy, Vol. 33, No. 4, 2008, pp. 1142-1152.

13. L. Solera, J. Macanása, M. Muñoza and J. Casado, "Electrocatalytic Production of Hydrogen Boosted by Organic Pollutants and Visible Light," International Journal of Hydrogen Energy, Vol. 31, No. 1, 2006, pp. 129-139.doi:10.1016/j.ijhydene.2004.11.001

14. M. Cooper and G. Botte, "Hydrogen Production from the Electro-Oxidation of Ammonia Catalyzed by Platinum and Rhodium on Raney Nickel Substrate," Journal of the Electrochemical Society, Vol. 153, No. 10, 2006, pp. A1894-A1901.

15. L. Chatbri, "Traitement des Eaux Usées et Recupération de l'Hydrogène," ISSATG, Gabes, 2010.

CITATION

CHAPTER 1

Sergei Grokhovsky, Irina Il'icheva, Dmitry Nechipurenko, Michail Golovkin, Georgy Taranov, Larisa Panchenko, Robert Polozov and Yury Nechipurenko (2012). Quantitative Analysis of Electrophoresis Data - Application to Sequence-Specific Ultrasonic Cleavage of DNA, Gel Electrophoresis - Principles and Basics, Dr. Sameh Magdeldin (Ed.), ISBN: 978-953-51-0458-2, InTech, DOI: 10.5772/37686.

CHAPTER 2

Martins Vanags, Janis Kleperis and Gunars Bajars (2012). Water Electrolysis with Inductive Voltage Pulses, Electrolysis, Dr. Janis Kleperis (Ed.), ISBN: 978-953-51-0793-4, InTech, DOI: 10.5772/52453.

CHAPTER 3

Meir A, Rubinsky B (2015) Electrical Impedance Tomography of Electrolysis. PLoS ONE 10(6): e0126332. doi:10.1371/journal.pone.0126332.

CHAPTER 4

Stehling MK, Guenther E, Mikus P, Klein N, Rubinsky L, Rubinsky B (2016) Synergistic Combination of Electrolysis and Electroporation for Tissue Ablation. PLoS ONE 11(2): e0148317. doi:10.1371/journal.pone.0148317.

CHAPTER 5

Ruyao Wang and Weihua Lu (2012). Direct Electrolytic Al-Si Alloys (DEASA) – An Undercooled Alloy Self-Modified Structure and Mechanical Properties, Electrolysis, Dr. Janis Kleperis (Ed.), ISBN: 978-953-51-0793-4, InTech, DOI: 10.5772/52962.

CHAPTER 6

Stephen J. Andersen, Pieter Candry, Thais Basadre, Way Cern Khor, Hugo Roume, Emma Hernandez-Sanabria, Marta Coma and Korneel Rabaey, "Electrolytic extraction drives volatile fatty acid chain elongation through lactic acid and replaces chemical pH control in thin stillage fermentation," Biotechnology for Biofuels20158:221, DOI: 10.1186/s13068-015-0396-7.

CHAPTER 7

Nong, G.; Zhou, Z.; Wang, S. Generation of Hydrogen, Lignin and Sodium Hydroxide from Pulping Black Liquor by Electrolysis. Energies 2016, 9, 13.

CHAPTER 8

Park, H.S.; Lee, J.S.; Han, J.; Park, S.; Park, J.; Min, B.R. CO2 Fixation by Membrane Separated NaCl Electrolysis. Energies2015, 8, 8704-8715.

CHAPTER 9

Ewan, B.C.; Adeniyi, O.D. A Demonstration of Carbon-Assisted Water Electrolysis. Energies 2013, 6, 1657-1668.

CHAPTER 10

Bamba, K. and Ziao, N. (2016) Cathodic Using of ZrB2-αSiC and TiB2-αSiC for PEM Electrolysis and Water Electrolysis at Low Temperature. American Journal of Analytical Chemistry, 7, 1-11. doi:10.4236/ajac.2016.71001.

CHAPTER 11

R. Slama, "Hydrogen Production by Water Electrolysis Effects of the Electrodes Materials Nature on the Solar Water Electrolysis Performances," Natural Resources, Vol. 4 No. 1, 2013, pp. 1-7. doi: 10.4236/nr.2013.41001.

INDEX

A

alloys 121, 122, 123, 124, 125, 126, 127, 129, 130, 131, 132, 133, 137, 138, 144, 147, 148, 149, 150, 151, 152, 154, 155, 156, 163

anode 243, 244, 245, 247, 249, 250, 252, 253, 254

B

biorefinery sidestreams 166, 196

C

carbon assisted a electrolysis cell (CAEC) 248

Carbon capture and storage (CCS) 223

Carbon capture and utilization (CCU) 223

castability 121, 130, 155

cationic exchange membrane (CEM) 203

chain elongation (CE) 178

circular dichroism (CD) 21

complete electrode model (CEM) 69

Cyclic voltammetric 261

cytoplasm 100, 101, 105, 110

D

dendrite arm space (DAS) 127, 144

diffractometer 261

direct carbon fuel cell (DCFC) 244

E

E.Coli 63, 68, 77, 79

Electrical Impedance Tomography (EIT) 63, 65

electrochemotherapy 80, 87, 88, 89, 97

electrodes 88, 90, 91, 92, 93, 95, 96, 98, 99, 100, 101, 102, 107, 108, 109, 110, 111, 112, 113

electrolysers 29, 31, 33, 277, 278, 280, 282, 287, 290

Electrolysis 63, 64, 67, 70, 74, 76, 85, 87, 88, 95, 96, 103, 119, 293

electrolysis cell (EC) 248, 251

electroporation 87, 88, 89, 90, 91, 92, 93, 95, 96, 97, 100, 101, 102, 104, 107, 108, 110, 111, 112, 113, 114, 118, 119

electropositivity 288

F

field emission scanning electron microscopy (FE-SEM) 223, 234